LECTURES ON

AIR POLLUTION MODELING

LECTURES ON

AIR POLLUTION MODELING

Akula Venkatram and John C. Wyngaard

Editors

American Meteorological Society
Boston
1988

ISBN 0-933876-67-X

American Meteorological Society
45 Beacon Street, Boston, MA 02108

We dedicate this volume to Hans Panofsky.

Contents

8. Concentration Fluctuations Within a Laboratory Convectively Mixed Layer, *James W. Deardorff and Glen E. Willis*

Preface

This volume is based on the "Short Course on Air Pollution Modeling" held in San Diego, CA, on March 17–21, 1986, and sponsored by the American Meteorological Society. The brainchild of my coeditor, Akula Venkatram, it attracted more than 80 attendees, including researchers from government laboratories and academic institutions; practitioners from the private sector; employees of regulatory agencies, public utilities, and energy companies; and graduate students.

Dr. Venkatram's idea for the course came during his experience as an air quality modeler and program manager for a consulting firm. A leader in applying the rapid advances in boundary layer meteorology during the 1970s and 1980s to the improvement of air quality modeling, he felt that his field as a whole was not advancing as it should. Outspokenly critical of what he saw as a failure of the air-quality modeling field to respond to the times, he reasoned that a short course, elucidating today's understanding of the physics of the lower atmosphere and of dispersion within it, and demonstrating how this understanding can be used to solve practical problems, could begin to make converts to a new practice of air quality modeling.

After the course he and I agreed to edit the lectures for publication, a task that has taken longer than we expected. We hope that this volume is worth the wait, and that it will indeed stimulate the thinking of a new generation of air quality modelers.

The diligent help of my NCAR colleagues Joann Van Dyke, Mildred Farnsworth, Hope Hamilton, Fran Huth, and Pat Waukau, in word processing, and Justin Kitsutaka, in graphics, was invaluable. Jeff Weil helped me in many ways during his long-term visit to NCAR, and Jim Adams skillfully prepared the camera-ready copy. The project was sponsored by the American Meteorological Society and the Electric Power Research Institute. Finally, I thank my coeditor Akula Venkatram for inspiring this effort.

John C. Wyngaard
National Center for Atmospheric Research
Boulder, Colorado

February, 1988

Introduction

John C. Wyngaard

This volume is concerned with the physics and application of air-pollution modeling on scales up to about 50 km. Its eight chapters, comprising the diverse points of view of seven authors, remain substantially in their original, lecture-note form. The result is not a smoothly flowing monograph but instead a richly textured, lively collection of the seasoned thoughts and perspectives of experienced researchers and practitioners. My coauthor, Akula Venkatram, and I hope that the volume conveys some of the excitement of a vital, rapidly evolving field.

In this Introduction I attempt to impart some order to the volume by giving some background history on the various topics within it. In the process I shall give some background on the various authors as well, since they are very much a part of that history.

Given our focus on relatively short spatial scales, our topic is dominated by the dispersive properties of the planetary boundary layer (PBL) into which the pollutant is emitted. For that reason, Ch. 1 is an up-to-date review of PBL structure and dynamics as it bears on air quality modeling.

Like any other turbulent flow, the PBL has yielded most readily to experimental, as opposed to theoretical, investigation. Theory has been very useful in designing experiments and in interpreting their results, however. For example, early micrometeorologists, studying turbulent exchange between the surface and the atmosphere, were guided by the Monin-Obukhov (M-O) similarity hypothesis for the structure of the atmospheric surface layer. Such studies reached a climax with the 1968 Kansas experiments, which revealed the structure of the first few tens of meters of the atmosphere in very complete detail. The 1973 AMS monograph *Workshop on Micrometeorology* edited by Duane Haugen thoroughly discusses these results.

By the early 1970s, however, it became apparent that the M-O hypothesis, which held that surface-layer structure was independent of the properties of the PBL as a whole (and specifically, independent of its depth) had some flaws. For example, it did not account for the daytime behavior of horizontal wind fluctuations near the surface, which are very important in dispersion applications. Researchers were limited in their ability to probe higher into the PBL however, and the next steps were actually made through unlikely means: experiments in convection tanks and on the computer.

As discussed in Ch. 1, Deardorff's numerical experiments with large-eddy simulation, and his convection tank experiments with Glen Willis, opened a new era in PBL understanding. Their work, begun at the National Center for Atmospheric Research in the late 1960s and continued at Oregon State University after their move there some ten years later, revealed clearly the scaling parameters for turbulence in the convective PBL and gave new insight into its dispersive properties. This allowed experimentalists to regroup. Soon, armed with a new conceptual framework for CBL structure, they began to rely heavily on aircraft and balloon-borne sensors in their experimental thrusts deep into the boundary layer. Before long, "mixed-layer scaling" became standard in presentations of PBL research findings.

During this period acoustic sounders revealed some intriguing aspects of the stably stratified nocturnal PBL. According to the sounder records, its fully turbulent layer could be quite thin; 100 m was not unusual. Its top was often marked by a layer of strong stability, which extinguished any turbulence and, hence, any acoustic scattering; thus, it often showed quite clearly in the returns. Although the records sometimes showed rich gravity-wave activity, which could at times complicate this simple picture, researchers nonetheless could see for the first time evidence of the dramatic diurnal swings in PBL depth and structure.

Meanwhile, the emergence of "second-order closure" was strongly influencing turbulent-flow modeling. It was particularly attractive to boundary-layer meteorologists because it seemed applicable, without ad hoc adjustments, through the diurnal cycle of the PBL; this was not true of simpler closures. Before long, modelers reproduced at least qualitatively the observed behavior of both the convective PBL and the nocturnal PBL. This work, together with further experimental programs, has given us much insight into the mechanisms controlling the structure and evolution of the stably stratified PBL.

In Ch. 2, "Analysis of Diffusion Field Experiments," Gary Briggs reviews changing perspectives on diffusion data gathered over the past 30 years. He shows how surface-layer similarity, mixed-layer scaling, and statistical theory can be used to interpret diffusion data, using a minimum of meteorological inputs, and he clearly delineates the strengths and weaknesses of each approach.

Briggs was introduced to diffusion in the early 1960s as a graduate student in meteorology at the Pennsylvania State University under thesis advisor Hans Panofsky. In 1964 he took a summer job at the Atmospheric Turbulence and Diffusion Laboratory (ATDL) in Oak Ridge, Tennessee. In so doing he missed fellow graduate student Wyngaard's introduction to the subject, a splendid summer course on atmospheric diffusion taught by Visiting Professor Frank Pasquill, but he soon received a permanent position at ATDL. Its director, Frank Gifford, encouraged Briggs' predilection for theoretical studies of plume rise and diffusion problems, followed by thoroughgoing data analyses to test the theories. Prior to receiving his Ph.D. from Penn State in 1970, Briggs became well known for his 1969 monograph *Plume Rise*. Since 1980 he has been with the Meteorology Laboratory of the Environmental Protection Agency in Raleigh, NC.

The buoyancy of many pollution sources makes plume rise an important factor in air quality modeling. Jeff Weil addresses this topic in Ch. 3, using the language and techniques of engineering fluid mechanics but setting his discussion firmly in the context of PBL physics. His experience stems from his own laboratory experiments, done in the course of his Ph.D. research at MIT under David Hoult, and more than a dozen years in air quality modeling research at Martin Marietta Laboratories in Baltimore.

By the close of the 1970s our knowledge of PBL structure and dynamics substantially exceeded that of the mid-1960s, say, when the pathbreaking Lumley-Panofsky monograph *The Structure of Atmospheric Turbulence* appeared. Although the general practice of air quality modeling seemed firmly rooted in the 1960s, in that it still strongly reflected the Pasquill-Gifford-Turner school, some practitioners in the air quality field were quick to use this new PBL understanding to improve their models. Weil, who has been a leader in this activity, meticulously discusses this new approach in Ch. 4.

In Chs. 5 and 6 the emphasis shifts to applications of air quality models. The author, Akula Venkatram, is in the unusual position of having extensive experience in both the research and applications sides of the air quality field. After receiving his Ph.D. in Mechanical Engineering in 1976 from Purdue University, where his thesis research concerned the radiative/dynamical coupling effects of pollution in an urban boundary layer, he did boundary layer research at the Atmospheric Environment Service and later at the Ontario Ministry of the Environment, both in Toronto. He joined Environmental Research and Technology, Inc., in 1981 where he has been heavily involved in air quality model development and application.

Our readers might notice that Venkatram's chapters have a distinctly "hands on" nature, compared with the preceding ones. Perhaps this reflects the values of one who understands and respects the challenges of dealing with the hard, and sometimes mutually exclusive, constraints of theory and practice. Such challenges are central to the field of engineering, of course, but also occur in other areas; for example, anyone who has actually developed and used a numerical model can readily appreciate them. What seems theoretically correct sometimes is not practical to implement in a numerical model; indeed, it may even lead to demonstrably incorrect results.

With regard to the last point, one of the editors remembers well his early experiences with second-order closure modeling, in which one carries a set of equations for second moments of turbulence, including variances. Variances, and the mean values of quantities such as concentrations, must of course be positive; in addition, constraints such as Schwarz's inequality must be satisfied. Early second-order models (and, indeed, most of today's versions) were not formulated in a way to guarantee this, a defect that did not attract much notice for several years. In the meantime, many modelers (including that editor) were sometimes frustrated by working hard on an improved parameterization, one with an obviously better basis, only to find that it could produce negative variances! That can be a humbling experience, but one that can also lead to a broader appreciation of the numerical modeling problem, and of those who can meet its challenges.

In recent years improved air quality data bases have allowed models to be tested more thoroughly. Broadly speaking, we have found that although improvements of the sort discussed by Weil and Venkatram do result in better model performance, there usually remains a good deal of scatter between model predictions and observations. We now feel that part of this scatter represents an irreducible uncertainty caused by the stochastic nature of the atmosphere. This greatly complicates the model evaluation process, as Venkatram discusses in Ch. 6, because we cannot attribute differences between predictions and observations entirely to model errors.

Air quality modeling is not only challenging, but also controversial. Consider, for example, the following quote by Scorer (1980), taken from his review of a monograph on the modeling of turbulent diffusion in the environment:

> Turbulent diffusion in the air and bodies of water is of fundamental importance to life on earth, and so it is natural that applied mathematics should attempt to say something about it; after all much of our modern way of life is based on a mathematical understanding of our physical environment.
>
> Because turbulence is a very complicated phenomenon simple mathematics is not adequate to the task. Because those who work in this field are more interested in the diffusion of pollution or other components of the air or water than in the mechanics of the fluid itself, and because turbulence makes itself more complicated all the time so that no full description is possible, most theories are not theories about the turbulence at all, but about the consequences....
>
> It is fashionable today to talk about "mathematical modelling", as if that were something rather new. Actually it is what mathematicians have done ever since they first drew a diagram and used Euclidean geometry to calculate things: but it has become rather fashionable and it is thought to give status to other sciences and technologies which have lacked it in the past. The art of modelling, of course, is not in making it sophisticated and difficult, although that is part of the game, but in finding the very simplest model which is as good in practice as the much more sophisticated ones.
>
> In the great outdoors we have, particularly in the atmosphere, an unending succession of different cases, whereas it is a characteristic of the models that they refer only to particular cases. One is always bound to wonder whether the models are relevant enough to be worth the bother.... Unfortunately...most of the papers in this volume are more concerned...with the properties of the models rather than the ways of nature. This is a valid exercise but it does not carry much conviction for the practical person who is really concerned with the behaviour of a fixed installation in an infinite variety of flow patterns.

Hunt (1981), in discussing what he sees as the practical benefits of the standardized methods now used in many countries to predict the diffusion of pollutants emitted into the atmosphere, questioned Scorer's views:

> Usually the source is a buoyant plume emitted from a chimney of height H which first mixes with the atmosphere under the action of its own thermal and mechanical energy, by the processes of "entraining" the surrounding fluid.... At some stage this process is overtaken by the diffusing action of the external turbulence and then the pollutant is assumed to diffuse like a passive scalar from a source at greater height ΔH above the original source. This transition process is not really understood at all. The standardized methods for predicting the rate of growth of plumes (pioneered by Pasquill in 1961) were developed in the framework of G. I. Taylor's...statistical theory of diffusion in *homogeneous* flow and Monin and Obukhov's...analysis of the diabatic surface layer. Despite the theoretical flaws in such a procedure, it is not so inaccurate that it cannot provide rough guidance. As the research on meteorological dispersion advances, so do these guidelines. This approach is dismissed as useless by Professor Scorer...; because nature is too complicated, he says, it cannot be codified. In a sense he is right, but if we are to follow his advice and discuss nature eddy by eddy, how is the government inspector to make his decision and how are others to argue with him?

Scorer's "unending succession of different cases" refers to the stochastic nature of the atmosphere, and reminds us again of the irreducible uncertainty in dispersion modeling. This is intimately related to the fluctuating concentration field, which has recently become an area of increased research interest. In Ch. 7 Ian Sykes presents his treatment of the concentration fluctuations in dispersing plumes. Making innovative use of simple physical models of the underlying processes, he nonetheless will undoubtedly persuade even the hardiest reader that the fluctuating concentration problem is at least as challenging as the mean one.

Sykes' research career began with the British Meteorological Office, where he did computational fluid mechanics, including flow over hills and boundary layer studies, from 1973 to 1980. He received his Ph.D. in mathematics from Imperial College in 1979. In 1980 he joined Aeronautical Research Associates of Princeton (now called ARAP) where he has done theoretical, computational, and numerical model studies of atmospheric turbulence and diffusion.

It is fitting that this volume should conclude with a contribution by Jim Deardorff and Glen Willis, who have perhaps done as much as anyone to advance the modern practice of air quality modeling. In Ch. 8 they discuss their convection tank measurements of concentration fluctuations, again in dispersing plumes. Their earlier measurements of mean concentration in the same tank provided the basis for the overhaul of the Gaussian-plume approach in the 1970s. Perhaps in their contribution here (which could be their last to the field since each has retired from Oregon State University to pursue other interests) we have the seeds of a fresh start on the challenging problem of irreducible uncertainty in air quality modeling.

References

Briggs, G. A., 1969: *Plume Rise*. USAEC Critical Review Series, TID–25075, NTIS, 81 pp.

Haugen, D. A., Ed., 1973: *Workshop on Micrometeorology*. Amer. Meteor. Soc., Boston, 392 pp.

Hunt, J. C. R., 1981: Some connections between fluid mechanics and the solving of industrial and environmental fluid-flow problems. *J. Fluid Mech.*, **106**, 103–130.

Lumley, J. L, and H. A. Panofsky, 1964: *The Structure of Atmospheric Turbulence*. Wiley Interscience, New York, 239 pp.

Scorer, R. S., 1980: Book review. *Clean Air*, **10**, 148–149.

CHAPTER 1

Structure of the PBL

John C. Wyngaard

1.1. Overview

Even the most casual observer notices the changes in the wind. Though it has a certain persistence in time, the details of its swirls and eddies seem infinitely variable. Its strength changes day to day as weather systems evolve, and day to night as the sun rises and sets. It is modulated strongly by terrain features and by urban architecture. How can we hope to describe such a complicated phenomenon?

The answer, of course, is that even though the wind fluctuates randomly, its behavior—even in its minuscule details—is not without order. We believe that it satisfies the Navier-Stokes equation, and for over 100 years we have used that equation, in conjunction with measurements, to study the behavior of the wind in the lower atmosphere. Although its instantaneous, local details are essentially unpredictable, as anyone who watches the continuously changing, meandering plume from a smokestack might conclude, we do understand the behavior of the average wind—its dependence on stability, distance from the surface, mean weather conditions, and the like—at least in uncomplicated terrain. We also understand much about its turbulent fluctuations. Today we can estimate the "most likely" behavior in many important turbulent dispersion problems in the planetary boundary layer (PBL), in spite of the chaos of the actual process.

I shall outline some of the knowledge about the PBL and its turbulence that allows us to deal successfully with applications such as dispersion. This knowledge rests principally on observations, but there is room for a good deal of individuality in interpreting these observations and fabricating useful conceptual models from them. Thus, the interpretations and explanations in this chapter reflect to some extent my own way of viewing the PBL. I shall concentrate on the physics and general structure; later chapters will cover in more detail the currently accepted forms for vertical profiles of mean wind and turbulence quantities important in dispersion applications.

1.2. What Is the PBL, and Why Is It Important?

By "planetary boundary layer" we mean the continuously and vigorously turbulent layer of the atmosphere adjacent to the earth's surface. Its upper edge can be made quite distinct by the turbulence-extinguishing effects of stable stratification, but is less clearly defined in other cases, as in the presence of large, active cumulus clouds. Other authors in this volume will at times refer to the PBL as the "mixed layer," since its turbulence often is sufficiently vigorous to mix contaminants as well as temperature, water vapor, and momentum very well in the vertical. This is often not the case, however; at night and during the day mean profiles within the PBL can be far from "well mixed," as we shall see later in this chapter.

Turbulent fluctuations of air temperature and humidity cause fluctuations in the refractive index for both acoustic and electromagnetic waves. This makes the PBL "visible" through backscattered radiation emitted from remote sensors. The 1970s saw exciting developments in boundary layer meteorology as acoustic sounders and other such devices revealed the locally sharp top and surprisingly shallow depth of the nocturnal PBL, and the rapidly deepening convective PBL in the morning. Figure 1.1 shows such acoustic sounder traces.

There are often substantial changes in wind speed, temperature, humidity, and scalar constituent concentration across the boundary layer. These changes, in the presence of a turbulent velocity field, ensure strong turbulent transfer in the vertical. The resulting turbulent fluxes of momentum, heat, and mass within the PBL have very important effects on larger-scale meteorology, since they represent a key link between the earth's surface and the atmosphere. For this reason the meteorological community has traditionally put a high value on PBL research.

The turbulence in the PBL is also responsible for the dispersion of smokestack effluents. In the absence of turbulence, their concentrations can decrease only through molecular diffusion, a very slow process. The turbulent diffusivity is larger by a factor of the Reynolds number (Tennekes and Lumley, 1972), which is why, for example, we stir our coffee after adding cream. The diffusivity of PBL turbulence

South Pole Acoustic Sounder

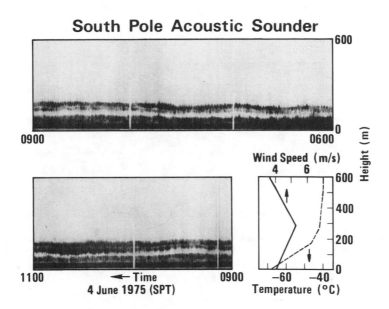

Figure 1.1a. An acoustic sounder record from a stably stratified PBL at the South Pole. The depth is steady at about 200 m, and turbulence level (indicated by echo intensity) is greatest near the surface, with a secondary maximum near PBL top, where the lapse rate seems to be less than at the surface (see profiles, lower right). Courtesy W. D. Neff, NOAA/ERL/WPL.

South Pole Acoustic Sounder

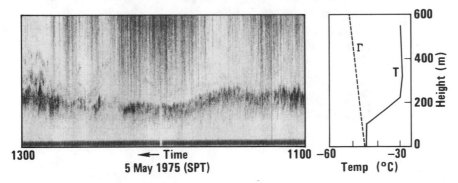

Figure 1.1b. An acoustic sounder record from the South Pole showing a weakly stable PBL capped by an inversion at 100–200 m height (see right panel). Courtesy W. D. Neff, NOAA/ERL/WPL.

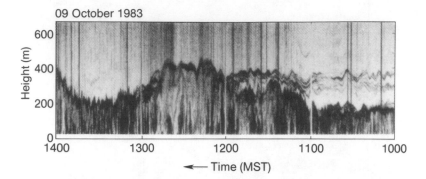

09 October 1983

Figure 1.1c. An acoustic sounder record from a convective PBL at the Boulder Atmospheric Observatory. The PBL is about 200 m deep until 1100 MST, when it begins growing to about 400 m at about 1230; its depth then falls to 200 m at 1330, before growing to 400 m again at 1400. Courtesy W. D. Neff, NOAA/ERL/WPL.

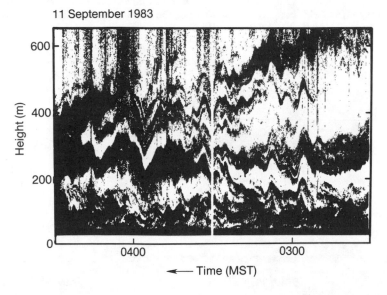

11 September 1983

Figure 1.1d. An acoustic sounder record from a nocturnal PBL at the Boulder Atmospheric Observatory. Note the rich wave activity. Courtesy W. D. Neff, NOAA/ERL/WPL.

can be impressively large; in the daytime with clear weather over land, it can exceed the molecular diffusivity by 7 orders of magnitude!

Airplane passengers flying over urban areas are apt to notice the PBL top during air pollution episodes, when it acts as a lid on the murky air. This reminds us that vertical transport out of the PBL can be quite weak; a principal mechanism is the venting by cumulus clouds, and in their absence pollutants can remain in the PBL for a long time.

It is clear, then, that important processes within the PBL are dominated by turbulence. We have learned much about PBL turbulence in the past fifteen years or so, thanks to new ways of observing it. Whereas we once were limited to measurements from towers, we now have remote sensors, laboratory simulations, and numerical simulations on the supercomputer to give us new insights.

1.3. The Ekman Spiral

The PBL bears a family resemblance to the viscous boundary layers of fluid mechanics, as we can see by examining Ekman's classical solution for the "wind profile," the distribution of mean wind with height. In the simplest case—steady flow over a horizontally homogeneous surface—the mean horizontal winds satisfy the following momentum balance (Panofsky and Dutton, 1984):

$$\frac{\partial}{\partial z}\tau_{xz} = f(V - V_g) \tag{1.1}$$

$$\frac{\partial}{\partial z}\tau_{yz} = f(U_g - U) . \tag{1.2}$$

Here τ_{xz} and τ_{yz} are components of the turbulent stress, (U, V) is the mean wind vector, f is the Coriolis parameter, z is distance from the surface and $(fV_g, -fU_g)$ is the mean pressure gradient, which in uncomplicated cases is determined by the large-scale conditions, i.e., by the weather map. As with any turbulent flow, we need a closure approximation—a relation between the turbulence terms (here τ_{xz} and τ_{yz}) and the mean fields—in order to be able to solve Eq. (1.1) and Eq. (1.2) for the mean wind profile.

The earliest closure for Eq. (1.1) and Eq. (1.2) assumed a constant "eddy viscosity" K that relates the turbulent stress to the mean velocity shear,

$$\tau_{xz} = -K\frac{\partial U}{\partial z} \tag{1.3}$$

$$\tau_{yz} = -K\frac{\partial V}{\partial z} \tag{1.4}$$

in analogy to the behavior in laminar flow. The wind vanishes at the surface, and becomes geostrophic (i.e., perpendicular to the pressure gradient) far above it; this gives the following analytical solution, called the "Ekman spiral" after the Swedish oceanographer who first derived it for the corresponding problem in the upper ocean (Holton, 1972):

$$U = U_g(1 - e^{-\gamma z}\cos\gamma z) \tag{1.5}$$

$$V = U_g e^{-\gamma z}\sin\gamma z . \tag{1.6}$$

Here we have aligned the axes with the geostrophic wind, with

$$\gamma = (f/2K)^{1/2} . \tag{1.7}$$

A plot of this solution, Fig. 1.2, shows that the mean wind direction rotates 45° across the Ekman layer. This characteristic and distinctive feature of the Ekman layer is due to the Coriolis force and is not often found in the boundary layers of engineering flows.

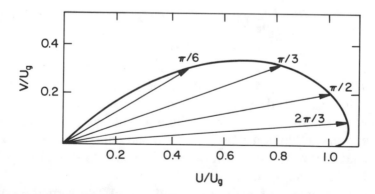

Figure 1.2. A plot of the Ekman solution for the PBL wind profile. The numbers on the curve are values of γz, the dimensionless distance from the surface.

According to Eqs. (1.5) and (1.6), the depth of the region over which the Ekman solution makes its transition to geostrophic flow is γ^{-1}; therefore, the Ekman layer depth is $h_e \sim \gamma^{-1} \sim (K/f)^{1/2}$. Note the subtle difference between h_e and PBL depth h defined in Sec. 1.2; h is the depth of the continuously turbulent layer, while h_e is the depth of the transition region in the mean wind profile.

Let's examine the consequences of assuming that h and h_e are of the same order. In any turbulent flow we expect the eddy diffusivity K to be of order $q\ell$, where q and ℓ are velocity and length scales of the energy-containing eddies (Tennekes and Lumley, 1972). Since we expect ℓ to be of the order of h, we can write

$$h \sim h_e \sim \left(\frac{K}{f}\right)^{1/2} \sim \left(\frac{qh}{f}\right)^{1/2} \qquad (1.8)$$

so that

$$h \sim \frac{q}{f} . \qquad (1.9)$$

One interpretation of this result is that $h \sim h_e$ implies that the steady Ekman layer adjusts its depth so that the eddy turnover time h_e/q and the Coriolis time $1/f$ are of the same order. Note also that since q is typically an order of magnitude smaller at night than during the day, as we will see later in this chapter, Eq. (1.9) implies that the steady-state Ekman layer depth varies similarly.

How well do the expressions Eq. (1.5) and Eq. (1.6) for the wind profile in the Ekman layer, and Eq. (1.9) for its depth, fit PBL observations? Let us survey the situation, remembering that we must restrict ourselves to steady, horizontally homogeneous conditions.

In the daytime over land, wind profiles in clear weather typically have very little speed or direction change within the PBL. Figure 1.3, for example, shows profiles measured in the 1973 Minnesota experiments; their flatness shows clearly why the daytime PBL, with its large, intense convective eddies, is often called the "mixed layer." To be sure, daytime PBL profiles with considerably more vertical structure are sometimes measured, but these deviations appear to be due to unsteadiness, mean advection, height-dependent mean pressure gradients, or other departures from the Ekman assumptions.

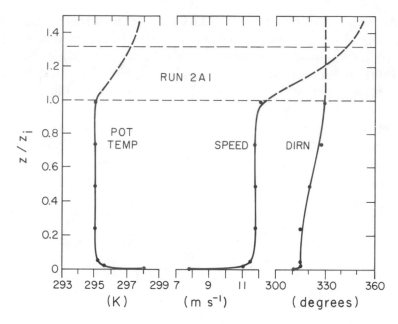

Figure 1.3. Early-afternoon profiles of mean wind and potential temperature in the 1973 Minnesota experiments. This moderately convective PBL had z_i = 1250 m, $-z_i/L \sim 30$. From Kaimal et al. (1976).

The abrupt change in turbulence properties at h makes it relatively easy to detect through remote sensing, and it is now clear that under convective conditions h bears no more than a coincidental relation to our Ekman prediction Eq. (1.9). The explanation for this seems equally clear. If $q \sim 1$ m s^{-1} and $f = 10^{-4}$ s^{-1}, typical values for midlatitude, convective conditions, then $q/f \sim 10^5$ m; if the proportionality factor implied in Eq. (1.9) is between 0.01 and 0.1, as seems likely, then h is predicted to lie between 10^3 and 10^4 m. Usually, however, there exists an inversion below this height, and this sets the convective PBL depth by acting as a "lid" on its turbulence. Thus, it is now generally accepted that in the convective case h is simply the distance to the first inversion base, frequently denoted by z_i. This distance usually evolves with time in response to the entrainment process (which tends to increase z_i as the turbulence erodes the inversion from below) and the mean vertical velocity there, which can increase it or decrease it depending on its sign. Clearly, the Ekman solution is not a good representation of the daytime, clear-weather PBL over land.

At neutral stability (which can be found under high wind speed, cloudy conditions) wind profiles are often at least qualitatively consistent with Eq. (1.5) and Eq. (1.6), although the observed direction change is usually distinctly less than the predicted 45°. Theory indicates the profile should be logarithmic near the surface (Lumley and Panofsky, 1964) and observations bear this out; Panofsky and Dutton (1984), for example, show neutral wind profiles with logarithmic behavior up to 100 m. Numerical solutions to the mean momentum equations with more realistic, z-dependent K values (e.g., Wipperman, 1973) also show smaller "cross-isobar" angles, also suggesting that the constant-K Ekman solutions exaggerate it.

Three-dimensional numerical simulations of the neutral PBL (e.g., Deardorff, 1973a) confirm this finding, but do indicate that its steady-state depth does scale with q/f as predicted. These simulations reveal two other important properties of the neutral case. First, it is very "slow," typically requiring several $(1/f)$ times to come to equilibrium with a new set of boundary conditions; this can mean on the order of one day. Second, it is very sensitive to heating or cooling at its top or bottom, which can induce stable or unstable stratification and radically change its structure. Thus, the truly neutral PBL is probably quite rare.

Finally, let us consider the stable PBL, commonly found over land at night or, at any time, over a surface colder than the overlying air. Observations (e.g., Fig. 1.4) and numerical model studies (e.g., Delage, 1974; Brost and Wyngaard, 1978; Nieuwstadt, 1984) indicate that the steady-state wind profile does turn through a larger angle than at neutral; the 45° Ekman prediction appears to fit this case best. Data also show that stably stratified PBLs are typically very shallow, perhaps an order of magnitude thinner than their neutral or convective counterparts. Their velocity fluctuation level is also reduced roughly in the same proportion. The associated decrease in the eddy diffusivity has important consequences for dispersion in the stable PBL, as we shall see clearly in Chs. 2 and 5.

We conclude that while the Ekman solution does represent qualitatively some of the features of the PBL, it is not an adequate description. In both convective and neutral conditions the PBL top is usually determined by the lowest inversion base, which typically occurs between a few hundred meters and a few kilometers; in practice this makes quite irrelevant the ability of the Ekman layer establish its own, and usually larger, depth proportional to q/f. Furthermore,

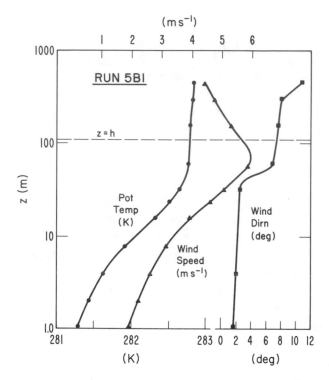

Figure 1.4. Early-evening profiles of mean wind and potential temperature in the 1973 Minnesota experiments. Transition occurred about 30 min after the start of this 90-min run. From Caughey et al. (1979).

neutral and convective PBL wind profiles are flatter than the Ekman spiral, with less turning. The best agreement with observations is likely to be found in a stable PBL of the type generated by a surface inversion (e.g., in the nocturnal PBL) evidently because the assumptions underlying the Ekman solution are most nearly met. Ekman predictions of both the wind turning and PBL depth have been fairly successsful in that case.

In summary, even in the simplest situation—steady, horizontally homogeneous flow—the wind profile in the lower atmosphere depends on stability and on the location of the capping inversion. The diurnal cycle makes true steadiness rare over land, of course, and terrain variations and moving weather systems combine to add inhomogeneity as well. Subtle temperature variations in the horizontal can cause the mean pressure gradient to vary significantly with height. These complications make a simple, universal description of the wind profile impossible. A good deal of progress has been made, however, by developing models for specific conditions. Before we discuss these "tailored" models, let us review in more detail the influence of stability on turbulence and the PBL.

1.4. Stability and PBL Structure

1.4.1. The Surface Layer

Perhaps the best-known success of micrometeorology is its description of the "surface layer," the lowest 10% or so of the PBL. Thanks to the similarity hypothesis of Monin and Obukhov, which provided a simple conceptual framework, and to extensive field measurements, which defined the range of applicability of the framework and yielded its various universal functions, we now have a vast store of useful knowledge about that part of the PBL closest to the surface.

M-O similarity, as we call it, is discussed in detail in numerous publications; good references are Monin and Yaglom (1971), Haugen (1973), McBean (1979), Nieuwstadt and van Dop (1982), and Panofsky and Dutton (1984). Because of this wide exposure I need not review the applications of M-O similarity in great detail. Instead, I will cover its underlying basis and physical interpretation.

The essence of M-O similarity is the hypothesis that mean-field and turbulence properties in the surface layer depend only on height z and three governing flow parameters: the buoyancy parameter g/T_0, the friction velocity $u_* = [(\tau_{x0}^2 + \tau_{y0}^2)^{1/2}/\rho]^{1/2}$ (the square root of the kinematic surface stress), and the surface temperature flux $\overline{w\theta_0}$. If water vapor is present, we use the surface *virtual* temperature flux. Defining a virtual temperature is a simple way to account for the effect of water vapor on buoyancy; by treating the mixture as a perfect gas, one finds (Lumley and Panofsky, 1964) that $\overline{w\theta_v} = \overline{w\theta} + 0.61T\overline{wq}$, where q is the fluctuation in specific humidity (mass water vapor/mass air).

These three governing flow parameters define a length scale $L = -u_*^3 T_0/(kg\overline{w\theta_0})$ (the M-O length), a velocity scale u_*, and a temperature scale $\theta_* = -\overline{w\theta_0}/u_*$. The M-O length L is negative under unstable conditions (positive surface temperature flux), infinite at neutral, and positive under stable conditions.

The M-O hypothesis attracted much interest when it became well known in the western world in the 1960s. Calder (1966) claimed to prove rigorously from the turbulent flow equations that it held for mean velocity and temperature gradients. Later, however, Monin and Obukhov commented on that effort (Calder, 1968):

K. L. Calder's paper is devoted to an attempt to demonstrate the similarity theory for the wind velocity and
temperature gradients (according to which the values of
dV/dx_3 and dT/dx_3 in the case of fully developed turbulence are uniquely determined by just the three parameters...) as *a legitimate consequence of dynamical equations
of turbulent flow*.... It is our belief that the similarity theory, even for the wind velocity and temperature gradients
only, *is not a strict consequence of dynamical equations*.... In
our opinion the basic assumption of the similarity theory
(the assumption of the existence of just the three determining parameters...) is but a statistical hypothesis based
upon the invariance of dynamical equations under certain transformations...and upon a specific viewpoint on
the mechanism of vertical turbulent mixing. The similarity theory needs observational justification even in the
case of wind velocity and temperature gradients.

The M-O hypothesis thus rests essentially on the physical and
mathematical intuition that a given dependent variable depends only
on the three M-O governing parameters. Given that assumption, dimensional analysis allows one to make a statement about the behavior
of that dependent variable: specifically, it says that the variable, when
made dimensionless by z, u_*, and θ_*, is a universal (i.e., the same in
all surface layers) function of z/L. The independent variable z/L is
a stability index.

Physically, this means that the surface layer has a velocity scale,
u_*, that sets the "level" of the velocity fluctuations; $|\theta_*|$ does the same
for the temperature fluctuations. If there is a passive, conservative,
horizontally homogeneous scalar field diffusing through the surface
(a trace constituent c, say) then its level is set by $c_* = C_0/u_*$, where
C_0 is the surface flux of c. Roughly speaking, the magnitude of
the length scale L corresponds to the height at which the shear and
buoyant production (or destruction, under stable conditions) rates of
turbulent kinetic energy are equal.

M-O similarity has been widely used in surface-layer research
for the past 20 years. It accounts well for the observed behavior of
wind, temperature, and water vapor profiles, and for the variances
and covariances of most turbulence quantities. We can demonstrate
its use with the mean potential temperature gradient $\partial\Theta/\partial z$. By the
M-O hypothesis, when we make $\partial\Theta/\partial z$ dimensionless with z and θ_*

(we traditionally also use von Kármán's constant $k \sim 0.4$, in analogy with the nondimensionalization of the mean wind shear) we have a universal function of z/L. This function is usually called ϕ_h:

$$\phi_h = \frac{kz}{\theta_*}\frac{\partial \Theta}{\partial z} . \tag{1.10}$$

Figure 1.5 shows data on the mean temperature gradient near the surface, plotted as suggested by the M-O hypothesis. The data were gathered over a period of several weeks from a 32-m tower in Kansas (Businger et al., 1971); the temperature gradient magnitudes varied widely. The nondimensionalization causes the data to fall along a universal curve, however, illustrating the power of M-O similarity.

As I shall explain later, M-O similarity fails for the horizontal wind components in unstable conditions. This has important implications for dispersion.

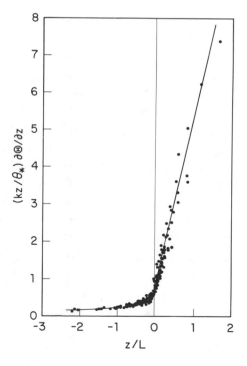

Figure 1.5. Data on the mean temperature gradient in the first 32 m, from the 1968 Kansas experiments, presented in the M-O similarity framework. The data, gathered over several weeks, collapse nicely. From Businger et al. (1971).

1.4.2. The Convective Boundary Layer (CBL)

A three-layer model

The large convective eddies driven by heating and/or evaporation from the surface tend to fill the vertical space available—up to the base of the inversion "lid." In some early morning situations the convection, after breaking through the remnants of the nocturnal, surface-based inversion, meets neutral air above; in such cases the entrainment of this essentially nonturbulent air by the convective turbulence is very fast. Deardorff (1974), for example, suggests the deepening rate of the convective boundary layer (CBL) in this case is of the order of 0.2 of the velocity characteristic of the convective eddies, or typically 0.2 m s^{-1} (700 m h^{-1}). Thus, in the absence of an inversion lid, the CBL deepens very rapidly until it does reach an overlying inversion that can act as the lid.

This inversion lid moves vertically because of entrainment from below but also because of the mean vertical velocity:

$$\frac{\partial z_i}{\partial t} = w_e + W(z_i) . \tag{1.11}$$

The entrainment velocity w_e is thus the time rate of change of the PBL depth z_i in coordinates moving with the mean vertical velocity at z_i, $W(z_i)$. In steady, clear conditions in the northern hemisphere, the large-scale mean flow around a high-pressure center is clockwise and the "Ekman-pumping" mechanism causes this to generate a negative (downward) mean vertical velocity at the top of the PBL (Holton, 1972). This subsiding motion is typically of the order of a few tenths of a centimeter per second, but can be quite effective in limiting PBL deepening and cloud growth. The vertical velocity at z_i is also influenced by terrain-induced divergence of the horizontal velocity, and in cloudy air is augmented locally by the compensating subsidence resulting from the mass flux out of the CBL and into clouds.

We see, then, that CBL growth results from the competition of small-scale (turbulent entrainment) and large-scale (vertical advection) processes. Over the ocean these can be evenly matched, and the CBL depth can remain steady for days at a time. In fair weather over land, on the other hand, the surface temperature flux increases strongly in the hours after sunrise, reaches a midday peak, and then decays. During most of this time entrainment usually dominates,

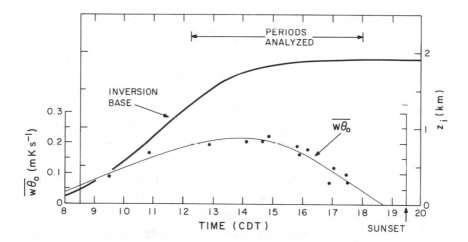

Figure 1.6. Typical daytime histories of surface heat flux and mixed-layer depth over land in clear weather during the warm season. From Kaimal et al. (1976).

deepening the CBL over the course of the day. Figure 1.6 shows the behavior observed during the 1973 Minnesota experiments (Kaimal et al., 1976).

This physical picture prompted Deardorff (1979) to propose a three-layer model of the CBL (Fig. 1.7). I reviewed this in detail earlier (Wyngaard, 1983) but briefly summarize it here.

The surface layer, nominally the lowest 10% of the CBL, is the region of M-O similarity of most (but not all) flow properties. The mixed layer extends roughly to the inversion base; a more precise definition is the crossover point of the heat flux profile (Fig. 1.7).

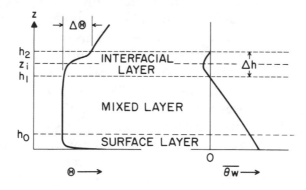

Figure 1.7. A three-layer model of the convective boundary layer. Adapted from Deardorff (1979).

The interfacial layer buffers the mixed layer from the free atmosphere, extending upwards to the level of deepest penetration of the convective eddies.

This model describes the area-averaged structure of the CBL, which is why Fig. 1.7 shows a smooth temperature profile in the interfacial layer. By contrast, individual, local temperature profiles from rawinsonde ascents often show very sharp inversions and an interfacial layer only meters thick. The convective eddies wrinkle and contort this inversion lid as they move it up and down, making the area-averaged interfacial layer fairly thick and smoothing the temperature profile. Deardorff (1979) indicated that $(h_2 - h_1)/h$ typically ranges from 0.3 to 0.6 or more.

Deardorff (1970) suggested that the parameters $g/T_0, z, z_i$, and $\overline{w\theta}_0$ alone determine the properties of turbulence in the mixed layer. In this mixed-layer scaling hypothesis the velocity scale w_* of the mixed-layer turbulence is therefore

$$w_* = \left(\frac{g}{T_0}\overline{w\theta}_0 z_i\right)^{1/3}. \tag{1.12}$$

As with M-O similarity, the additional property needed for scalars other than temperature (e.g., trace constituents) is taken to be the surface flux C_0. Thus, the scale for mixed-layer fluctuations of a scalar constituent c is $c_* = C_0/w_*$; one must be careful not to confuse it with the analogous M-O scale, which shares the same symbol.

The eddy integral scale is hypothesized to be of order z_i, so that the "usual" eddy diffusivity K (i.e., the negative of the turbulent flux/mean gradient ratio for the horizontally homogeneous vertical diffusion of temperature, moisture, trace constituents, or momentum through the PBL) should scale with $w_* z_i$. This is the meaning of the eddy diffusivity I use in this chapter; later authors will introduce a different one that applies to the special case of diffusion from discrete sources such as effluent stacks. Csanady (1973) has discussed in detail the differences between these eddy diffusivities.

In general, the mixed layer has vertical fluxes of horizontal momentum, temperature, moisture, and scalar constituents throughout its depth. These are turbulent fluxes (i.e., covariances of fluctuating vertical velocity and the other quantity) and originate from interactions with the surface and with the free atmosphere; thus, the fluxes can be nonzero at both the bottom and top. Figure 1.8 shows the

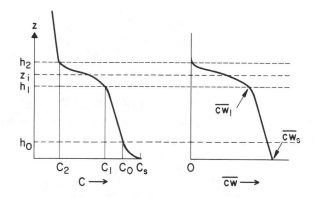

Figure 1.8. A schematic of the profiles of the mean (left) and turbulent flux (right) of a conservative scalar in the convective boundary layer.

mean value and turbulent flux of a conservative scalar (potential temperature or water vapor mixing ratio, for example).

The mean momentum balance takes a special form in the steady, horizontally homogeneous mixed layer. It is simplest here to align the axes with the mixed-layer mean wind, so that $V = 0$ (Fig. 1.9). If the flow is barotropic (i.e., U_g and V_g are independent of z) then Eq. (1.1) and Eq. (1.2) become

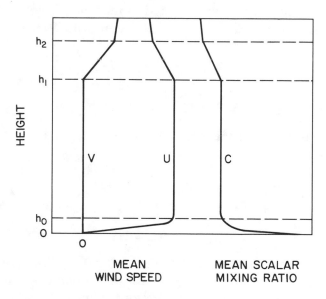

Figure 1.9. A schematic of the mean wind and scalar mixing-ratio profiles in the convective boundary layer. Axes are such that $V = 0$ in the mixed layer.

$$\frac{\partial}{\partial z}\tau_{xz} = -fV_g \tag{1.13}$$

$$\frac{\partial}{\partial z}\tau_{yz} = f(U_g - U) . \tag{1.14}$$

Since the right sides are independent of z, the left sides are as well; thus, the friction terms in the mixed layer are (replacing h_1 with the nearly equivalent and more familiar height z_i)

$$\frac{\partial \tau_{xz}}{\partial z} \cong \frac{\tau_{xz}(z_i) - \tau_{xz}(h_0)}{z_i} \tag{1.15}$$

$$\frac{\partial \tau_{yz}}{\partial z} \cong \frac{\tau_{yz}(z_i) - \tau_{yz}(h_0)}{z_i} , \tag{1.16}$$

where we take $h_0 \ll h_1$ (Fig. 1.9). At h_0 we use the usual "drag-law" assumption that the stress and wind are parallel. We take the stress at h_0 as the surface stress, so we write

$$\tau_{xz}(h_0) = -u_*^2 \tag{1.17}$$

$$\tau_{yz}(h_0) = 0 . \tag{1.18}$$

The stress at mixed-layer top is related to the velocity jump through the entrainment expression (Lilly, 1968; Deardorff, 1973b)

$$\tau_{xz}(z_i) = -w_e(U_g - U) \tag{1.19}$$

$$\tau_{yz}(z_i) = -w_e(V_g - V) = -w_e V_g , \tag{1.20}$$

where w_e (Eq. 1.11) is the entrainment velocity. The momentum balance then becomes

$$-w_e(U_g - U) + u_*^2 = -fz_i V_g \tag{1.21}$$

$$-w_e V_g = fz_i(U_g - U) . \tag{1.22}$$

If we denote by e the small parameter $w_e/(fz_i)$ (typical values are $w_e = 0.01$ m s^{-1}, $f = 10^{-4}$ s^{-1}, $z_i = 10^3$ m, so that $e = 10^{-1}$) the solution to Eqs. (1.21)–(1.22) is, to first order in e,

$$\frac{U_g - U}{u_*} = \frac{u_*}{fz_i}e \tag{1.23}$$

$$\frac{(V - V_g)}{u_*} = \frac{u_*}{fz_i}. \tag{1.24}$$

If e is negligible, Eq. (1.21) and Eq. (1.22) show that in this idealized mixed layer $U_g = U$ and $V_g = -u_*^2/fz_i$. If $U = 10$ m s^{-1}, $u_* = 0.5$ m s^{-1}, $f = 10^{-4}$ s^{-1} and $z_i = 10^3$ m we have $U_g = 10$ m s^{-1}, $V_g = -2.5$ m s^{-1}, so that the wind turns about 14° in the interfacial layer as it becomes geostrophic (Fig. 1.9).

Using this mixed-layer model, Garratt et al. (1982) derived general expressions for the mean momentum balance for the general case with time changes, mean advection, and baroclinity (i.e., z-dependent U_g and V_g) and found that their results agreed well with observations. The full expressions are

$$(\overline{U}_g - U)/u_* = \frac{e}{1 + e^2}\left(\frac{u_*}{fz_i}\right)$$
$$- \left(\frac{0.5e^2}{1 + e^2}M_x + \frac{0.5e}{1 + e^2}M_y\right) \tag{1.25}$$
$$+ \frac{1}{fu_*}\left(\frac{e}{1 + e^2}\frac{DU}{Dt} + \frac{1}{1 + e^2}\frac{DV}{Dt}\right),$$

$$\overline{V}_g/u_* = -\frac{u_*}{fz_i} + \frac{e^2}{1 + e^2}\left(\frac{u_*}{fz_i}\right)$$
$$- \left(\frac{0.5e^2}{1 + e^2}M_y - \frac{0.5e}{1 + e^2}M_x\right) \tag{1.26}$$
$$+ \frac{1}{fu_*}\left(\frac{e}{1 + e^2}\frac{DV}{Dt} - \frac{1}{1 + e^2}\frac{DU}{Dt}\right),$$

where

$$M_x = \frac{z_i}{u_*}\frac{\partial U_g}{\partial z} \cong \frac{-z_i}{u_*}\frac{g}{fT_0}\frac{\partial T}{\partial y}$$
$$M_y = \frac{z_i}{u_*}\frac{\partial V_g}{\partial z} \cong \frac{z_i}{u_*}\frac{g}{fT_0}\frac{\partial T}{\partial x} \tag{1.27}$$

and

$$\frac{D}{Dt} = \frac{\partial}{\partial t} + U\frac{\partial}{\partial x} + V\frac{\partial}{\partial y}$$

and the overbar on U_g and V_g means the average over the mixed layer. Joffre (1985a) applied the Garratt et al. model to the advective, baroclinic PBL over the sea with good results.

One interesting feature of this solution is that since baroclinity (i.e., M_x and M_y) appears multiplied by entrainment effects (e), and since e is typically small, the effects of baroclinity are relatively weak. Thus, although a horizontal gradient of mean temperature of 3 K per 100 km (not unusual in coastal areas) creates a geostrophic wind shear of 10 m s^{-1} per kilometer, according to Eq. (1.27), the results Eq. (1.25) and Eq. (1.26) indicate this does not have strong effects on the mean momentum balance if e is of the order of 0.1, say. By contrast, some mean advection effects are felt directly, according to Eq. (1.25) and Eq. (1.26), and can easily influence the mean momentum balance quite strongly.

One might now ask: Does our shear-free model (Fig. 1.9) of the mixed layer remain valid under baroclinic conditions? Not always, say the observations (Fig. 1.10). Unfortunately, our knowledge of PBL physics does not yet allow us to calculate the wind profile from first principles, but let us speculate a bit about how it behaves.

If the mixed-layer wind profile is to remain flat under baroclinic conditions, the mean momentum balance [Eqs. (1.1)–(1.2)] requires that the turbulent friction vary with height to balance the variation in the pressure gradient (i.e., in the geostrophic wind). If the geostrophic wind varies linearly with height, the stress profile must be parabolic, which means that the mid-layer stress can differ substantially from its barotropic value. A K-closure implies that wind shear is needed to support this additional stress, and, thus, that the wind profile can only be approximately flat. In this way I estimated (Wyngaard, 1985) the wind shear in the mixed layer as

$$\frac{\partial U}{\partial z} \simeq \left(\frac{1}{m^2 + 1}\right)\frac{\partial U_g}{\partial z} + \left(\frac{m}{m^2 + 1}\right)\frac{\partial V_g}{\partial z} \qquad (1.28)$$

$$\frac{\partial V}{\partial z} \simeq -\left(\frac{m}{m^2 + 1}\right)\frac{\partial U_g}{\partial z} + \left(\frac{1}{m^2 + 1}\right)\frac{\partial V_g}{\partial z} . \qquad (1.29)$$

This indicates that the wind shear decreases as the parameter $m = fz_i/w_*$ decreases. Whether a wind profile is "well-mixed" is a subjective issue, of course, but perhaps most observers would agree that it depends on an index such as the wind change across the layer di-

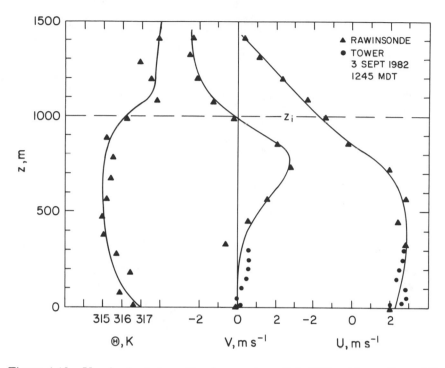

Figure 1.10. Unmixed wind profiles from a baroclinic PBL with $-z_i/L \sim 250$, measured at the Boulder Atmospheric Observatory. From Wyngaard (1985).

vided by the mean wind averaged over the layer. If so, a CBL with given turbulence propertics and given baroclinity would be judged "unmixed" if the mean wind speed were sufficiently small, and "well-mixed" if it were sufficiently large.

Turbulence structure

Progress in understanding turbulence structure in the CBL has been rapid since Deardorff's (1970) introduction of mixed-layer scaling. Using w_* as a velocity scale and z_i as a length scale does collapse data on velocity variances and their spectra, variance and covariance budgets, and many aspects of scalar statistics (Panofsky and Dutton, 1984; Caughey, 1982; Joffre, 1985b). Figure 1.11, for example, shows a collection of vertical velocity fluctuation data from the AMTEX and GATE experiments (Lenschow et al., 1980; Nicholls and LeMone, 1980), and from a laboratory convection tank (Willis and Deardorff, 1974; Deardorff and Willis, 1985). Mixed-layer scaling

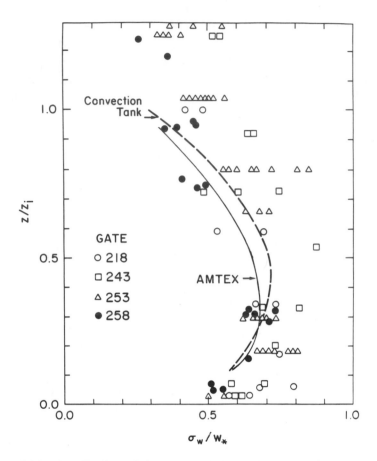

Figure 1.11. A collection of data on vertical velocity fluctuations in convective mixed layers. From Nicholls and LeMone (1980), with AMTEX data (Lenschow et al., 1980) added. Mixed-layer scaling collapses data from these disparate sources fairly well.

has brought these disparate data together, even though the convective mixed layers in these cases have the real-world complications of advection, baroclinity, and growth by entrainment. Simple parameterizations of the horizontal and vertical velocity variances useful in dispersion applications are discussed in Chs. 2 and 4.

At the root of this success is the fact that convective eddies are often "fast" compared with the time scale of CBL evolution. Hence, we can often treat the CBL as quasi-steady, in a state of "moving equilibrium" as its depth evolves with time and its surface heat flux rises to its midday peak and falls again (Fig. 1.6). Similar arguments hold for the effects of sufficiently slow changes in surface or upper-air properties. For this reason, observations and numerical simulations of steady, horizontally homogeneous CBLs are often justifiably and usefully applied to real-world cases.

Nonetheless, mixed-layer scaling has limitations. Principal among these is its neglect of properties at mixed-layer top in the list of governing parameters. This is a particularly poor assumption for the statistics of scalars (e.g., variances of temperature and water vapor). As we saw earlier, mixed-layer scaling implies that their fluctuations scale with surface scalar flux, but data make it clear (Deardorff, 1974; Caughey, 1982) that they also can be generated by the entrainment process at mixed-layer top. Figure 1.12 shows the spectacular failure of mixed-layer scaling for water vapor fluctuations.

In an attempt to remedy this deficiency, Moeng and Wyngaard (1984) developed an extension of mixed-layer scaling for conservative scalars. They used the Wyngaard-Brost (1984) decomposition of the fluctuating scalar field in the mixed layer into "bottom-up" and "top-down" parts,

$$c = c_b + c_t . \tag{1.30}$$

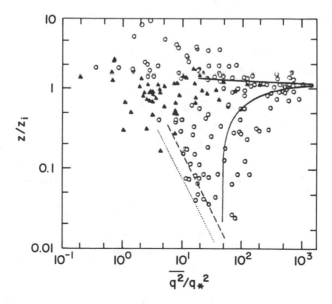

Figure 1.12. Mixed-layer scaling fails for humidity fluctuations because their source is often the entrainment process at PBL top. From Tuzet (1982).

They assumed that c_b scales with the usual scale (i.e., C_0/w_*), but that c_t scales with C_e/w_*, where C_e is the entrainment-induced scalar flux at mixed-layer top. Using large-eddy simulation (fine-mesh, time-dependent, three-dimensional numerical modeling) they explored the properties of these entrainment-induced scalar fluctuations.

It seems likely that velocity statistics within the mixed layer also depend on properties at mixed-layer top, including the structure of the interfacial layer. Little is known about these dependencies, however.

Some implications for dispersion

Let's now discuss some properties of the CBL that are relevant to diffusion applications. First, its large eddies are vigorous and highly diffusive, and yet slow when scaled by the averaging times we would like to use in field measurements. If we take typical midsummer values over land as $\overline{w\theta}_0 = 0.2$ m s^{-1} K and $z_i = 1500$ m, we find $w_* = 2$ m s^{-1}. If we estimate an eddy diffusivity as $0.1w_*z_i$ (Lamb and Durran, 1978), we find $K = 300$ m^2 s^{-1}, more than 10^7 times the molecular value. However, the "eddy turnover time" z_i/w_* is nearly 15 minutes, and the "eddy passage time" z_i/U will not be much less in light winds. If we need to average for, say, 10 eddy turnover or eddy passage times in order to obtain statistically stable mean concentrations in a diffusion problem, then we must average for a few hours. This is seldom practical, and as a result diffusion measurements in the CBL are often plagued by scatter. Put another way, concentration fluctuations can be very important in CBL diffusion problems! This has strong implications for field experiments and model evaluation, as we shall see in later chapters.

These large, convective eddies extend into the surface layer, where they have purely horizontal motions. Thus, the horizontal wind fluctuations in the unstable surface layer should scale in part with w_*, and as a result σ_u and σ_v should not follow M-O similarity. Fig. 1.13 shows the behavior of σ_v and σ_w observed at 5.7, 11.3 and 22.6 m height in the 1968 Kansas experiments. The only hint of the failure of M-O similarity for σ_v is that its plot has much more scatter than that for σ_w. If we replot the σ_v data we find more evidence, however. The top panel of Fig. 1.14 suggests that within a given CBL σ_v does not depend on height, but does depend on L. Although the CBL depth z_i was not measured in the Kansas experiments, we can presume it did not vary greatly for these runs since

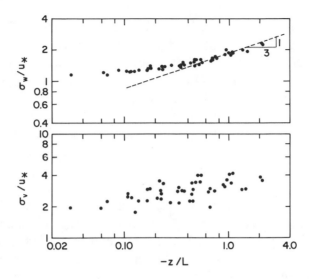

Figure 1.13. The standard deviation of vertical (upper panel) and lateral (lower panel) wind fluctuations at 5.7, 11.3, and 22.6 m height in the convective surface layer of the 1968 Kansas experiments. The scatter in the σ_v plot is due to the failure of M-O similarity, but σ_w seems to follow it quite well, even showing indications of the $(-z/L)^{1/3}$ regime predicted by "local free convection" scaling (Wyngaard et al., 1971).

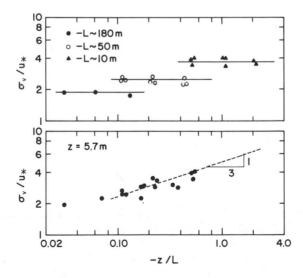

Figure 1.14. Replots of some of the data of Fig. 1.13. The upper panel, based on data from the three heights and three different, narrow stability ranges, suggests that σ_v depends on $-L$ but not on z. The lower panel shows that at a given z (here 5.7 m) σ_v may seem to have a "local free convection" regime like that of σ_w (Fig. 1.13). However, this is more likely due to the influence of w_*.

all were made in the afternoon under similar, fair-weather conditions. If in the very unstable surface layer $\sigma_v = Aw_*$, where A is a constant, then one can show from the definitions of L and w_* that $\sigma_v/u_* = C(-z/L)^{1/3}$ where $C = (A^3 z_i k^{-1} z^{-1})^{1/3}$. If we take $A = 0.5$ (Willis and Deardorff, 1974; Lenschow et al., 1980), $z_i = 1500$ m, $k = 0.35$ and $z = 5.7$ m, we find $\sigma_v/u_* = 4.5(-z/L)^{1/3}$. This does describe the behavior of the very unstable Kansas data from 5.7 m height (Fig. 1.14). Thus, σ_v data from a single height can erroneously give the appearance of following M-O similarity.

This sharply contrasting behavior of the horizontal and vertical velocity fluctuations is demonstrated dramatically in Fig. 1.15, which shows time histories of near-surface velocity variances measured in a growing CBL by Banta (1985). The boundary layer broke through the morning inversion and became an order of magnitude deeper between 1100 and 1200 local time. Note that the horizontal variances

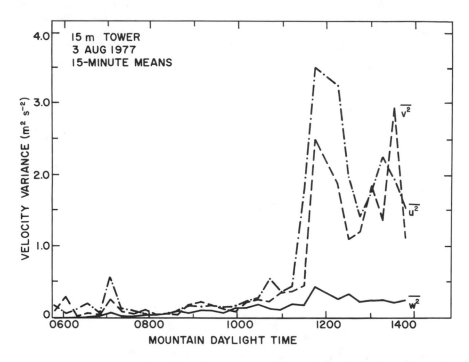

Figure 1.15. Time histories of the variance of horizontal and vertical wind fluctuations measured at 15 m height by Banta (1985). Shortly after 1100 the mixed layer began a period of rapid growth, which increased its depth by about an order of magnitude. The horizontal wind fluctuations increased as predicted by mixed-layer scaling, but the vertical ones, being M-O similar, did not.

increased strongly (their change scaled with the change in $w_*^2 \sim z_i^{2/3}$, as expected) but the vertical variance, being M-O similar and hence not depending on z_i, changed only slightly.

It has taken a number of years for this simple model of the behavior of the horizontal wind fluctuations in the unstable surface layer to emerge. In their 1964 monograph, Lumley and Panofsky wrote on p. 147:

> "...it might be tempting to follow similarity theory and write $\sigma_v = Cu_* f(z/L)$ for diabatic conditions. However, this formulation is plainly incorrect. For, whereas σ_v is remarkably insensitive to changes in height...this quantity is so sensitive to changes in stability that, for example, Cramer (1959) has used $\sigma_A = \sigma_v/V$ as predictor for other characteristics of turbulence...."

The top panel of Fig. 1.14 demonstrates clearly the insensitivity of σ_v to changes in height and its sensitivity to changes in stability (i.e., to changes in L). We can also see that the authors were getting close to the idea that σ_v scales with w_*. Calder (1966), in his attempt to put M-O similarity of mean wind and temperature gradients on a solid theoretical foundation, argued that it should not hold for horizontal velocity fluctuations; however, in their rebuttal, Monin and Obukhov (see Calder, 1968) disputed this. Busch (1973), citing new data, repeated the claim that horizontal wind fluctuations are not M-O similar. Shortly after that, Wyngaard and Coté (1974), citing Deardorff's LES studies, his convection tank simulations with Willis, and the 1968 Kansas data, suggested w_* scaling for the horizontal velocity fluctuations in the very unstable surface layer. Panofsky et al. (1977) showed that it indeed did account for the behavior of σ_u and σ_v in several experiments, and they proposed interpolation formulas with u_* scaling in the neutral limit and w_* scaling in the convective limit. These are presented as Eqs. (4.1) and (4.2) in Ch. 4, and explained in detail there.

The vertical velocity field also has distinctive features. The rate of buoyant production of turbulence (which, according to the budget of turbulent kinetic energy, is proportional to the vertical heat flux) in the convective PBL is not uniformly distributed in the vertical (Fig. 1.7). Cloud processes, including cloud-top cooling, can modify the heat flux profile but in general the buoyant production is largest near the lower surface. As a result, the vertical velocity fluctuations

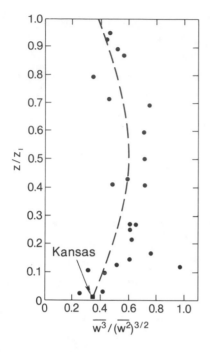

Figure 1.16.   Profile of the skewness of vertical velocity fluctuations measured in the 1973 Minnesota experiment.

are skewed (Fig. 1.16). This strongly affects plume dispersion; one example is the elevated locus of maximum concentration downwind of a surface source of nonbuoyant material, which invalidates the Gaussian vertical profile assumption. In Ch. 2 Briggs discusses these effects in detail.

Clearly, the vertical velocity field in the CBL has profound effects on its diffusivity properties, and some of these properties might seem quite unusual. Wyngaard and Brost (1984), for example, found that the diffusivities for horizontally homogeneous scalar fields entering the PBL through the surface and through the top (i.e., the "bottom-up" and "top-down" diffusivities) are substantially different. This means that the concept of eddy diffusivity is not strictly applicable to the CBL.

The results just cited also imply that Smith's "Reciprocal Theorem" (1957) is invalid. This theorem states that

> The concentration at x' due to a source at x'', with the flow in the positive x_1 direction, is equal to the concentration at x'' due to an identical source at x' when the direction of the flow is reversed.

Earlier workers used this theorem to calculate the groundlevel effects of dispersion from elevated sources, since it says they are equivalent to the effects at that height from surface sources. We now know that this is not generally true.

1.4.3. The Stable PBL

Transition

Over land, the CBL begins to decay in midafternoon to late afternoon, when the decreasing angle of the sun has reduced the surface heat flux to a small fraction of its midday peak. Their energy supply severely restricted, the kinetic energy of the large eddies, and the fluxes they carry, fall off quickly. The surface stress and wind speed decrease (Fig. 1.17). If skies are clear radiative cooling will set in at the surface, driving the turbulent heat flux negative shortly before sunset.

Several field studies have shown the qualitative features of CBL decay, but none has yet documented it in detail. Perhaps the most informative work to date is that of Nieuwstadt and Brost (1986), who used large-eddy simulation to study the CBL response to the sudden removal of surface heat flux. They found that the time scale of turbulent kinetic energy decay is z_i/w_*, as one would expect, but

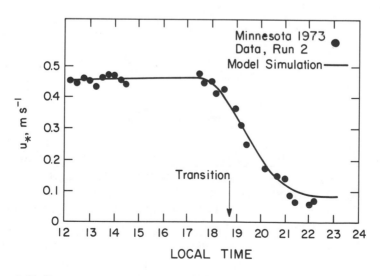

Figure 1.17. The late-afternoon decay of friction velocity u_* in the 1973 Minnesota experiments. From Wyngaard (1975).

that the decay begins only about one z_i/w_* time after the heat flux shutdown. Temperature variance decay, by contrast, begins almost immediately.

The transition to a negative surface heat flux triggers profound changes in the structure of the overlying PBL, as the study by Caughey et al. (1979) shows. The negative heat flux propagates upward through the diffusive action of the turbulence, but at the same time this negative flux extracts kinetic energy from the turbulence. Since the largest eddies are affected the most, the surviving turbulence is of smaller scale than turbulence in the unstable or neutral surface layer.

Meanwhile, the convective turbulence aloft decays, and the virtually shear-free wind profile prevents subsequent regeneration of turbulence by shear production, at least initially. Above the growing, ground-based inversion layer, therefore, the flow soon approaches laminar. The turbulent flux divergence terms in the mean-field conservation equations fall to zero there, throwing these equations out of balance. As a result, the mean fields begin to evolve with time.

Let us use the mean momentum equation to illustrate this initial evolution of the wind field aloft. After transition the turbulent friction quickly decays, but let us assume that the horizontal pressure gradient remains steady so that the wind field does not change appreciably during this time. The wind acceleration is then due to the imbalance between Coriolis and pressure-gradient forces; from our afternoon solution Eq. (1.23) and Eq. (1.24) this gives initially

$$\frac{\partial U}{\partial t} \simeq f(V - V_g) \simeq \frac{u_*^2}{z_i} \tag{1.31}$$

$$\frac{\partial V}{\partial t} \simeq f(U_g - U) \simeq \frac{u_*^2}{z_i}e \; , \tag{1.32}$$

where e is the entrainment parameter $w_e/(fz_i)$, which is usually small. Thus, U begins to increase before V changes. Eq. (1.31) indicates that U typically changes by about 1 m s^{-1} per hour; when it exceeds U_g in magnitude, V grows negative. This inertial oscillation leads to what is called the nocturnal jet (Blackadar and Buajitti, 1957; Garratt, 1985).

We saw earlier that if the afternoon CBL has the real-world complications of baroclinity, entrainment, and advection, its geostrophic departure could differ from Eq. (1.23) and Eq. (1.24); in such cases the subsequent evolution of the nocturnal jet could vary substantially.

Speeds just above the top of the nocturnal PBL can be difficult to predict as a result, but observations show they can be supergeostrophic. They also can differ substantially in direction from the surface winds, due to the Ekman spiral. The resulting large changes in mean wind speed and direction across the nocturnal PBL can have strong effects on long-range transport of pollutants.

Within a few hours of the sign change in surface heat flux the remnants of the CBL are usually gone and the nocturnal PBL is well established. Early acoustic sounder observations (McAllister et al., 1969) showed that it was typically no more than a few hundred meters thick, the flow above that being laminar or at most only intermittently turbulent. Zilitinkevich (1972) argued on theoretical grounds that its equilibrium depth is proportional to $(u_*L/f)^{1/2}$, but this estimate is not always applicable to the evolving nocturnal PBL over real terrain, as we shall see shortly.

Structure and dynamics

By contrast with convective turbulence, the turbulence in the stable PBL is in a delicate dynamical balance; it is produced by interactions with the mean wind shear but destroyed by the downward buoyancy flux and viscous dissipation. It is also found mixed together with wave motion, some of which seems to transfer heat, according to Hunt (1985), which as he said contradicts some simple ideas about wave motion.

Progress in understanding the stable PBL has come more slowly than for its convective counterpart. There are several reasons why this is so:

(1) Much of the stable PBL is near the critical Richardson number, beyond which turbulence cannot be sustained against the loss to buoyancy and viscous dissipation; this is the "delicate dynamical balance" we just referred to. However, the unstratified, shear-driven turbulence ubiquitous in engineering flows has a balance between shear production and viscous dissipation, and there seems nothing delicate about that balance. One might ask: Why does the addition of stable stratification change things?

The answer lies in the completely different mechanisms of viscous dissipation and buoyant extraction. Viscous forces, being proportional to velocity gradients, are strongest at the smallest scales of motion (Tennekes and Lumley, 1972). In shear flows the turbulent kinetic energy input mechanism, shear production, acts at the much larger, energy-containing scales. Nonlinear, eddy-eddy interactions transfer energy to smaller scales; this "pipeline" terminates in the viscous range (of the order of 1 mm in the PBL). The magnitude of this energy transfer (and, hence, of the viscous dissipation) is determined simply by the shear production rate.

By contrast, the downward buoyancy flux in the stable PBL extracts kinetic energy from the vertical velocity field in the energy-containing range. The vertical velocity field also has a loss to viscous dissipation; in steady conditions this total loss is balanced by transfer from the streamwise fluctuations, which receive the input from shear production. This transfer is done by the fluctuating pressure field. Thus, the dynamical path here is more complex than in shear flow. Experiments in stably stratified shear flows, including the nocturnal PBL, suggest that when the buoyant destruction rate exceeds about 1/4 of the shear production rate the turbulence can no longer sustain itself.

As a result, nocturnal PBL turbulence tends to be intermittent, even within the surface layer (Kondo et al., 1978). Furthermore, key parameters (e.g., the surface energy budget, the wind aloft, the PBL depth, its mean momentum balance) evolve with time, making the nocturnal PBL less likely than the CBL to be in a quasi-steady state. These properties make the measurement and parameterization of nocturnal PBL turbulence relatively difficult.

(2) The drainage forces typically experienced by cooler fluid over even weakly sloping terrain can be quite important in the mean momentum balance. Brost and Wyngaard (1978) showed, for example, that the ratio (S) of mean drainage and turbulent friction forces is

$$S = (g/T_0)T'Bh/u_*^2 , \tag{1.33}$$

where T' is the mean temperature deficit of the cooler air, B is the terrain slope, and h is the PBL depth. Note that if $B = 0.001$ (i.e., a slope of 5 ft per mile), $T' = 3\,\text{K}, h = 100\,\text{m}$, and $u_* = 0.1\,\text{m s}^{-1}$, then $S = 1$; that is, drainage forces are equal to the turbulent frictional forces. Since the earth's surface is rarely this level, we must expect that the structure and evolution of the nocturnal PBL are sensitive to the local terrain. This makes it difficult to extract generalizations about its behavior from the results of individual experiments, which, as we saw earlier, we can do for the CBL.

(3) The nocturnal PBL is a rich medium for the growth and propagation of internal gravity waves, which can interact with the turbulence in ways that are not yet completely understood. This can complicate its structure and dynamics.

(4) In comparison with the CBL, turbulence levels in the stably stratified PBL are relatively low and integral scales are small—so that experimentalists need more sensitive, lower-noise, faster turbulence instrumentation.

Our knowledge of the CBL grew through observations in the atmosphere, in the laboratory (e.g., Willis and Deardorff, 1974) and on the computer (through large-eddy simulation); the last have been particularly valuable, yielding mixed-layer scaling, for example (Wyngaard, 1984). Furthermore, even a CBL with the complications of time evolution and inhomogeneity can give useful data, since its large, fast eddies tend to keep it near a locally homogeneous, quasi-steady state. The stable PBL has yet to yield to either laboratory or large-eddy computer simulation, to my knowledge, and the observations we have come mainly from nocturnal cases which are often evolving significantly with time.

Nonetheless, second-order closure model studies (Delage, 1974, Brost and Wyngaard, 1978; Nieuwstadt, 1984) seem to have shed a good deal of light on the internal dynamics of the nocturnal PBL. Although these models use a seemingly complex set of equations, the underlying physics can be illustrated quite simply.

If we average over an ensemble of realizations (or over a large horizontal plane) the horizontal mean momentum and potential temperature equations are

$$\frac{\partial U}{\partial t} + \frac{\partial \overline{uw}}{\partial z} = f(V - V_g) \tag{1.34}$$

$$\frac{\partial V}{\partial t} + \frac{\partial \overline{vw}}{\partial z} = f(U_g - U) \tag{1.35}$$

$$\frac{\partial \Theta}{\partial t} + \frac{\partial \overline{\theta w}}{\partial z} = 0. \tag{1.36}$$

We have ignored radiative flux divergence, which is not essential to the dynamics, at least early in the evening (Garratt and Brost, 1981). We also assume level terrain and horizontal homogeneity.

These mean equations are unsteady because their turbulent flux terms evolve strongly with time after sunset. A simplified form of the conservation equations (Wyngaard, 1983) for these fluxes is

$$\frac{\partial}{\partial t}\overline{uw} = -\overline{w^2}\frac{\partial U}{\partial z} + \frac{g}{T_0}\overline{\theta u} - \frac{\overline{u\partial p}}{\partial z} - \frac{\overline{w\partial p}}{\partial x} \qquad (1.37)$$

$$\frac{\partial}{\partial t}\overline{vw} = -\overline{w^2}\frac{\partial V}{\partial z} + \frac{g}{T_0}\overline{\theta v} - \frac{\overline{v\partial p}}{\partial z} - \frac{\overline{w\partial p}}{\partial y} \qquad (1.38)$$

$$\frac{\partial}{\partial t}\overline{\theta w} = -\overline{w^2}\frac{\partial \Theta}{\partial z} + \frac{g}{T_0}\overline{\theta^2} - \frac{\overline{\theta\partial p}}{\partial z} , \qquad (1.39)$$

where we have neglected the third-moment divergence terms, which experimental data (Wyngaard and Coté, 1971; Caughey et al., 1979) indicate are often quite small. This set, in turn, involves other second moments. This process does not converge; that is, there is no set of exact equations where the number of unknowns equals the number of equations. Thus, one needs "closures," approximations for certain unknowns so the set can be solved.

Brost and Wyngaard carried 7 second-moment equations plus the 3 mean-field equations, for a total of 10. Since that is too many to discuss in detail here, we illustrate their closure by discussing the turbulent kinetic energy equation. Again ignoring the third-moment divergence term, that equation is

$$\frac{1}{2}\frac{\partial}{\partial t}\overline{q^2} = \underbrace{-\overline{uw}\frac{\partial U}{\partial z} - \overline{vw}\frac{\partial V}{\partial z}}_{\substack{\text{shear} \\ \text{production}}} + \underbrace{\frac{g}{T}\overline{\theta w}}_{\substack{\text{buoyant} \\ \text{production}}} - \underbrace{\epsilon}_{\substack{\text{viscous} \\ \text{dissipation}}} . \qquad (1.40)$$

We have indicated the physical meaning of the terms. The shear production term would be expected (through a K-closure for $\overline{uw}$ and $\overline{vw}$) to be positive, and hence is a source term. The buoyant production term is negative in stable conditions, and thus represents destruction, or a kinetic energy sink. Viscous dissipation is always a sink. If ℓ and σ_w are the length and velocity scales of the energy-containing eddies, we can write for the order of each term

$$\text{shear production} \sim \sigma_w^2 \frac{\partial U}{\partial z} \qquad (1.41a)$$

buoyant production $\sim -K\dfrac{g}{T_0}\dfrac{\partial\Theta}{\partial z} \sim -\sigma_w\ell\dfrac{g}{T_0}\dfrac{\partial\Theta}{\partial z}$ (1.41b)

viscous dissipation $\sim \sigma_w^3/\ell,$ (1.41c)

where we have used standard turbulence scaling arguments (Tennekes and Lumley, 1972). Ignoring constants of order one, we can now write the turbulent kinetic energy equation as

$$\sigma_w^2\frac{\partial U}{\partial z} = \sigma_w\ell\frac{g}{T_0}\frac{\partial\Theta}{\partial z} + \frac{\sigma_w^3}{\ell} .$$ (1.42)

Examination of Eq. (1.42) shows that near the surface, where from M-O similarity we know that $\sigma_w \sim u_*$ and the mean wind and temperature gradients are inversely proportional to z, the buoyancy term is quite small and the shear production is balanced by viscous dissipation. This implies that $\ell \sim z$ there.

Now consider the behavior in the very stable outer region. Let the length scale there be ℓ_B, and write $(g/T_0)d\Theta/dz = N^2$, the square of the Brunt-Väisälä frequency. If we assume that the flux Richardson number, the negative of the ratio of the buoyant and shear production terms in Eq. (1.42), approaches a constant (i.e., its critical value) in this limit, then the terms in Eq. (1.42) approach a fixed ratio. Then it follows that

$$\ell_B \sim \sigma_w/N .$$ (1.43)

This means that in the limit of very stable stratification the largest eddy that can survive has a scale of the order of ℓ_B; larger eddies are extinguished by working against the buoyancy forces. Much smaller eddies, by contrast, hardly feel the direct effects of buoyancy.

The Brost-Wyngaard model uses the closure

$$1/\ell = 1/z + 1/\ell_B ,$$ (1.44)

which interpolates between these two limits. Thus, it determines its own closure length scale; given initial and boundary conditions, this allows the nocturnal PBL to evolve and determine its own depth and internal structure. One of the first findings from the model was that the Zilitinkevich depth relation could be maintained only if the cooling rate at the lower surface (assumed level) remained constant. If instead the cooling rate evolved according to a surface energy budget, the usual case over land, no equilibrium height relation was found.

One of the most complete studies to date of the structure of the nocturnal PBL is that of Nieuwstadt (1984). He analyzed a large number of runs from the Cabauw 200-m mast, but excluded those that were dominated by gravity waves. This was done by using only cases in which the vertical velocity variance decreased continuously with height. Using a derivative of the Brost-Wyngaard stably stratified turbulence closure, he produced predictions of turbulence profiles in good agreement with the Cabauw observations (Fig. 1.18). Parameterizations of σ_w based on Nieuwstadt (1984) or Caughey et al. (1979) are used in dispersion applications in Ch. 5.

These studies show that at least in quasi-steady, horizontally homogeneous conditions a relatively simple model of stably stratified turbulence dynamics accounts well for the observed behavior of the nocturnal PBL. What can we say about its behavior in more typical conditions, when it is apt to be evolving in time, for example?

Caughey et al. (1979) analyzed nocturnal PBL data taken within a few hours of transition under far from steady conditions, and found that the M-O similarity functions for the surface layer agreed well with those from earlier experiments. They also found that by filtering their data over the frequency band 0.001–5 Hz (which for a mean wind speed of 5 m s^{-1} corresponds to a streamwise wavelength band of 5000–1 m, through Taylor's hypothesis) they were able to find considerable order in the data throughout the depth of the PBL. Turbulence profiles were well behaved, decreasing monotonically from the surface to the PBL top.

Nieuwstadt (1981) analyzed nocturnal observations over flat terrain in Cabauw, and found that the PBL depth (measured by acoustic sounder) did not correlate well with the equilibrium prediction of Zilitinkevich. Nieuwstadt and Tennekes (1981) subsequently used dynamical arguments to derive a rate equation for PBL depth,

$$\frac{dh}{dt} = T^{-1}(h_e - h) , \qquad (1.45)$$

where T is a time scale dependent on the mean temperature field and h_e is the equilibrium depth of Zilitinkevich. Their data indicate that predictions based on this rate equation can give reasonable agreement with observed depths.

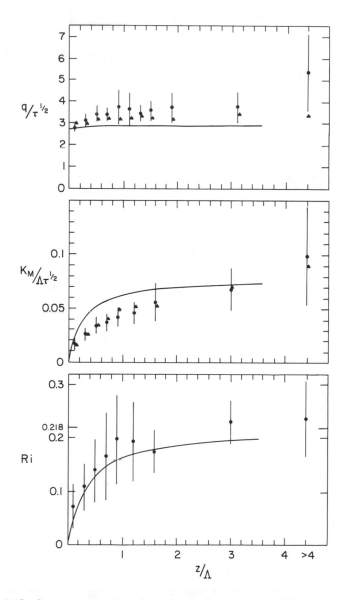

Figure 1.18. Some observed statistics from the nocturnal PBL (data points) and theoretical predictions (solid curves) from Nieuwstadt (1984). Solid triangles are high-pass filtered data (cutoff 0.1 Hz); solid circles are unfiltered; vertical bars are $\pm 1\sigma$ for the data, which are averaged over stability ranges. Abscissa is z/Λ, where Λ is a local M-O length. Upper panel: rms turbulent kinetic energy, scaled with the local friction velocity; middle: dimensionless eddy diffusivity; lower: Richardson number.

As we mentioned, Brost and Wyngaard (1978) found that the nocturnal PBL is extremely sensitive to terrain slope. In their simulations, as the wind direction varied, the ratio of equilibrium PBL depth to the Zilitinkevich depth scale varied by a factor of 4 for a slope of 0.002 and a surface cooling rate of 2 K h^{-1}. This is simply the result of the downslope force on the cooler, denser air near the surface. Caughey et al. (1979) found that the early-evening evolution of the PBL in the 1973 Minnesota experiments was also clearly influenced by the terrain slope, which was only 0.0014. They showed that the drainage forces evidently contributed to the "backward" Ekman spiral observed in some of the runs.

This suggests that in the general case with sloping terrain and time changes, we probably cannot reliably "diagnose" the structure of the nocturnal PBL simply from knowledge of gross conditions, as we can do for the CBL. Both the mean and turbulent structure tend to evolve continuously in response to the delicately balanced turbulence dynamics and the inevitably changing boundary conditions. Thus, a simple, diagnostic similarity framework for the *structure* of the nocturnal PBL seems not to exist. What we have described instead is a relatively simple model of its *dynamics*, a set of rate equations for its mean structure plus a simple turbulence closure, also in the form of rate equations. The solution of this set gives the structure, which then depends on the history of the flow and boundary conditions.

One meets other complications in the nocturnal PBL. For example, Hunt (1985) claimed that it was only the high-pass filtering that allowed Caughey et al. (1979) to find structure in the Minnesota data. The discarded data at lower frequencies are typically rich in wave motion, he pointed out, and he contended that these frequencies are important for heat and mass transfer. In a later study at the Cabauw mast, de Baas and Driedonks (1985) showed that the measured velocity and temperature statistics in the nocturnal PBL did have identifiable contributions from wave motions. They carried out a linear stability analysis on the measured mean wind and temperature profiles, calculating the properties of the most unstable disturbances. The profiles of vertical velocity and temperature statistics at low frequencies were explained well by their calculated vertical structure of the unstable Kelvin-Helmholtz waves.

Finally, we probably should not expect the horizontal velocity fluctuations in the surface layer over typical terrain to be well behaved, e.g., to follow M-O similarity. As later chapters will indicate, the evidence suggests that the strong sensitivity of the stably stratified PBL to downslope drainage forces, and to gravity wave motions, makes the horizontal wind fluctuations site dependent. Unfortunately this limits the universality of the statements we can make about diffusion in the stable surface layer, as we shall see later.

1.5. Surface Effects

The interactions between the air and the underlying surface are the dominant influence on the PBL. Even over a homogeneous surface these interactions control most of the parameters that govern PBL structure.

1.5.1. The Energy Budget

We have seen that the surface heat flux usually determines the stability state of the PBL [one exception is the cloud-topped PBL, whose convection can be driven by cloud-top cooling (Lilly, 1968); another is a "neutral" PBL (i.e., one with no surface heat transfer) made stably stratified by entrainment from above]. This heat flux, in turn, is closely tied to the surface energy budget, which is a first-law energy balance applied to the surface-air interface. Surface properties that are important in the energy budget include albedo, emissivity, roughness, conductivity, heat capacity, permeability, and vegetative resistance to water losses through evapotranspiration. As that list suggests, the physics of the surface energy budget could fill a monograph.

The control the surface energy budget exerts over the stability of the PBL can have a strong effect on dispersion. A cover of snow can greatly reduce the daytime surface heat flux, preventing the PBL from becoming more than weakly convective even under clear skies. This snow cover can have an equally dramatic effect on the nocturnal PBL, according to the model study of Brost and Wyngaard (1978). They found that its low conductivity and heat capacity caused the surface temperature to fall sharply, so stabilizing the PBL that it virtually collapsed within a few hours, leaving laminar, nondiffusive flow. By contrast, a high-conductivity, high-heat-capacity surface such as rock cooled at a more nearly constant rate and generated a quasi-steady, turbulent PBL.

Real land surfaces inevitably have spatial variations in their properties, and hence in their surface energy budget. Key contributors here are natural variations in terrain slope, soil composition and vegetative cover, and of course the man-made effects of farming and urbanization. As a result, surface temperature typically is spatially nonuniform. The recent study of Hechtel and Stull (1985) shows that these temperature variations can be substantial over agricultural land. Figure 1.19 shows two of their surface temperature records from flights over Oklahoma in 1983; note that variations of several kelvins occur between adjacent fields. Hechtel and Stull suspected that these temperature variations are the cause of the observed variations in the lifting condensation level and the height of cloudbase.

The finding that apparently homogeneous terrain can have strong variations in surface temperature probably has implications for PBL structure. These have yet to be explored in any detail, it seems, but let us speculate about some possible effects.

For simplicity imagine that the surface temperature varies sinusoidally with wavelength λ along the mean wind direction, and is

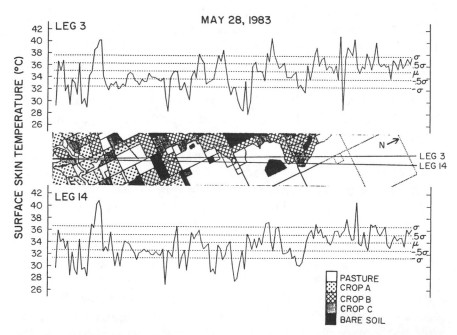

Figure 1.19. Surface temperature records (and the surface pattern) from flights near Chickasa, Oklahoma, at an altitude of about 50 m above ground level. On the temperature trace for each flight leg, the leg average (μ), the standard deviation (σ), and one-half the standard deviation (0.5σ) are indicated with dashed lines. The large boxes marked out in a north-south direction are 1-mile-wide section lines. From Hechtel and Stull (1985).

uniform in the lateral direction. The mean air temperature Θ then adjusts according to a balance between mean advection and flux divergence:

$$U\frac{\partial \Theta}{\partial x} + \frac{\partial}{\partial z}\overline{w\theta} = 0 \ . \tag{1.46}$$

If the time required to advect the PBL air through one cycle of the surface-temperature wave is short compared with the time required to diffuse the wave to the top of the PBL, the temperature wave will not penetrate far into the PBL. A sufficiently-long-wavelength temperature variation, on the other hand, will be felt throughout the PBL.

An eddy-diffusivity closure allows one to express this mathematically by solving Eq. (1.46). One finds that the important parameter is $r = Uz_i^2/(K\lambda)$, the ratio of advection and diffusion time scales. When r is large the air temperature response extends to the top of the PBL and has the same amplitude throughout. Through this mechanism, long-wavelength variations in surface temperature can generate pressure variations extending to the top of the PBL. The resulting pressure gradients would be expected to generate flow circulations of that wavelength, thereby adding "standing" eddy structures not present over a uniform-temperature surface. These eddies could be an important influence on dispersion.

1.5.2. The Marine PBL

Recent reviews of the marine PBL by LeMone (1980), Businger (1985), and Joffre (1985b) discuss the similarities and differences between the marine and continental PBLs, and I shall briefly highlight some of these aspects here.

The marine PBL is typically quite moist and very nearly neutrally stratified. In part because of the large heat capacity of water, its diurnal changes are usually small. Its lifting condensation level is relatively low; it is typically 550–700 m over the tropical ocean, for example, perhaps half the usual value over land. If the air-sea temperature difference is unusually large, which occurs in cold air outbreaks, much of the buoyancy is due to the water vapor flux and as a result the PBL tends to be cloud topped. Such cases were studied in the AMTEX experiments (Lenschow et al., 1980).

Much of the thermal energy transfer at the sea surface takes the form of latent heat flux, rather than sensible heat flux; the opposite situation usually prevails over land. The ratio of sensible to latent heat fluxes is called the Bowen ratio.

The entrainment process at the upper edge of a cloud-topped PBL brings in dry air, and if during mixing and evaporation a parcel's density increases, it is unstable to further downward movement. Deardorff (1980) referred to this as "cloud-top entrainment instability" and showed that it occurs when the jump in equivalent potential temperature (which decreases through cloud top) exceeds about 1 K to 3 K in magnitude. (Equivalent potential temperature is essentially the potential temperature that dry air would need in order to have the same enthalpy as a given air-water vapor mixture). Deardorff showed that the entrainment rate increases decisively when the jump exceeds this threshold, triggering processes that can lead to breakup of the stratocumulus. Rogers et al. (1985) showed how this can occur when a marine PBL passes over an increasingly warmer ocean.

The sea surface interacts with the air above, causing the roughness length and drag coefficient to increase with wind speed. The details of this dependence are difficult to determine experimentally, but the consensus is that the drag coefficient (the ratio of u_*^2 and the square of the 10-m mean wind speed) varies from about 0.001 at 5 m s^{-1} to about 0.002 at 25 m s^{-1} (Joffre, 1985b). Surface temperature variations, although usually smaller than over land, can be quite important. Joffre (1985b) cited evidence that frontal structure and mesoscale eddy motion can produce a patchy distribution of surface temperature, with local horizontal temperature differences of 1 K over 30 km. Larger gradients can occur in enclosed seas, in the vicinity of currents or upwelling areas, and in coastal regions.

Given the same synoptic conditions, therefore, the structural differences between the continental and marine PBLs will make their dispersive properties different.

1.5.3. The Urban PBL

Air flowing over an urban area encounters strong changes in its lower boundary conditions. The surface elevation, roughness, and radiative, thermal, and moisture properties are not only changed from the rural environment, but are also apt to vary spatially and temporally within the city. As a result the urban PBL is apt to be inhomogeneous, and even in its homogeneous limit is apt to be structurally different from the rural case.

Hildebrand and Ackerman (1984) used PBL data taken from research flights over St. Louis to document some of the differences in turbulence structure in the daytime urban PBL. They found, for example, that the vertical heat flux was 2–4 times the rural value. The inversion base z_i was slightly greater, meaning that the convective velocity scale w_* (and therefore the diffusivity) was 25–60% larger. The pressure perturbations created by the warmer urban PBL air create an urban circulation, the maximum updraft in the St. Louis data being 0.1 m s^{-1}. Such circulations can be very important in dispersion applications.

Using data from the same study, Ching (1985) documented some of the spatial variability of the St. Louis PBL. He showed that the Bowen ratio ranged from a maximum of about 1.5 over the city to less than 0.2 in surrounding nonurban areas; other turbulence properties varied significantly, if less dramatically. Ching stressed the difficulty of making representative measurements from towers in an urban environment.

1.6. Modeling and Measurements

In many ways the most difficult part of air quality modeling is comparing predictions and observations. Let's discuss some of the reasons for this.

An air quality model predicts dispersion, using as inputs the atmospheric conditions and the characteristics of the site. We can never specify the state of the atmosphere completely, however; the best we can typically do is give PBL depth, a measure of stability, and perhaps the mean wind profile. The crudeness of these descriptors means that any number of different realizations of the atmosphere will, in principle, be judged to be identical. As a result, apparently identical atmospheric conditions can produce different dispersion patterns. This is a very confusing state of affairs.

Let's revise what we just wrote, then, and say that an air quality model predicts the "most likely" dispersion—that is, the dispersion that we would obtain if we averaged a large number of realizations of a given dispersion problem, each realization having the same meteorology as judged by our crude descriptors but differing in the details of its eddies. The dispersion will vary from realization to realization, of course, because of these detail differences in the meteorology.

We conclude that to evaluate a dispersion model we should compare its predictions with ensemble-averaged observations. A "perfect" model will agree only with the observed dispersion averaged over the ensemble; it will invariably differ from the dispersion observed in individual realizations.

The key to model evaluation, then, is generating a large ensemble of realizations of the given problem. In a statistically stationary case we can do this by generating a long time series, and doing time averaging; this is often possible in laboratory experiments. By contrast, the atmosphere is often nonstationary over the time scales of interest in many dispersion applications, making it all but impossible to approximate the ensemble average with a time average. This means that evaluating a dispersion model in the atmosphere invariably requires doing an experiment many times, averaging each realization over no more than a few hours to avoid nonstationarity effects. This is usually very expensive, unfortunately, and as a result we have very little atmospheric data with which to do proper model evaluations.

Nonetheless, over the past decade there have been many attempts to verify simple diffusion models with data bases. Such exercises can give startling results, with plots of (time-averaged) observations vs. predictions showing much scatter. The first reaction has been to conclude that the model is poor, but more recently the community has come to suspect that some of the scatter could be due to "inherent uncertainty"—the inevitable difference between the ensemble-averaged behavior and the behavior in any given realization. Hanna (1982), for example, found roughly a factor-of-2 scatter in the hourly averaged SO_2 concentrations measured in the St. Louis study, due primarily to the inherent uncertainty in the 1-h averages.

There are averaging techniques to minimize the inherent uncertainty in PBL data (Wyngaard, 1986), but in general it remains a pressing issue in dispersion applications (Fox, 1984). There is a growing feeling that air-quality models should predict the inherent uncertainty as well as the mean dispersion, but as yet I do not sense much movement in this direction. Later chapters will cover this important topic in more detail.

1.7. Dealing With Reality

A reader might complain, "Your models seem highly idealized; the wind and turbulence fields that I meet in many of my dispersion problems seem much more complex than anything you've described. Furthermore, in my real world I often must do without the sophisticated flux measurements needed as inputs for your similarity theories of turbulence and dispersion. How do I deal with this?" Let's broaden our discussion in order to examine some of these issues.

1.7.1. Indirect Measurements of Fluxes

Not long ago, direct measurements of turbulent fluxes in the surface layer were a central research goal in micrometeorology. Dramatic improvements in sensor and data-handling technology have since made flux measurements routine in research. Nonetheless, they are still not practical in many dispersion applications.

This is not all bad, however. The last 20 years' experience in micrometeorology has shown us that while flux measurements are possible, they are notoriously difficult. At the conclusion of the 1968 Kansas experiments the investigators mounted their three 3-component sonic anemometers at the same height (5.6 m) less than 1 m apart in order to do a 1-h comparison of outputs. They were startled to find that while the means and standard deviations of the wind components agreed quite well, the covariance $\overline{uw}$ did not; the 1-h averages were 1.50, 0.93, and 0.57 times the overall mean. They traced the problem to small (of the order of 1 degree) errors in instrument alignment (Kaimal and Haugen, 1969). Subsequent work has shown that flow distortion by the sensor array itself can cause comparably large errors in measurements of fluxes of momentum and scalars (Wyngaard, 1986, 1987). Thus, it is very difficult to make *accurate* measurements of turbulent fluxes.

It is also difficult to make *representative* measurements of turbulent fluxes. In an early experiment comparing 30-min direct measurements of temperature fluxes from two instruments separated horizontally by 5 m at 4 m height, Businger et al. (1967) found that the two agreed only within a factor of 2 for individual runs. Averaged over all runs the agreement was very much better, leading them to conclude that "...the Eulerian point average does not provide an adequate statistical sample of the heat flux."

Similar results, for both momentum and heat fluxes from instruments separated vertically, were found in the 1968 Kansas experiments. Fig. 1.20 shows the ratio of $\overline{uw}$ from 22.6 m and 5.66 m; we know that for quasi-steady flow over a homogeneous site, this stress ratio should typically be within a few percent of 1.0. Although the alignment errors mentioned earlier were accounted for in data processing (so the ratio averaged over the experiment is indeed close to

1.0), there is a good deal of scatter within a given hour. This scatter is due to the failure of the time average to converge to the ensemble average; i.e., to insufficient averaging time (Wyngaard, 1986).

This means that a direct measurement of momentum or heat flux, however *accurate*, may not be *representative*.

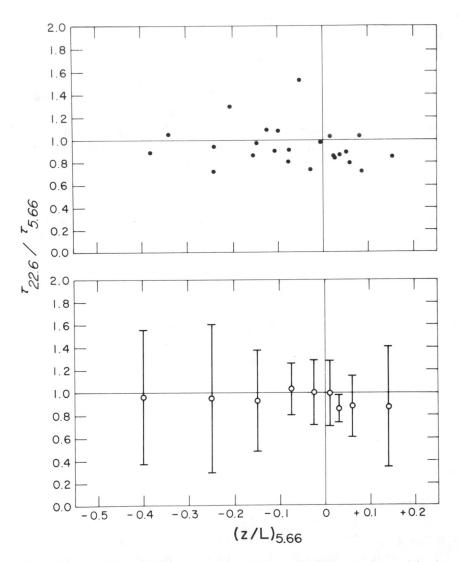

Figure 1.20. Ratio of shearing stresses at 22.6 m and 5.66 m measured in the 1968 Kansas experiments. Upper plot, 15-min data in 1-h blocks. Lower plot, data grouped into z/L intervals, with circle indicating mean, and bars plus and minus one standard deviation. Adapted from Haugen et al. (1971).

For these reasons—i.e., the expense, difficulty, and nonrepresentativeness of direct flux measurements—indirect techniques for inferring turbulent fluxes in the surface layer are of great interest in both research and applications. Strange as it may seem, they can be better than direct techniques in each of these respects. Indirect techniques relevant to dispersion problems are discussed in Chs. 4 and 5.

1.7.2. Local Meteorology and Terrain Effects

The mean wind and turbulence profiles within the boundary layer have a profound influence on the transport and dispersal of pollutants. In this chapter I have summarized current thinking about these profiles, and the physics behind them. That thinking rests in part on observations in ideal situations—flat, homogeneous terrain, fair weather, quasi-steady conditions—and from numerical model and convection tank studies, also under idealized conditions. In many situations the PBL profiles do not behave as simply as we implied earlier in this chapter, however.

I live on the eastern side of the Rocky Mountains. During the summer the daytime winds near the surface, being dominated by the upslope buoyancy force, are usually easterly. At the same time, the winds above the boundary layer, responding principally to the larger-scale synoptic forcing, can be westerly. Thus, the wind direction change across the boundary layer can be nearly 180° (Fig. 1.10)! In other situations when our surface winds are southeasterly a boundary-layer circulation called the "Denver cyclone" can form; it is apparently due to interactions with the local topography. This large eddy has a convergence zone that generates a strong mean vertical velocity northeast of Denver that can trigger severe storms in the warm season (Szoke et al., 1984).

Our mountain valleys provide another example of the strong influence of topography on boundary layer winds. Using data from a pulsed, infrared Doppler lidar, Post and Neff (1986) showed the structure of the nocturnal drainage winds within a narrow mountain valley in western Colorado, their decay in the morning, and the formation of thermally driven up-valley flow later in the day. Their study was conducted as part of the U.S. Department of Energy's Atmospheric Studies in Complex Terrain (ASCOT) program, which has as a goal the development of models for predicting dispersion in complex terrain.

When dealing with pollution transport and dispersion in such flows one must take into account such interactions between meteorology and the local topography. Given sufficient resources, one can now do this through detailed, three-dimensional numerical modeling of the wind field and the resulting dispersion. Such methods bring along a host of new issues, including description of the unresolvable subgrid-scale processes, incorporation of real data on initial and boundary conditions into the model, and model validation. This type of wind field modeling is in the mainstream of mesoscale meteorology today (Ray, 1986), although its applications to dispersion are not as common. Given the increasing availability of supercomputers this approach might become more accessible as time goes on.

In the meantime, those dealing with dispersion problems having important meteorology-terrain interactions will most likely have to assess their effects with less sophisticated tools in conjunction with an understanding of the underlying meteorological principles (see Randerson, 1984). In some problems of sufficiently simple geometry (e.g., isolated hills, simple valleys) one can gain a good deal of insight by using analytical approaches, as Hunt (1981) lucidly discussed. Finally, it is well to remember that in any fluid mechanics problem there is no substitute for good observations.

References

Banta, R. M., 1985: Late-morning jump in TKE in the mixed layer over a mountain basin. *J. Atmos. Sci.*, **42**, 407–411.

Blackadar, A. K., and K. Buajitti, 1957: Theoretical studies of diurnal wind variations. *Quart. J. Roy. Meteor. Soc.*, **83**, 486–500.

Brost, R. A., and J. C. Wyngaard, 1978: A model study of the stably stratified planetary boundary layer. *J. Atmos. Sci.*, **35**, 1427–1440.

Busch, N. E., 1973: On the mechanics of atmospheric turbulence. *Workshop on Micrometeorology*, D. A. Haugen, Ed., Amer. Meteor. Soc., Boston, 1–65.

Businger, J. A., 1985: The marine boundary layer, from air-sea interface to inversion. NCAR Technical Note, NCAR/TN–252 + STR, National Center for Atmospheric Research, Boulder, CO, 84 pp.

Businger, J. A., M. Miyake, A. J. Dyer, and E. F. Bradley, 1967: On the direct determination of the turbulent heat flux near the ground. *J. Appl. Meteor.*, **6**, 1025–1032.

Businger, J. A., J. C. Wyngaard, Y. Izumi, and E. F. Bradley, 1971: Flux-profile relationships in the atmospheric surface layer. *J. Atmos. Sci.*, **28**, 181–189.

Calder, K. L., 1966: Concerning the similarity theory of A. S. Monin and A. M. Obukhov for the turbulent structure of the thermally stratified surface layer of the atmosphere. *Quart. J. Roy. Meteor. Soc.*, **92**, 141–146.

Calder, K. L., 1968: Concerning the similarity theory of A. S. Monin and A. M. Obukhov for the turbulent structure of the thermally stratified surface layer of the atmosphere (Discussion). *Quart. J. Roy. Meteor. Soc.*, **94**, 108–113.

Caughey, S. J., 1982: Observed characteristics of the atmospheric boundary layer. *Atmospheric Turbulence and Air Pollution Modelling*, F. T. M. Nieuwstadt and H. van Dop, Eds., Reidel, Dordrecht, 107–158.

Caughey, S. J., J. C. Wyngaard, and J. C. Kaimal, 1979: Turbulence in the evolving stable boundary layer. *J. Atmos. Sci.*, **36**, 1041–1052.

Ching, J. K. S., 1985: Urban-scale variations of turbulence parameters and fluxes. *Bound.-Layer Meteor.*, **33**, 335–362.

Cramer, H. E., 1959: Engineering estimates of atmospheric dispersal capacity. *Am. Ind. Hyg. J.*, **20**, 183–188.

Csanady, G. T., 1973: *Turbulent Diffusion in the Environment.* Reidel, 248 pp.

Deardorff, J. W., 1970: Convective velocity and temperature scales for the unstable planetary boundary layer and for Raleigh convection. *J. Atmos. Sci.*, **27**, 1211–1213.

Deardorff, J. W., 1973a: Three-dimensional numerical modeling of the planetary boundary layer. *Workshop on Micrometeorology*, D.A. Haugen, Ed., Amer. Meteor. Soc., Boston, 271–311.

Deardorff, J. W., 1973b: An explanation of anomalously large Reynolds stresses within the convective planetary boundary layer. *J. Atmos. Sci.*, **30**, 1070–1076.

Deardorff, J. W., 1974: Three-dimensional numerical study of turbulence in an entraining mixed layer. *Bound.-Layer Meteor.*, **7**, 199–226.

Deardorff, J. W., 1979: Prediction of convective mixed–layer entrainment for realistic capping inversion structure. *J. Atmos. Sci.*, **36**, 424–436.

Deardorff, J. W., 1980: Cloud top entrainment instability. *J. Atmos. Sci.*, **37**, 131–147.

Deardorff, J. W., and G. E. Willis, 1985: Further results from a laboratory model of the convective planetary boundary layer. *Bound.-Layer Meteor.*, **32**, 205–236.

de Baas, A. F., and A. G. M. Driedonks, 1985: Internal gravity waves in a stably stratified boundary layer. *Bound.-Layer Meteor.*, **31**, 303–323.

Delage, Y., 1974: A numerical study of the nocturnal atmospheric boundary layer. *Quart. J. Roy. Meteor. Soc.*, **100**, 351–364.

Fox, D. G., 1984: Uncertainty in air-quality modeling. *Bull. Amer. Meteor. Soc.*, **65**, 27–36.

Garratt, J. R., 1985: The inland boundary layer at low latitudes. I. The nocturnal jet. *Bound.-Layer Meteor.*, **32**, 307–328.

Garratt, J. R., and R. A. Brost, 1981: Radiative cooling effects within and above the nocturnal boundary layer. *J. Atmos. Sci.*, **38**, 2730–2746.

Garratt, J. R., J. C. Wyngaard, and R. J. Francey, 1982: Winds in the atmospheric boundary layer—prediction and observation. *J. Atmos. Sci.*, **39**, 1307–1316.

Hanna, S. R., 1982: Natural variability of observed hourly SO_2 and CO concentrations in St. Louis. *Atmos. Environ.*, **16**, 1435–1440.

Haugen, D., Ed., 1973: *Workshop on Micrometeorology*. Amer. Meteor. Soc., Boston, 392 pp.

Haugen, D. A., J. C. Kaimal, and E. F. Bradley, 1971: An experimental study of Reynolds stress and heat flux in the atmospheric surface layer. *Quart. J. Roy. Meteor. Soc.*, **97**, 168–180.

Hechtel, L. M, and R. Stull, 1985: Statistical measures of surface inhomogeneity, and its potential impact on boundary-layer turbulence. *7th Symposium on Turbulence and Diffusion*, Amer. Meteor. Soc., Boston, 144–146.

Hildebrand, P. H., and B. Ackerman, 1984: Urban effects on the convective boundary layer. *J. Atmos. Sci.*, **41**, 76–91.

Holton, J. L., 1972: *An Introduction to Dynamic Meteorology*. Academic Press, New York, 319 pp.

Hunt, J. C., 1981: Some connections between fluid mechanics and the solving of industrial and environmental fluid-flow problems. *J. Fluid Mech.*, **106**, 103–130.

Hunt, J. C. R., 1985: Diffusion in the stably stratified atmospheric boundary layer. *J. Climate Appl. Meteor.*, **24**, 1187–1195.

Joffre, S. M., 1985a: Effects of local accelerations and baroclinity on the mean structure of the atmospheric boundary layer over the sea. *Bound.-Layer Meteor.*, **32**, 237–255.

Joffre, S. M., 1985b: The structure of the marine atmospheric boundary layer: A review from the point of view of diffusivity, transport and deposition processes. Technical Report No. 29, Finnish Meteorological Institute, Helsinki.

Kaimal, J. C., and D. A. Haugen, 1969: Some errors in the measurement of Reynolds stress. *J. Appl. Meteor.*, **8**, 460–462.

Kaimal, J. C., J. C. Wyngaard, D. A. Haugen, O. R. Coté, Y. Izumi, S. J. Caughey, and C. J. Readings, 1976: Turbulence structure in the convective boundary layer. *J. Atmos. Sci.*, **33**, 2152–2169.

Kondo, J., O. Kanechika, and N. Yasuda, 1978: Heat and momentum transfers under strong stability in the atmospheric surface layer. *J. Atmos. Sci.*, **35**, 1012–1021.

LeMone, M. A., 1980: The marine boundary layer. *Workshop on the Planetary Boundary Layer*, J. C. Wyngaard, Ed., Amer. Meteor. Soc., Boston, 182–234.

Lenschow, D. H., J. C. Wyngaard, and W. T. Pennell, 1980: Mean-field and second-moment budgets in a baroclinic convective boundary layer. *J. Atmos. Sci.*, **37**, 1313–1326.

Lilly, D. K., 1968: Models of cloud-topped mixed layers under a strong inversion. *Quart. J. Roy. Meteor. Soc.*, **94**, 292–309.

Lumley, J. L., and H. A. Panofsky, 1964: *The Structure of Atmospheric Turbulence*. Wiley Interscience, New York, 239 pp.

McAllister, L. G., J. R. Pollard, A. R. Mahoney, and P. J. R. Shaw, 1969: Acoustic sounding—A new approach to the study of atmospheric structure. *Proc. IEEE*, **57**, 571–578.

McBean, G., Ed., 1979: *The Planetary Boundary Layer*. Technical Note No. 165, WMO, Geneva, Switzerland, 201 pp.

Moeng, C. H., and J. C. Wyngaard, 1984: Statistics of conservative scalars in the convective boundary layer. *J. Atmos. Sci.*, **41**, 3161–3169.

Monin, A. S., and A. M. Yaglom, 1971: *Statistical Fluid Mechanics*. MIT Press, Cambridge, 769 pp.

Nicholls, S., and M. A. LeMone, 1980: The fair weather boundary layer in GATE: The relationship of subcloud fluxes and structure to the distribution and enhancement of cumulus clouds. *J. Atmos. Sci.*, **37**, 2051–2067.

Nieuwstadt, F. T. M., 1981: The steady-state height and resistance laws of the nocturnal boundary layer: Theory compared with Cabauw observations. *Bound.-Layer Meteor.*, **20**, 3–17.

Nieuwstadt, F. T. M., 1984: The turbulent structure of the stable, nocturnal boundary layer. *J. Atmos. Sci.*, **41**, 2202–2216.

Nieuwstadt, F. T. M., and R. A. Brost, 1986: The decay of convective turbulence. *J. Atmos. Sci.*, **43**, 532–546.

Nieuwstadt, F. T. M., and H. Tennekes, 1981: A rate equation for the nocturnal boundary-layer height. *J. Atmos. Sci.*, **38**, 1418–1428.

Nieuwstadt, F. T. M., and H. van Dop, Eds., 1982: *Atmospheric Turbulence and Air Pollution Modelling*. Reidel, Dordrecht, 358 pp.

Panofsky, H. A., and J. Dutton, 1984: *Atmospheric Turbulence*. Wiley, New York, 397 pp.

Panofsky, H. A., H. Tennekes, D. H. Lenschow, and J. C. Wyngaard, 1977: The characteristics of turbulent velocity components in the surface layer under convective conditions. *Bound.-Layer Meteor.*, **11**, 355–361.

Post, M. J., and W. D. Neff, 1986: Doppler lidar measurements of winds in a narrow mountain valley. *Bull. Amer. Meteor. Soc.*, **67**, 274–281.

Randerson, D., Ed., 1984: *Atmospheric Science and Power Production*. U. S. Dept. of Energy DOE/TIC–27601. (Available from NTIS as DE84005177.)

Ray, P. S., Ed., 1986: *Mesoscale Meteorology and Forecasting*. Amer. Meteor. Soc., Boston, 793 pp.

Rogers, D. P., J. A. Businger, and H. Charnock, 1985: A numerical investigation of the JASIN boundary layer. *Bound.-Layer Meteor.*, **32**, 373–400.

Smith, F. B., 1957: The diffusion of smoke from a continuous elevated point-source into a turbulent atmosphere. *J. Fluid Mech.*, **2**, 49–76.

Szoke, E. J., M. L. Weisman, J. M. Brown, F. Caracena, and T. W. Schlatter, 1984: A subsynoptic analysis of the Denver tornadoes of 3 June 1981. *Mon. Wea. Rev.*, **112**, 790–808.

Tennekes, H., and J. L. Lumley, 1972: *A First Course in Turbulence*. MIT Press, Cambridge, 300 pp.

Tuzet, A., 1982: Contribution a l'étude des lois de similitude dans la couche limité planetaire en regime convectif. Thesis, Université de Clermont II, France.

Willis, G. E., and J. W. Deardorff, 1974: A laboratory model of the unstable planetary boundary layer. *J. Atmos. Sci.*, **31**, 1297–1307.

Wipperman, F., 1973: *The Planetary Boundary Layer of the Atmosphere*. Deutscher Wetterdienst, Offenbach, 346 pp.

Wyngaard, J. C., 1975: Modeling the planetary boundary layer—extension to the stable case. *Bound.-Layer Meteor.*, **9**, 441–460.

Wyngaard, J. C., 1983: Lectures on the planetary boundary layer. *Mesoscale Meteorology—Theories, Observations, and Models*, D. K. Lilly and T. Gal-Chen, Eds., Reidel, Dordrecht, 603–650.

Wyngaard, J. C., Ed., 1984: *Large-Eddy Simulation: Guidelines for its Application to Planetary Boundary Layer Research*. Available from DTIC, AD–A146381.

Wyngaard, J. C., 1985: Structure of the planetary boundary layer and implications for its modeling. *J. Climate Appl. Meteor.*, **24**, 1131–1142.

Wyngaard, J. C., 1986: Measurement physics. *Probing the Atmospheric Boundary Layer*, D. H. Lenschow, Ed., Amer. Meteor. Soc., Boston, 5–18.

Wyngaard, J. C., 1987: Flow-distortion effects on scalar flux measurements in the surface layer: Implications for sensor design. To appear, *Bound.-Layer Meteor.*.

Wyngaard, J. C., and R. A. Brost, 1984: Top-down and bottom-up diffusion of a scalar in the convective boundary layer. *J. Atmos. Sci.*, **41**, 102–112.

Wyngaard, J. C., and O. R. Coté, 1971: The budgets of turbulent kinetic energy and temperature variance in the atmospheric surface layer. *J. Atmos. Sci.*, **28**, 190–201.

Wyngaard, J. C., and O. R. Coté, 1974: The evolution of a convective planetary boundary layer–A higher-order-closure model study. *Bound.-Layer Meteor.*, **7**, 289–308.

Wyngaard, J. C., O. R. Coté, and Y. Izumi, 1971: Local free convection, similarity, and the budgets of shear stress and heat flux. *J. Atmos. Sci.*, **28**, 1171–1182.

Zilitinkevich, S. S., 1972: On the determination of the height of the Ekman boundary layer. *Bound.-Layer Meteor.*, **3**, 141–145.

CHAPTER 2

Analysis of Diffusion Field Experiments

Gary A. Briggs

2.1. Theoretical Frameworks for Analysis of Diffusion Data

There now exist a multitude of theoretical methods for making diffusion predictions for comparison with results of field and laboratory diffusion experiments. These include gradient transfer, spectral diffusivity, second-order closure, large-eddy simulation, and random perturbation models (Briggs and Binkowski, 1985). These models generally require either detailed meteorological and turbulence measurements or many assumptions about these quantities, and may require considerable computational effort. In this chapter we seek more general theoretical frameworks that allow considerable ordering of diffusion data without requiring complex meteorological information or computational efforts.

It is not only impossible, but also undesirable to specify entire boundary layer flow fields for each diffusion trial. On the other hand, one wants to carry the measured data beyond the stage of huge lists of diffusion and meteorological variables. Ideally, we would like to reduce these data to a few neat correlations between the diffusion variables and key meteorological variables. Such correlations may be imperfect, but are still of great utility in developing practical diffusion models.

Three theoretical methods have been of great service in digesting measurements from field experiments. Two of the methods, surface layer similarity and convective scaling, are forms of dimensional analysis (Barenblatt, 1979). They nondimensionalize the characteristic lateral and vertical diffusion scales, such as σ_y and σ_z, and the diffusion time or distance traveled by scales thought to be the most important in characterizing the turbulence responsible for diffusion. This reduces the number of variables to be correlated, and in turn greatly reduces the number of variable combinations that need to be considered in further analysis. A theoretical model can often be expressed in terms of the same nondimensionalizations; by reducing the number of independent variable combinations in the model, this may help avoid redundant computations. The most obvious rewards of this approach occur in diffusion situations where the salient effects are contained within the chosen scales, so that a very usable correlation exists in simple nondimensional plots of measured σ_y versus x, for instance.

The third analysis method also begins with a nondimensionalization. The diffusion coefficients σ_y and σ_z are assumed to be proportional to the statistical theory predictions for very short ranges, based on turbulent velocity or wind direction and inclination variances. Thus, σ_y or σ_z measurements can be divided by these predicted values and then plotted against distance, time or nondimensional time. Recent analyses favor the latter, using some assumption about the Lagrangian time scale of turbulence. This method may use the same scales used in surface layer similarity and/or convective scaling analyses; in such cases it differs from them mainly in its starting point, the use of measured or estimated turbulent variances.

2.1.1. Surface Layer Similarity

The "similarity theory" of Monin and Obukhov (1954) has proved a very useful tool for describing the profiles of wind speed, temperature, humidity, and some turbulence quantities in the surface layer, approximately the lowest tenth of the boundary layer. Its success became most evident following the 1968 Kansas surface layer experiment (Businger et al., 1971; Kaimal et al., 1972), which used much improved instrumentation compared with earlier meteorological experiments. Ch. 1 covers this similarity framework in detail.

Any theoretical framework that successfully describes turbulence should describe turbulent diffusion as well. Thus, surface layer similarity has been used in analyses of diffusion field experiment data since 1973, shortly after publication of the first Kansas results. It has been quite successful for particular cases, especially for vertical diffusion from near-surface sources.

If we apply the surface-layer similarity of Monin and Obukhov (M-O) to diffusion, we expect the rate of diffusion to be u_* times a function of z/L, where z may be a source height, z_s, a mean particle height, $\bar{z}$, a fraction of σ_z, or a combination of such height variables; u_* and L are the friction velocity and M-O length (Ch. 1). For instance, for a surface source, one way to apply surface layer similarity assumptions is

$$d\sigma_z/dt = u_* f(\sigma_z/L) \ . \tag{2.1}$$

The time-to-distance conversion is usually provided by $dx/dt = U$ evaluated at height z proportional to σ_z or $\bar{z}$ or, in the case of an elevated source, at $z = z_s$. Conveniently, $U(z)$ is provided by the same theoretical framework, $\partial U/\partial z = u_*/(kz)$ times $\phi_m(z/L)$, where the dimensionless function ϕ_m has been determined empirically from field studies such as the Kansas experiment (Businger et al., 1971). We review this and other past applications of surface layer similarity to diffusion field data in Sec. 2.3.

Although we shall see that surface-layer similarity gives good results in analyses of near-surface vertical diffusion data, it has quite a few limitations. These are discussed in general terms in Ch. 1 and in some detail by Briggs and Binkowski (1985), but can be briefly stated here. It does not work as well for lateral diffusion as for vertical. In unstable (convective) conditions, σ_v, as well as σ_y, scales with the convective scaling velocity w_* rather than with u_* (see Sec. 2.1.2). In neutral conditions it works fairly well, but σ_v/u_* reportedly ranges from 1.3 to 2.6 at different sites (Lumley and Panofsky, 1964); this might be due to effects of terrain and surface roughness inhomogeneities on horizontal eddies, or again due to some w_* influence. In stable conditions, at many sites there are large horizontal eddies caused by drainage flows or inhomogeneous conditions upstream; presumably only the smaller, more isotropic eddies scale with u_*, so only short-term averages of σ_y follow M-O similarity.

For vertical diffusion the theory works best, naturally, in about the lowest tenth of the PBL. Above this height in unstable conditions we have the option of convective scaling. In near-neutral conditions, there is little information above the surface layer. Numerical models of neutral boundary layers indicate that turbulence velocities drop off very slowly with height, with $\sigma_w/u_* \simeq 1.3 \exp(-2zf/u_*)$, where f is the Coriolis parameter ($\simeq 10^{-4}\text{s}^{-1}$ at middle latitudes) (Hanna,

1981). However, in nature the neutral boundary-layer depth is usually determined by an overlying inversion rather than the value of u_*/f. Thus, aircraft measurements during the morning in neutral boundary layers capped by stable air at h indicate $\sigma_w \propto (1 - z/h)$(Yokoyama et al., 1977); the data reported by Brost et al. (1982) behave similarly.

The greatest problem in describing diffusion in stable conditions is determination of h, the height of continuous turbulent mixing, which we take as the PBL depth. The mean temperature or wind speed profiles taken alone are usually poor indicators of h, but the bulk Richardson number profile is an effective indicator (Briggs and Binkowski, 1985). The diagnostic equation $h \propto (u_*L/f)^{1/2}$ has worked well at some sites, but as we saw in Ch. 1 the constant of proportionality can be very sensitive to slight slopes and their angle with the horizontal pressure gradient (Garratt, 1982). In pure drainage flow, $h \propto L$, approximately (Briggs and Binkowski, 1985). The upper limit to h in very stable conditions is roughly $10L$. (Figure 2.2 in Sec 2.1.2 gives a schematic view of the various regimes governing turbulence and diffusion in atmospheric boundary layers.)

2.1.2. Convective Scaling

As we saw in Ch. 1, in unstable conditions warmer-than-average parcels of air tend to rise until they encounter a "capping" layer of stable, higher-potential-temperature air just above height h, the mixed-layer depth. Their momentum tends to carry them somewhat above h, where they mix a little with the base of the stable air and then are returned below h by their negative buoyancy. The upward acceleration of "thermals" is reduced somewhat by the need to push previously risen air trapped beneath the capping layer out of the way; this continuity requirement generates a slightly positive pressure gradient that pushes previously risen air downward, forming "downdrafts." These tend to extend to the surface (Willis and Deardorff, 1976b). Thus, h is the length scale of the largest eddies in the convective boundary layer (CBL). It applies to horizontal motions as well, as typical downdraft diameters are $1.3h$ (Willis and Deardorff, 1976b). Unless generated over "hot spots" due to uneven surface characteristics, thermals tend to gather at the intersections of adjacent downdrafts, as divergence from the center of each downdraft near the ground moves surface-heated air in those directions.

Naturally, this surface divergence tends to move passive material near the surface toward the thermals, the "up elevators" of the CBL.

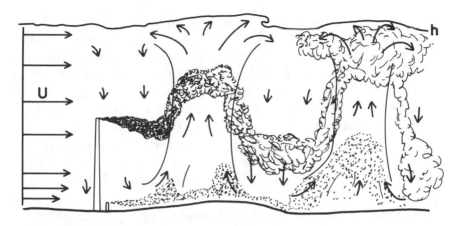

Figure 2.1. A schematic view of large-eddy diffusion processes in a convective boundary layer.

Another important aspect of CBL diffusion is the tendency of the buoyant thermals to accelerate and elongate in the vertical direction during the active stages of CBL development. This causes them to be relatively narrow, owing to internal horizontal convergence; downdrafts, which are driven more by pressure than by negative buoyancy, tend to be wider. The result is that the ratio of downdraft and updraft areas through most of the CBL is typically about 60%/40%. This means that material released from an elevated source has about a 60% chance of being released into a "down elevator" in the form of a downdraft [there is some evidence now that in late afternoon, as surface heating drops off and convective motions become partly driven by cooling from above, owing to radiation flux divergence, the downdraft/updraft ratio is closer to 50/50 (Briggs and Binkowski, 1985)]. These large-eddy phenomena in the CBL are not accommodated well by classical Gaussian plume models, and were the basis of much controversy in the late 1970s. Figure 2.1 gives a simplified overview of the large eddies and their salient effects on diffusion in the CBL.

There are two main scales characterizing turbulence in the bulk of the CBL, away from the surface and the entrainment zone near the top: w_* and h. These are applied to diffusion by nondimensionalizing vertical and lateral length scales with h and time scales, including time of travel, with h/w_* (this is proportional to the Lagrangian time scale). The travel time t can be accurately approximated by $x/\overline{U}$, where $\overline{U}$ is the wind speed averaged through the mixing depth. As we saw in Ch. 1, all quantities, including momentum, tend to become well mixed above the surface layer and the wind speed shear is usually quite small. In practice, U evaluated at $0.1h$ or $0.2h$ usually provides an adequate approximation to $\overline{U}$. Most convective scaling analyses of diffusion data consist of plots of σ_y/h, σ_z/h, or $h \int \chi dy/(Q/U)$, where χ is the ambient tracer concentration and Q/U is source strength per downwind length of air advected past a continuous source or line source strength. These ratios are usually plotted against $X = tw_*/h$ or $X = (x/U)w_*/h$, where w_* is the convective velocity scale (Ch. 1).

Both surface layer similarity and convective scaling utilize the surface flux of buoyant accelerations, $H^* = (g/\Theta_0)\overline{w\theta_0}$. The main difference between them is in the length scales, h versus L. The velocity scales relate to the length scales in the same way: $w_* = (H^*h)^{1/3}$ and $u_* \propto (H^*L)^{1/3}$. Convective scaling has no validity in neutral or stable boundary layers, of course, but in unstable boundary layers, how does one choose between the two different scalings? One approach is to use both, by including $h/|L|$ as an additional parameter. This tends to complicate the analysis, particularly when the data scatter is large; as we saw in Ch. 1, this is often the case in convective conditions. However, we can define regimes in which one scaling or the other is more effective, and even find a regime in which *both* are effective, if $h/|L|$ is large enough.

For horizontal diffusion it turns out that for large enough $-h/L = k(w_*/u_*)^3$ convective scaling is better regardless of source height, because σ_u and σ_v are proportional to w_* even a few meters from the surface. Earlier attempts to order horizontal turbulent velocity spectra in terms of u_* and L, as in Kaimal et al. (1972), failed in unstable conditions. According to a recent survey of atmospheric measurements, w_* is the dominant scale characterizing σ_v when $w_* > 3.2u_*$, or $-h/L > 13$ (Hicks, 1985). For typical h values of the order of 1000 m, this transition corresponds roughly to Pasquill's "C" conditions. (Very roughly, the dividing point between "D" and "C" stability classes is around $-h/L \sim 5$, between "C" and "B" is ~ 20, and between "B" and "A" is ~ 100). In the "C" stability class both u_* and w_* make significant contributions to σ_v; there are empirical fits to σ_v measurements that combine these contributions, such as Panofsky et al. (1977) and Hicks (1985). However, it is more difficult to assess how the Lagrangian time scale for diffusion transforms from surface layer to convective scaling behavior; several guesses have been used in proposed analytical equations for σ_y (Briggs, 1985).

For vertical diffusion the height of diffusing material relative to $-L$ and to h is the determinant. Strictly speaking, surface layer similarity theory is only valid near the surface (below $0.1h$, say) where fluxes are nearly equal to their surface values. Above $0.1h$, $U(z)$ is nearly constant and $\sigma_w \simeq w_*$ times a function of z/h (Irwin, 1979; Hicks, 1981) in a sufficiently convective boundary layer. Roughly, $\sigma_w \simeq 0.6w_*$ in mid-layer, and $\sigma_w \simeq 1.3u_*$ in neutral surface layers (Panofsky et al., 1977); therefore, transition to convective scaling of vertical turbulent velocities occurs at about $w_* > 2.2u_*$ or $h/|L| > 4$, near the transition between "D" and "C" Pasquill categories. This criterion is easily met on any sunny or partly sunny day, except during vigorous winds (> 7 to 10 m s^{-1}) or in moderate winds over snow cover, when H^* is greatly diminished.

When $h/|L| > 10$, there is a "free convection" regime above $|L|$ and below $0.1h$ in which $\sigma_w \simeq 1.3(H^*z)^{1/3}$; it is convective, but vertical velocities are more limited by the $w = 0$ effect of the surface than by the capping at $z = h$. In this regime, $\sigma_w \propto u_*|z/L|^{1/3} \propto w_*(z/h)^{1/3}$, so both scalings apply. For the vertical diffusion from a surface source, free convective growth ($\sigma_z \propto H^{*1/2}t^{3/2}$) fits laboratory observations well when $X \leq 0.5$ or $\sigma_z/h \leq 0.3$ (Willis and Deardorff, 1976a). Beyond this point, the maximum concentrations lift from the surface as most material feeds into thermals and lofts to the upper part of the CBL; in this stage, only convective scaling can give a satisfactory description of diffusion. In absolute distance, this transition is at $x = 0.5Uh/w_*$, which ought to be considered the maximum distance for applicability of surface layer similarity scaling; $0.5h/w_*$ is typically about 300 s, and U is more variable. Provided that $h/|L| > 4$, convective scaling is fully valid when $z_s > |L|$ or when $x > (U/u_*)L \sim 10L$ (based on transition at $\sigma_z \simeq 0.8|L|$ for a surface source, and using $\sigma_z \simeq 0.8u_*t$, the neutral asymptote from Briggs [1982b]).

Figure 2.2 gives a schematic view of the various regimes affecting diffusion in atmospheric boundary layers, including stable ones.

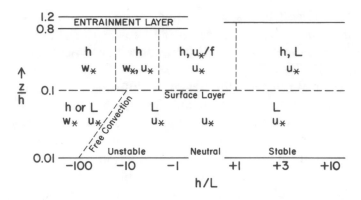

Figure 2.2. A schematic view of scaling regions of atmospheric boundary layers with principal length and velocity scales affecting turbulence and diffusion; z is also important, especially for vertical diffusion. Adapted from Sivertsen et al. (1987).

2.1.3. Statistical Theory Analyses Using Wind Fluctuations

The two scaling approaches just discussed are used to analyze diffusion data in terms of the fundamental parameters influencing turbulence in the atmospheric boundary layer—surface buoyancy flux, surface friction, and boundary layer depth. The statistical approach aims to be a little more direct, by using at least a single-point measurement of turbulent velocity. However, this is not sufficient information for accurate description of whole diffusion patterns, so often surface layer or convective scales are used in refinements of this approach.

Even in "looping" conditions, when plumes meander up and down, one can note that individual flow features, when followed by eye, tend to move in nearly a straight line for considerable distances. Atmospheric vertical accelerations seldom exceed $(1 \text{ K}/300 \text{ K})10$ m s^{-2}, or 1 m s^{-1} change in vertical velocity in 30 s. Thus, within the first few tens of meters of a source we can assume essentially straight-line travel for each plume element; the plume goes in the direction of the wind at the source position at the release time. At short distances from a continuous release the lateral and vertical concentration distributions should be almost the same as the distributions of occurrence of the tangent wind angles at the source times x. We can make the small-angle approximation $\tan\beta \simeq \beta$, where the wind angle is in radians, to get the result

$$\sigma_y \simeq \sigma_a x \quad \text{and} \quad \sigma_z \simeq \sigma_e x , \tag{2.2}$$

where σ_a and σ_e are the standard deviations of the azimuth and the elevation wind angles. Alternatively, we approximate the tangent of azimuth and elevation angles to get

$$\sigma_y \simeq (\sigma_v/U)x \quad \text{and} \quad \sigma_z \simeq (\sigma_w/U)x . \tag{2.3}$$

Equation (2.3) is the near-source asymptote of the classic statistical theory prediction (Taylor, 1921). Written for lateral diffusion, it is

$$d\sigma_y^2/dt = 2\sigma_v^2 \int_0^t R(\xi)d\xi , \tag{2.4}$$

where $R(\xi)$ is the Lagrangian (i.e., following a fluid particle) correlation coefficient of flow velocities separated by a time lag ξ. At small ξ, there is practically no change in particle velocities; i.e., plume features tend to move in a straight line, so $R \simeq 1$. Then the integral in Eq. (2.4) is equal to t, and another integration over t gives $\sigma_y \simeq \sigma_v t$, which is approximated in Eqs. (2.2) and (2.3) using $t \simeq x/U$ and $\sigma_a \simeq \sigma_v/U$. The classical theory assumes homogeneous and stationary turbulence, but Eqs. (2.2) and (2.3) are also valid for nonstationary conditions if σ_v or σ_a (or σ_w or σ_e) are measured over the same sampling time as σ_y (or σ_z).

As the lag time increases, R becomes less than unity and σ_y becomes less than $\sigma_v t$. In stationary conditions, R can be expressed as a function of t/T_L, where the Lagrangian time scale T_L characterizes the decay of velocity correlation. In this case, $\sigma_y/(\sigma_v t)$ begins at unity and also falls off as some function of t/T_L. Writing this result for both lateral and vertical diffusion, we have

$$\sigma_y/(\sigma_v t) = f_1(t/T_L)$$
$$\sigma_z/(\sigma_w t) = f_2(t/T_L) . \tag{2.5}$$

These equations, along with variants based on $t \simeq x/U$, $\sigma_v \simeq \sigma_a U$, and $\sigma_w \simeq \sigma_e U$, have been the basis of much practical analysis of diffusion data since Draxler's 1976 paper. Diffusion scientists have often suggested that an approach like this, beginning with direct measurements of turbulence near the source, is the most reliable way to model diffusion (Hanna et al., 1977); this is especially true in cases that are strongly influenced by site characteristics, such as lateral diffusion at night. The analyses vary in their treatment of the unknown functions f_1 and f_2; these range from simple empirical functions of x to assuming that T_L is related to similarity scales. These applications of the statistical approach are reviewed mostly in Sec. 2.5.

2.2. Early Diffusion Experiment Analyses and Resulting Sigma Curves

Convective scaling was unknown and surface layer similarity just beginning to attract theorists' attention when Pasquill and Brookhaven stability classification schemes and sigma curves were developed (see Pasquill, 1962, Sec. 2.1). Statistical theory had long been familiar (Taylor, 1921), but its application to diffusion was sometimes handicapped by the wind measurement instruments of the time, which were often either too sluggish or underdamped. Nonetheless, diffusion data analyses of this period did make some use of statistical theory or of variables that correlate strongly with L. It is interesting to review these efforts in the light of the developments that came later with improved meteorological instrumentation and better knowledge of PBL structure.

2.2.1. Early Analyses of the Round Hill and Prairie Grass Data Sets

In the mid-1950s substantial efforts were made to carry out upgraded diffusion experiments with quantitative sampling and, for that time, high quality meteorological measurements. To quote from the introduction to an experimental study at Round Hill, MA (Cramer and Record, 1957):

> It is felt that improved understanding of the physics of diffusion requires empirical studies of the structure of atmospheric turbulence. The current research program has a two-fold objective: simultaneous field measurements of diffusion and meteorological parameters, including fluctuations in wind velocity as well as the vertical gradients of mean wind speed and air temperature; establishment of empirical relations between diffusion and meteorological factors.

The tracer gas used in these experiments was SO_2, released near the surface and collected by midget impingers and bubbled through dilute hydrogen peroxide solution. The SO_2 reaction with this solution increases its conductivity, providing a rather convenient measure of dosage at each sampler. Samplers were placed at 3° spacing in arcs at x = 50, 100, and 200 m at Round Hill (RH). For the Prairie Grass (PGr) experiment in O'Neill, Nebraska, 1956, there were additional arcs at 400 and 800 m, with 2° or 1° spacing (Barad and Haugen, 1959). The PGr site was quite flat and much smoother than the RH site (roughness lengths 0.6 cm and >10 cm, respectively). In both experiments, the release duration was only 10 min.

Measurements of peak concentration on an arc, cross-arc integrated concentration, or plume width were compared with several meteorological indices. One of these was the standard deviation of wind azimuth, σ_a, measured over the release period and near the height of the samples (1.5 or 2 m). Another index was the "stability ratio" SR = $\Delta T/U^2$, with ΔT a mean temperature difference. At Round Hill, ΔT was taken between 1.5 and 6 m with U at 3 m (Cramer and Record, 1957); at O'Neill, ΔT was taken between 0.5 and 4 m with U at 2 m (Barad and Haugen, 1959). This close to the ground, where temperature gradients are usually strong, ΔT is very nearly equal to the potential temperature difference $\Delta \Theta$. Thus, SR is essentially a bulk Richardson number, which, according to surface-layer similarity theory, is monotonically related to the Obukhov length, L. Although the significance of L was not widely recognized in the United States circa 1955, the Richardson number had long been accepted as indicative of the ratio of buoyancy and shear forces in flows, and hence as a stability index. In the Barad and Haugen (1959) study of PGr data, comparisons were also made with an index n based on the exponent in z best fitting the wind profile: $n = 2p/(1 + p)$, where $U \propto z^p$, approximately. This was done to test a version of Sutton's (1947) diffusion theory in vogue at the time, in which it was hypothesized that $\sigma_y^2 \propto \sigma_z^2 \propto x^{2-n}$. They concluded that n did not give a useful correlation with the diffusion observations.

The only direct measure of σ_z was in the PGr experiment at the 100-m arc, where there were six 18-m towers carrying samplers. Barad and Haugen (1959) showed a clear relationship between this σ_z and SR, which implies a good correlation with L. (Several later studies, discussed in Sec. 2.3, reanalyzed the PGr data in terms of surface-layer scaling with u_* and L.) All other σ_z estimates were made on the basis of *surface* concentrations by assuming no loss of tracer and a normal distribution of concentration in the vertical dimension, with total reflection at the surface. Problems with this assumption are discussed in Sec. 2.2.4.

Table 2.1. Absolute Correlation Coefficients × 100, for Prairie Grass (PGr) and Round Hill (RH) Experiments.

		σ_a (PGr)	$1/\sigma_a$ (RH)	SR (PGr)	SR (RH)
Plume Width:	Day	85	91	55	78
	Night	85	52	12	43
$\int \chi dy$:	Day	81	90	84	54
	Night	7	71	84	77
Peak χ:	Day	91	93	76	60
	Night	76	72	38	49

Some linear correlations between plume measures and meteorological measures for the PGr and RH data sets are summarized in Table 2.1. These are from Cramer (1957) and Cramer and Record (1957). The PGr analysis used σ_a, and the RH analysis used its reciprocal, so moderate differences are to be expected. The stability ratio (SR) definitions are more similar, with slight height differences. For plume widths, both studies showed the expected high correlation with σ_a; the decrease in correlation with $1/\sigma_a$ at RH at night may arise from use of the reciprocal, which has a larger geometric range at night. The correlation with SR showed considerable site differences, being higher at RH and in the daytime; the SR essentially failed to correlate with σ_y at night at O'Neill. For $\int \chi dy$, which is inversely proportional to the "apparent" σ_z if a Gaussian vertical distribution is assumed, we see that σ_a gave a surprisingly good correlation in the day, considering that it is a measure of lateral turbulence, not vertical. Furthermore, σ_a failed to correlate with $\int \chi dy$ at O'Neill at night, but correlated well for RH data; in contrast, the SR correlated well day and night, with somewhat poorer performance for RH daytime data. This comparison agrees with later conclusions that σ_a is the best simple plume width indicator, that the SR, Richardson number or L is most reliable for short-range $\int \chi dy$ or σ_z from surface sources, and that large site differences can occur at night (Weber et al., 1977; Hanna et al., 1977).

For the centerline value of χ, both studies indicated somewhat better correlations with σ_a than with the SR, with somewhat degraded performance of either index for nighttime (stable) cases.

2.2.2. Pasquill's Use of Prairie Grass Data

Pasquill's (1961) σ_y curves were not based on diffusion measurements. They were derived from an adaptation of statistical theory developed by Hay and Pasquill (1959), which uses filtered values of σ_a to predict σ_y ; this method was applied to wind direction measurements at 16 m over rather smooth terrain (the roughness length $z_0 \simeq 3$ cm). The σ_z for "D" (neutral) stability was based on Calder's (1949) diffusivity theory, supported by some field data, applied to terrain of the same roughness, but Pasquill noted that a sixfold increase in z_0 produces only about 25% increase in σ_z. Pasquill's σ_z curves for all other stability classes and for x $\leq$ 1 km were based on the PGr data, using the behavior of σ_z relative to that for neutral conditions. The Prairie Grass site was very smooth, $z_0 \simeq 0.6$ cm, and the neutral σ_z values at $x = 100$ m in this experiment were about 25% smaller than Pasquill's D curve. Except at the arc with samplers on towers, at $x = 100$ m, these σ_z values were not direct measurements, of course. They were taken from Cramer's (1957) analysis, which used the peak concentration, the plume width, and the Gaussian vertical distribution assumption to infer σ_z.

Pasquill ordered these σ_y and σ_z values according to a stability classification scheme that was based on wind speed at $z = 10$ m, insolation, and cloudiness. The scheme was largely intuitive, and was intended to classify diffusion rates broadly, using the most important meteorological variables that are also easy to measure and commonly available. The latter requirement ruled out quantities such as h, H^*, u_*, and L. Yet at least since Golder (1972), it has been noted that there is a rough correspondence between Pasquill's categories and L values. As Golder suggested, this correspondence is affected somewhat by surface roughness; the larger is z_0, the larger are u_*/U_{10} and $L \propto u_*^3$ for a given category. Briggs (1982a) also demonstrated a strong correspondence between L values and "psuedo-L" values defined by U^3 divided by insolation or net radiation; these can be regarded as quantified forms of Pasquill's stability classes.

Consistent with this relationship between Pasquill's stability and L, Briggs and Binkowski (1985) found that this scheme performs best under the same restrictions for which surface-layer similarity gives a good description of turbulence; i.e., it is most effective for vertical diffusion from surface and near-surface sources. It is reasonably effective for lateral diffusion over short averaging times, because this diffusion is caused by smaller eddies that are much less influenced by h, terrain slopes, or surface inhomogeneities than are larger eddies. That is why Pasquill's σ_y curves are for only 3-min averages. Pasquill (1961) preferred σ_y and σ_z estimates based on wind direction fluctuation measurements and statistical theory, but he offered his σ_y and σ_z curves with simply determined stability classes as a backup scheme, with many stated limitations, for use when the preferred measurements are not available.

2.2.3. Brookhaven Diffusion Experiments Related to σ_a

Investigators at Brookhaven National Laboratory also wanted to order their diffusion experimental data in terms of a simple measurement of stability. Their index of choice was the strip chart record of wind vane fluctuations (Singer and Smith, 1966). Stability classification by such means goes back at least to 1932 (Pasquill, 1962, p. 26), and σ_a measurements were included with the first oil fog diffusion experiments at Brookhaven in 1949 (Lowry et al., 1951). Of course, plume width at small distances is directly proportional to σ_a, which is both intuitively evident and mathematically supported by statistical diffusion theory. Strip chart examples of Brookhaven's "gustiness" classifications are shown here in Fig. 2.3. In applying σ_a as an index for lateral plume growth out to $x \sim 50$ km, Singer and Smith assumed, in effect, that the dropoff of $\sigma_y/(\sigma_a x)$ versus x at medium

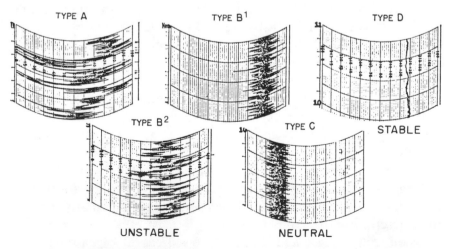

Figure 2.3. Brookhaven gustiness classifications and typical wind direction traces. Adapted from Singer and Smith (1966).

and long distances would also correlate with σ_a itself. This will be consistent with statistical theory in the form of Eq. (2.5) if effective eddy diameters D_e are correlated with σ_a. (The dropoff is a function of t/T_L; if T_L is proportional to D_e/σ_v and we use the approximations $t = x/U$ and $\sigma_v/U = \sigma_a$, we find that t/T_L is proportional to $x\sigma_a/D_e$.) At a specific site, it is quite possible that a good correlation exists between D_e and σ_a, but because D_e is strongly influenced by h in the day and by topographic factors at night, generalization to other sites might not be warranted.

The Brookhaven investigators and others have also used σ_a as an index of *vertical* diffusion, a proposition more difficult to justify theoretically because σ_a is a measure of *lateral* turbulence. However, it is justifiable when turbulence is roughly isotropic in the vertical and lateral directions, as in neutral conditions and also in unstable conditions, above $0.1h$ or so (recall that below $0.1h$ in strongly convective conditions we have $\sigma_v \propto (H^*h)^{1/3}$ whereas $\sigma_w \propto (H^*z)^{1/3}$). In stable conditions, at many sites, horizontal eddies or meanders are often much larger than the stability-restricted vertical turbulent motions, so σ_a and σ_z fail to correlate (e.g., see Table 2.1).

For the Brookhaven site, on the basis of both wind direction fluctuation and diffusion measurements, Singer and Smith (1966) suggested that the atmosphere seems to have "preferred states" that tend to cluster around their four stability categories (there was also a fifth category, with wind direction fluctuations exceeding 90°—extremely unstable—for which they were unable to provide σ_y and σ_z curves). They simplified determination of the categories by showing characteristic strip chart segments for each one; clearly these relate directly to σ_a, but visual inspection served in place of the calculation of σ_a, which was tedious in pre-microcomputer times.

Diffusion data of three types were gathered at Brookhaven over many years. Uranine dye from a near-surface (2 m) source provided a small amount of data at $x < 100$ m. Radiation from the A^{41} emissions from the research reactor provided some information on plume dimensions in the 10 to 60 km range, particularly during stable conditions. The great bulk of the data was from oil fog experiments, with concentration measured only near the surface, mostly in the 0.4 to 10 km range. The hourly mean concentrations of oil droplets were obtained using the fluorescent properties of the oil. These experiments contrasted greatly with the PGr measurements used by Pasquill in that the oil fog and A^{41} sources were considerably elevated, at $z = 108$ m, and the site was very much rougher, with $z_0 \simeq 100$ cm. As in the PGr and most other experiments, σ_z was inferred from surface concentrations by assuming a Gaussian plume and conservation of tracer (i.e., $\int\int \chi U dy dz$ constant with distance).

2.2.4. Problems Affecting Early σ_z Estimates

As is evident in the foregoing discussion, the only direct measurements of σ_z in these early experiments were at $x = 100$ m during the PGr experiment. It is very expensive and difficult to suspend a sufficiently dense vertical array of samplers to intercept plumes at larger distances and through much greater depths. Hence, vertical sampling of plumes has been extremely limited, even to the present time (see Table 1 of Briggs and Binkowski, 1985). In the last few years this experimental limitation has been largely overcome by making χ measurements in three dimensions using lidar and radar (Eberhard et al., 1985). Except for PGr at 100 m, on occasions when the six towers adequately caught the plume, the investigators analyzing the PGr and Brookhaven measurements were (practically) forced to estimate σ_z from surface χ by assuming (1) no loss of tracer, i.e., no deposition, evaporation, or chemical transformation, and (2) a known vertical profile shape (Gaussian was assumed). Unfortunately, there are difficulties arising from both these assumptions.

It has gradually come to light that none of the tracers used in diffusion experiments before 1970 was conservative (Briggs and Binkowski, 1985). This problem was not overcome until gas chromatography techniques were sufficiently developed to allow use of gases such as SF_6. Before then, the primary tracers were SO_2 gas, oil fog, and fluorescent particles, which tend to deposit on the surface. The PGr experiment used SO_2, which is now known to be taken up by vegetation. Gryning et al. (1983) estimated for this experiment a 20% to 25% reduction in surface χ due to deposition at $x = 200$ m and a 25% to 45% reduction at $x = 800$ m. This would make the previously inferred σ_z values at the latter distance 30% to 80% too large. Most Brookhaven experiments used oil fog, which was also one of the tracers used in the recent CONDORS (Convective Diffusion Observed with Remote Sensors) experiment (Eberhard et al., 1985). For CONDORS, significant evaporation of oil fog droplets with distance is strongly suspected, in spite of the use of a low-volatility oil (Eberhard, 1985: personal communication). Lidar scans indicate a

50% loss in the total return signal from the plume, after correction for attenuation and background haze, at travel distances of 1.4 to 2.8 km for most cases. This suggests a 50% loss of droplet volume at $x = 1$ to 2 km. Thereafter, evaporation probably slows down because of smaller droplet size and early loss of the more volatile components, but there are no direct measures of this. The degree to which oil fog evaporation would have affected σ_z assessments at Brookhaven is extremely difficult to estimate now because it depends on the oil used, the droplet size spectrum, and the air temperature, at least, but I suspect that it often caused a factor-of-2 error in inferred σ_z.

The second assumption made to get σ_z from surface concentrations, that of a Gaussian vertical profile within the plume (but with reflection at the surface), has also been called into question, especially for convective conditions. Even in neutral and stable conditions, vertical plume profiles systematically deviate from Gaussian—they could be truly Gaussian only in vertically homogeneous turbulence, which is practically nonexistent— but the deviations are not so serious that Gaussianity fails as a working assumption. However, in convective conditions the phenomena described at the opening of Sec. 2.1.2 cause some strong departures from the Gaussian profile shape. For a near-surface source, this marked departure was first demonstrated by Willis and Deardorff's (1976a) tank experiment. The plume remained at the surface for small times, with the $\int \chi dy$ profile growing upwards but maintaining an approximate half-Gaussian shape until $t \simeq 0.5z_i/w_*$. The maximum concentration then lifted from the surface, rising to the upper half of the mixing depth while surface concentrations dropped rapidly. The PGr field data, when analyzed in terms of convective scaling, are consistent with this result (Nieuwstadt, 1980). The implication for estimates of σ_z based on surface concentrations and the Gaussian shape assumption, such as those behind Pasquill's A and B curves, is that the apparent plume growth rate will be considerably exaggerated (the calculated σ_z is inversely proportional to surface $\int \chi dy$). The ultimate consequence, in applied diffusion modeling, is that one may still get the right answer if these σ_z values are inserted into Gaussian models for near-surface releases, but if applied in that manner to elevated releases one may overestimate vertical diffusion by factors up to 4.

For elevated sources in convective conditions about 60% of continuous emissions are released into passing downdrafts, so after plume rise is complete the centerline of averaged concentration descends towards the surface at a rate $\simeq 0.5w_*$. This tends to increase the maximum surface $\int \chi dy$ over that calculated by assuming a Gaussian plume centered at the effective source height, z_s, by a factor approximating $(1 + 2z_s/h)$ (Briggs, 1985). The Gaussian plume model predicts a maximum $\int \chi dy$ at the surface equal to $0.48\,Q/(Uz_s)$, which occurs when $\sigma_z = z_s$, regardless of how fast or slow σ_z grows. Thus, there is no way to manipulate σ_z artificially to get the right answer for convective conditions; one must include plume descent or non-Gaussianity in a realistic model (for a review of suggested models, see Briggs, 1985). In the Brookhaven cases, the surface concentration enhancement by dominance of downdrafts might have been masked by oil droplet evaporation since the two effects tend to counteract each other. It is difficult to estimate the net effect on the derived Brookhaven σ_z values; suffice it to say that the B1 and B2 curves are reasonably consistent with more recent convective scaling analyses of vertical plume growth from elevated sources.

Many uncertainties arise when σ_z is estimated only from surface concentrations. Many have mistakenly assumed that σ_z curves such as Pasquill's are based on direct measurements; that assumption has perhaps led to too much confidence in applying such curves to diverse sources and situations. It would be better to carry out data analyses in terms of more directly measured quantities like peak surface concentration versus distance or surface $\int \chi dy$; this does not assume the use of a particular model or shape distribution that is unsupported by measurements.

2.3. Surface-Layer-Similarity (SLS) Analyses

The next three sections give a number of examples of diffusion data analyses, mostly published in the last decade. Although some of these overlap several categories, they are divided here into the three analysis methods covered in Sec. 2.1, according to the emphasis. We begin with those data analyses primarily using SLS scaling, including L and Richardson number. These have been largely restricted to vertical diffusion from near-surface sources.

Pasquill (1966) made one of the first attempts to analyze diffusion experiment data in terms of SLS scaling. He showed how the directly determined PGr σ_z values at $x = 100$ m systematically vary with a close relative of L which is easier to determine, $L' = (u_*/k)(T_0/g)(\partial U/\partial z)/(\partial \Theta/\partial z)$. The ratio L'/L equals the ratio of eddy diffusivities of heat and momentum, which was presumed to be a function of z/L but was subject to some uncertainty at the time. Klug (1968) made an attempt to fit the PGr peak surface concentration χ_p versus distance with a form of SLS theory. In order to get good results, he had to abandon the idea that $U\bar{z}^2\chi_p/Q$ is a function of $\bar{z}/L$, where $\bar{z}$ is the (inferred) mean particle height; he concluded instead that it was correlated with σ_v/u_* in the PGr experiment. From his Fig. 3, we see that it is approximately proportional to $(\sigma_v/u_*)^{-1}$; since $\chi_p \propto \sigma_y^{-1}$, this suggests that $\sigma_y \propto \sigma_v$, but that σ_v/u_* did not conform to the SLS theory prediction that it should be a function of z/L.

The best results using SLS and related variables such as the Richardson number Ri $= (g/T_0)(\partial\Theta/\partial z)/(\partial U/\partial z)^2$ have been obtained in comparison with *vertical* measures of diffusion such as σ_z or $\int \chi dy$. Weber et al. (1977) computed linear correlations of these variables at fixed distances with various single-parameter indices for three different field experiments. For unstable conditions, they found very good correlation coefficients, $r = 0.75$ to 0.91, between $(\int \chi dy)^{-1}$ at various distances in the PGr experiment and Ri or L^{-1}, where L was inferred from detailed temperature and wind speed profiles. For the National Reactor Testing Station (NRTS) data (Islitzer and Dumbauld, 1963) the results were less clear, with $r \simeq 0.65$ at three distances (0.2, 1.6, and 3.2 km) and $r < 0.5$ at the remaining three distances (0.1, 0.4, and 0.8 km). However, overall correlations with σ_e and σ_a were poorer and even less consistent at NRTS; in the PGr data, these correlations were mostly near 0.4, σ_e being slightly superior to σ_a; correlations may have been adversely affected by the poorer response of wind vanes and bivanes of this period. Both these experiments showed negligible correlation between $(\int \chi dy)^{-1}$ and the mean temperature gradient in unstable conditions.

For stable conditions, Weber et al. (1977) found good correlations, $r = 0.63$ to 0.85, between PGr-measured $(\int \chi dy)^{-1}$ values and Ri, L^{-1}, σ_e, and even ΔT (temperature difference). At $x = 100$ m, the correlations with directly measured σ_z values from tower-mounted samplers was notably higher than with $(\int \chi dy)^{-1}$ measured at $z = 1.5$ m, perhaps in part because of the 1 m vertical offset from the release height (0.5 m). The correlations with σ_a were negligible. At NRTS, correlations with Ri, L^{-1}, σ_e , and ΔT were lower and more variable: $r = 0.39$ to 0.84, except for near-zero correlation with ΔT at the largest distances, $x > 1$ km.

Correlations with σ_a ranged from 0.24 to 0.59. In the Green Glow experiment (Fuquay et al., 1964), $r \simeq 0.63$ for both Ri and L^{-1} in the neighborhood of $x \sim 1$ km and was even better, $r \simeq 0.8$, for ΔT measured in the most ideal height range (0.6 to 6 m); however, all correlations with $(\int \chi dy)^{-1}$ degenerated at the largest distance, $x = 3.2$ km. These results are given in graphical form in Fig. 2.4.

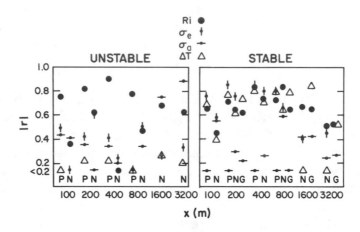

Figure 2.4. Correlation coefficients between $(\int \chi dy)^{-1}$ at the surface and meteorological indices, from Weber et al. (1977) tables. P = Prairie Grass, N = NRTS, G = Green Glow experiments.

The PGr experimental data set from 1956, because of its relative completeness in terms of number of releases, dense array of samplers, multilevel U and T measurements, and aircraft soundings, has been used many times since 1978 for analyses in terms of surface layer similarity and convective scalings. One of the first such analyses was by Briggs and McDonald (1978). A "scale height" of the cloud was defined using the crosswind-integrated concentration at the surface and U at $z = 8$ m: $\Delta z = Q/(U \int \chi dy)$. Then u_* and L were determined from the U and Θ profiles measured at factor-of-2 increments from $z = 0.25$ to 16 m. Particularly simple algorithms were developed for doing this, using published profile equations in terms of u_* and L. This method was applied to four different combinations of measurement levels with good consistency of results except for one run, subsequently eliminated. For convective scaling, a systematic method was applied to the aircraft soundings of temperature and specific humidity to determine h; only 3 of the 33 unstable runs had to be omitted because of ambiguous results for h. A number of plots were made of Δz at fixed distances (100 m and 800 m) versus single parameters, namely L, $U^2/\Delta\Theta$, U^2 , and $\Delta\Theta$. These plots showed the superiority of the Obukhov length L and its more easily measured relative $U^2/\Delta\Theta$ for characterizing vertical diffusion; U^2 and $\Delta\Theta$ worked nearly as well in stable conditions, but in unstable conditions $\Delta\Theta$ was "worthless."

Briggs and McDonald included a number of dimensionless plots. The most straightforward in terms of surface-layer similarity was $|\Delta z/L|$ versus $|tu_*/L|$, with t approximated by z/U_8. This is reproduced in Fig. 2.5. Defining $u_*x/|U_8 L| = X_L$, the unstable data points are fit nicely by the equation $\Delta z/|L| = X_L(1 + 5X_L^2)^{1/2}$. The large-$X_L$ asymptote is steeper than the "free convection" prediction $\Delta z/L \propto (tu_*/L)^{3/2}$, probably because of "sweepout" of near-surface material into thermals, a phenomenon better represented by convective scaling. Some of the stable data at large x are notably higher

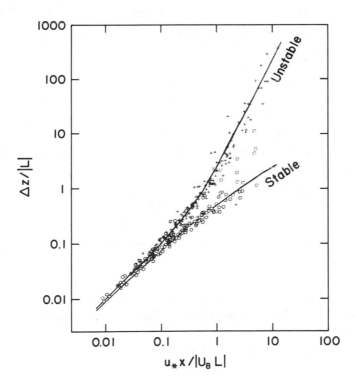

Figure 2.5. Surface layer similarity plot of Prairie Grass $\Delta z/|L|$ versus $u_*x/|U_8 L|$, where Δz is a plume thickness scale defined from surface concentrations: $\Delta z = Q/(U_8 \int \chi dy)$. From Briggs and McDonald (1978).

than the general cluster of points; this was thought to be due to vertical displacement of maximum $\int \chi dy$ values from the sampler heights,
especially in the most stable cases (this problem was addressed in
Briggs, 1982b). The general cluster of points was fit by the equation
$\Delta z/L = X_L/(1+X_L^{1/2})$. Realizing that L is generally difficult to determine reliably, the authors tried an alternative, nondimensionaliz-
ing both Δz and x with the length $U_8^2\Theta/(g\Delta\Theta)$, with $\Delta\Theta = (\Theta_8-\Theta_1)$.
According to SLS theory, this length should be proportional to L,
but is also a function of the measurement heights and z_0. This plot
was nearly as effective as Fig. 2.5, in terms of limited scatter, but the
result would be harder to transfer to sites having much different z_0
(the PGr site was extraordinarily smooth, with $z_0 \simeq 0.6$ cm).

Further testing of the PGr data in terms of SLS theory, under the
name of "Lagrangian similarity modeling," was carried out by Horst
(1979). To determine L from U and Θ profiles, he first calculated Ri
at several heights and then fit the SLS relationship $z/L = \phi_m^2 \text{Ri}/\phi_h$
with the ϕ_m and ϕ_h (the dimensionless forms of $\partial U/\partial z$ and $\partial\Theta/\partial z$)
of Businger et al. (1971). Since $\Delta z/L$ is a function of X_L, according
to SLS theory, then so is $X_L/(\Delta z/L) = xu_*\int \chi dy/Q$. Horst applied
this format at fixed values of x by simply plotting $u_*\int \chi dy/Q$ versus
L^{-1} (the ratio U/u_* that appears in X_L is also assumed to be a function of $\Delta z/L$). He compared the PGr data plotted in this way with
the theoretical prediction due to Chaudhry and Meroney (1973) that
$d\bar{z}/dt = ku_*/\phi_h$. An example of his result, for $x = 200$ m, is shown
here in Fig. 2.6 (the open circles indicate cases where computed L
values for the run differed by more than a factor of 2). The theoretical curve fits the PGr data rather well, except at $x = 800$ m for
the most unstable cases ($|L| < 10$ m). In these cases, Δz was of
the order of 1 km, i.e., of h, so the conditions were beyond those
for SLS and convective scaling would be the logical choice. Horst
also applied this treatment to the NRTS data at $x = 3200$ m (Islitzer
and Dumbauld, 1963), which present a similar situation for the most
unstable cases. Furthermore, very significant deposition losses were
estimated for the NRTS experiments (Islitzer and Dumbauld, 1963),
10% to 54% at $x = 400$ m; this casts much uncertainty over comparisons of χ at 3200 m with theoretical predictions that assume no
losses.

The effect of deposition losses in the PGr experiment was addressed by Gryning et al. (1983). For computational convenience

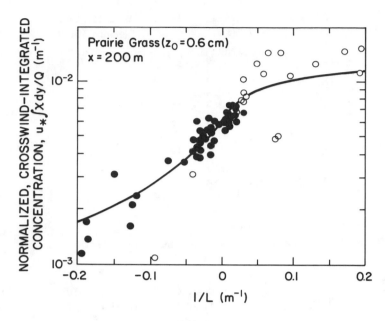

Figure 2.6. Horst's (1979) plot of $u_* \int \chi dy/Q$ (PGr surface values) versus $1/L$ at $x = 200$ m. Open circles indicate greater uncertainty in L values, inferred from U and Θ profiles. Solid line is theoretical prediction using $K = K_h$.

they assumed that the deposition velocity v_g was $0.05u_*$. The constant was chosen on the basis of estimates of transfer resistance through the laminar sublayer and typical surface resistance of medium-height grass to SO_2, evaluated at typical values of u_* for the PGr experiment (actually, the constant could be about 1.4 times as large for the most unstable runs and 2 times as large for the most stable runs, as it is larger for small u_*). Similar to Horst's (1979) presentation, plots of $u_* \int \chi dy/Q$ versus L^{-1} were shown for fixed distances and compared with theoretical predictions. The theory was more elaborate than Horst's, with the effects of the variation of U and eddy viscosity K with height on the vertical concentration profile being considered, as well as deposition effects. Examples of calculated profiles for neutral conditions with and without deposition are shown in Fig. 2.7. Deposition had little effect at the sampler height (1.5 m) at $x = 50$ m, but at 800 m it reduced χ by about 45% in very stable conditions and by about 30% in neutral and unstable conditions. With deposition considered, the fit to the PGr data is substantially improved, especially for neutral and unstable cases at $x = 800$ m.

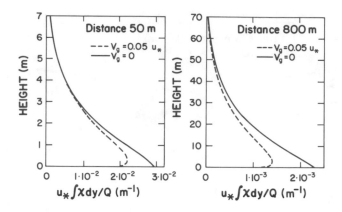

Figure 2.7. Gryning et al. (1983) $\int \chi dy$ profile predictions for PGr releases at $x = 50$ and 800 m, neutral conditions, with and without SO_2 deposition.

Attempts to simplify analysis of PGr data by eliminating U were made by both Venkatram (1982) and Briggs (1982b); because U is a function of z, it grows as σ_z or $\bar{z}$ grows, making computations more complex. Both authors noted that both the neutral and extremely stable asymptotes of $z^* \equiv Q/(u_* \int \chi dy)$ depend only on L and x, and not on U, using SLS theory. [The neutral asymptote is given by $d\Delta z/dt \propto u_*$; because $\Delta z = (u_*/U)z^*$ and $dt = dx/U$, U cancels out if we neglect dU/dx, and $z^* \propto x$ results. The stable asymptote is given by $K \propto K_m \propto u_* L$, which yields $\Delta z \propto (u_* Lx/U)^{1/2}$ and $\bar{u} \propto u_* \Delta z/L$, so $\Delta z \propto (L^2 x)^{1/3}$ and $z^* \propto (Lx^2)^{1/3}$; however, this theoretical asymptote for $\Delta z \gg L$ is not actually found in nature because the stable boundary layer depth seldom exceeds $10L$.] Briggs further noted that outside of these asymptotes z^* is weakly dependent on surface roughness; $z^* \propto z_0^n$, approximately, with $n < 0.04$ for stable and $n < 0.1$ for unstable cases. The analysis then neglected z_0 effects and fitted curves to PGr data plotted in the form z^*/x versus x/L. Six runs that give anomalous results were identified and possible reasons for their misbehavior were given (these included extremely low U, extreme stability, a terrain drop at $x = 800$ m, and anomalous σ_a). For the first time, a simple correction was made for the difference between the release and sampler heights, 1 m for most runs. This had the effect of reducing the neutral asymptote (small x/L) by 20% compared with Briggs and McDonald (1978) and Venkatram (1982), to $z^* = 0.8x$. However, the effect of SO_2 deposition was not taken into account; according to the Gryning et al. (1983) analysis, this would have the effect of reducing the large $|x/L|$ asymptotes of Fig. 2.8 by about 30% for unstable conditions and 45% for stable conditions. Note that the half-circled and circled unstable data points indicate that $\Delta z > 0.5h$ and $> 1.0h$, respectively, which means that these are more properly described using convective, rather than SLS, scaling.

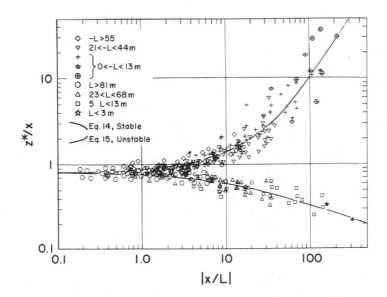

Figure 2.8. Surface layer similarity plot of Prairie Grass z^*/x versus $x/|L|$, where $z^* = Q/(u_* \int \chi dy)$ (Briggs, 1982b); Eq. (14) from this reference is $z^*/x = 0.8/(1 + 0.13x/L)^{1/3}$ and Eq. (15) is $z^*/x = 0.8[1 - 0.19x/L - 0.00014(x/L)^3]^{1/2}$.

2.4. Convective-Scaling Analyses

Convective scaling can be profitably applied to almost any diffusion data collected in the daytime, unless conditions were near neutral. However, substantial source buoyancy is a complicating factor that can affect a number of aspects of diffusion besides plume rise. For convenience, this discussion is split into analyses of nonbuoyant and buoyant sources.

2.4.1. Nonbuoyant Sources

Field data were first analyzed in terms of convective scaling by Briggs and McDonald (1978), who plotted PGr unstable values of $\Delta z/h$ versus $(x/U)w_*/h = X$. These points were compared with a curve derived from near-surface values of $\int \chi dy$ measured in the first laboratory tank diffusion study of Willis and Deardorff (1976a). The small-X field values lined up well with the curve, but the peak at $\Delta z/h \simeq 1.4$ was reached sooner in the field data, by roughly a factor of 2. Nieuwstadt (1980) carried out a similar analysis on the PGr data, but extended it to include σ_y. For $Uh \int \chi dy/Q$ (the inverse of $\Delta z/h$), he found a broad regime that is approximated by $0.9X^{-3/2}$, in agreement with the free-convection prediction $\sigma_z \propto \Delta z \propto hX^{3/2} \propto H*^{1/2}t^{3/2}$. Nieuwstadt's σ_y/h-versus-X plot revealed σ_y/h to be about 40% larger for the PGr data than the Deardorff and Willis (1975) result. This discrepancy may be in part due to the constraining effect of the tank walls on horizontal turbulent motions in the laboratory experiment.

The PGr experimental data did not go far downwind; only a few points were beyond $X \simeq 0.5$. Furthermore, since it was limited to near-surface measurements of only 10-min releases from near the surface, its usefulness is rather limited for testing diffusion models in the convective boundary layer, as vertical diffusion occurs through 1000 m or more and highly non-Gaussian behavior has been suggested by laboratory and numerical models (Sec. 2.2.4).

To expand the scope of field data for convective conditions, the CONDORS experiment was carried out near Boulder, CO, in 1982 and 1983 (Moninger et al., 1983; Eberhard et al., 1985). Although there was some very limited surface sampling of conservative tracers in 1983, the main tracers used were oil fog, scanned by lidar, and aluminum-coated "chaff," scanned by Doppler radar. This allowed the simultaneous mapping of both oil fog and chaff plumes in three dimensions, at distances up to 4 km. Both surface and elevated releases were made, sometimes with the sources collocated and sometimes with the oil fog released at the surface while the chaff was released from high up on the 300-m meteorological tower. Only some preliminary data analyses have been carried out at this time, but data processing is now complete and a data report is available (Kaimal et al., 1986).

In analyzing the CONDORS data, it cannot be assumed that either the chaff or oil fog is a conservative tracer. As mentioned in Sec. 2.2.4, there is strong evidence that significant evaporation of oil fog droplets took place (Eberhard et al., 1985), perhaps 50% in 1 or 2 km of travel. The smallest chaff available is known to have a settling velocity of about 0.3 m s^{-1}, so it deposits with distance, as the dropoff of measured $\int\int \chi dy dz$ with x indicates (beyond 1 or 2 km of travel). The settling velocity also distorts the vertical profile of χ in comparison with passive tracers; an effort will be made to correct this defect, using the oil fog scans during periods of collocated release as a standard. Since neither tracer is conservative, χ is nondimensionalized using the *local* downwind flux of tracer, $\int\int U\chi dy dz$, in place of the release rate, Q. This procedure distorts the results compared with those for passive tracers only to the extent that the percentage depletion of material is nonuniform.

Some preliminary analyses of CONDORS data from both surface and elevated chaff releases are shown here in Figs. 2.9 and 2.10. These convectively scaled plots include a variety of field data, including PGr, some power plant plume measurements (Weil, 1979), chaff measurements from CONDORS (labeled "BAO"), and SF_6 measure-

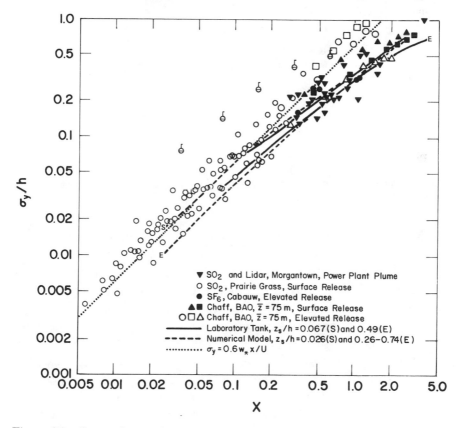

Figure 2.9. Convective scaling plot of σ_y/h versus $X = (x/U)w_*/h$ from Briggs (1985). "E" identifies best-fit lines for elevated sources.

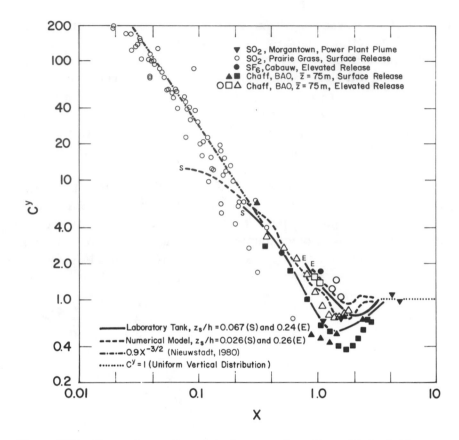

Figure 2.10. Convective scaling plot of near-surface $C^y = Uh \int \chi dy/Q$ versus $X = (x/U)w_*/h$. "E" and "S" identify best-fit lines for elevated and surface sources, respectively.

ments from elevated releases at Cabauw, The Netherlands (Agterberg et al., 1983). Figure 2.9 shows σ_y/h, versus X. In the case of σ_y, the chaff settling velocity is not expected to alter the results significantly, especially since the turbulence velocity σ_v is nearly uniform with height. The flagged run from PGr was anomalous in that the χ-versus-y distributions had a double peak, which suggests that the wind direction changed during the run. Most other points cluster around the small-X prediction assuming that $\sigma_y \simeq \sigma_v t$ and $\sigma_v \simeq 0.6w_*$, or the Willis and Deardorff (laboratory) and Lamb (numerical) curves, which drop off very gradually from the small-X prediction.

The laboratory and numerical experiments could not include σ_y enhancement due to mesoscale eddies or wind direction shear, both of which occur often in the real atmosphere. For instance, it is known that the two elevated chaff runs showing the largest σ_y/h values in CONDORS were made on a day with unusually large wind direction shear. Such occurrences are due to upslope effects on the eastern slope of the Rocky Mountains, which are just west of the BAO site.

Figure 2.10 shows $Uh \int \chi dy/Q$ versus X for the same data sets, with $\int \chi dy$ measured near the surface. Although the BAO chaff data fall reasonably near the Willis and Deardorff tank curves for both surface and elevated releases, there is more uncertainty here because of the chaff settling effect and because the measurement was made at $z = 75$ m, owing to radar blockage and ground clutter problems below 50 m (the bottom of the grid cells used here). Figure 2.11 shows a direct comparison between an elevated chaff release and laboratory tank results (Willis and Deardorff, 1981) for contours of C^y distribution in z/h and X (Moninger et al., 1983). The field data have more "lumpy" contours, in part because of shorter averaging in terms of numbers of eddies sampled, but there are some remarkable similarities with the contours from the tank experiment.

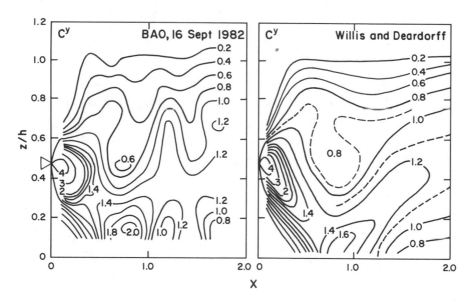

Figure 2.11. Convective scaling plot of C^y isopleths versus z/h and X. Chaff from CONDORS field experiment on left and laboratory tank experiment on right. From Moninger et al. (1983).

Systematic effects of chaff fall velocity can be seen in comparisons with oil fog distributions of C^y versus z/h during collocated releases; e.g., in Fig. 2.12 we see that during this period with a release height $\simeq 0.3h$, the maximum chaff impact at the surface occurred at $X \simeq 0.34$, whereas the maximum oil fog impact occurred at $X \simeq 0.46$. This difference is about what would be expected on the basis of a 0.3 m s^{-1} settling velocity. However, there is more variability in the behavior of individual averaging periods of field observations during CONDORS compared with the well-averaged and smoothed tank

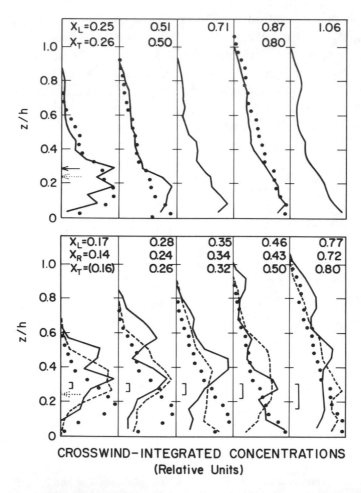

CROSSWIND–INTEGRATED CONCENTRATIONS
(Relative Units)

Figure 2.12. Comparison of vertical profiles $\int \chi dy$ for laboratory tank experiments (dots) and two CONDORS periods, oil fog measured with lidar (solid lines) and chaff measured with radar (dashed lines). X_L, X_R, and X_T are X values for lidar, radar, and tank data, respectively. From Eberhard et al. (1985).

data. Figure 2.12 shows a period with oil fog C^y profiles in very good agreement with tank results for $z_s/h = 0.24$, and also shows, from a run on the following day, oil fog profiles that initially do not descend. However, the scan at maximum surface impact ($X = 0.46$) shows a good comparison with Willis and Deardorff's (1978) tank result.

Further analyses of the CONDORS data are presented in Briggs et al. (1986). We hope to carry out much more analysis in the near future. These efforts will include attempts at corrections for blockage and settling velocity in the chaff distributions and extrapolations of both chaff and oil fog distributions to the surface.

2.4.2. Buoyant Sources

Convective scaling appears to be a useful tool for doing analyses of diffusion from buoyant, as well as passive, sources. Buoyancy may have significant effects on the diffusion, however. The degree of these effects is indicated mainly by a single dimensionless parameter, $F^* = F_b/(Uw_*^2h)$, where F_b is the source buoyancy flux; F_b is usually defined by $g(\rho_a - \rho_o)/\rho_a$ times the volume flux of source gases divided by π, where ρ_a and ρ_o are the ambient and source gas densities; (F_b/U) is the buoyancy per unit length in a bent-over plume (plumes are almost always bent over). Since the buoyancy contained in a segment of bent-over plume caught in a single thermal or downdraft is $\propto F_b(h/U)$ and the buoyancy contained in an ambient thermal is $\propto H^*h^2(h/w_*) = w_*^2h^2$, F^* can be thought of as the ratio of these buoyant forces, i.e., plume versus ambient.

The parameter F^* first appeared implicitly in forms for σ_y suggested by Weil (1979). These equations assumed vectorial addition of growth caused by ambient convective eddies and by entrainment due to buoyant plume rise. The latter is given by the "2/3 law." Weil assumed $0.9(F_b/U)^{1/3}(x/U)^{2/3}$ for this component of σ_y; this can also be expressed as $0.9h F^{*1/3}X^{2/3}$. He used Lamb's (1979) σ_y formulation for growth due to ambient, convective turbulence. Lamb's formulation is in two stages, so there are two results. At $X < 1$, the ambient-induced $\sigma_y = (1/3)hX$, so enhancement of σ_y due to buoyancy is proportional to $(1 + 7.3(F^*/X)^{2/3})^{1/2}$. For the power plant data shown in Figs. 2.9 and 2.10, F^*/X was used to screen out significant buoyancy effects. Data were excluded from Fig. 2.9 if $F^*/X > 0.2$ and from Fig. 2.10 if $F^*/X > 0.032$; however, larger ambient-induced growth and smaller buoyancy-induced growth, compared with those of Weil, were assumed. At $X > 1$, the ambient-induced $\sigma_y = (1/3)hX^{2/3}$, so the enhancement due to buoyancy was just $(1 + 7.3F^{*2/3})^{1/2}$ in Weil's formulation. Venkatram (1980) suggested some rather similar forms for σ_y but with different numbers, e.g., 1.57 in place of 7.3; Briggs (1985) suggested that both of these formulations underestimate σ_y somewhat. The above analyses suggest that $F^*/X = F_b/(w_*^3 x)$ is the primary index of the importance of buoyancy effects at small X, and F^* alone is the main index at large X.

Weil (1979) presented the first analysis of power plant diffusion data in terms of convective scaling, using lidar and a van-mounted correlation spectrometer to vertically integrate SO_2 from the Morgantown, MD, power plant. Both σ_y/h and σ_z/h were plotted against X, for comparison with Lamb's (1978) results from numerical modeling and with Weil's suggested forms for "2/3 law" enhancement. Although the latter gave a better fit to σ_y, there are other passive plume σ_y formulations that give a larger σ_y than Lamb's equations (Briggs, 1985) and would therefore require less buoyancy-induced enhancement to give a good fit. Weil (1983) also showed a plot of σ_y/h versus X for three power plants, plus PGr passive surface release data. This plot looks much like Fig. 2.9, with a little overlap and remarkable continuity between the PGr and power plant data. As in Fig. 2.9, the scatter increased somewhat at $X > 1$, with most data points clustering between the small-X asymptote ($\sigma_y \propto x$) and a curve based on tank-experiment results from Deardorff and Willis (1975).

Carras and Williams (1983) presented similar plots for vertical diffusion of plumes from several isolated smelter stacks in Australia. They determined a vertical plume width W_z from the envelope of 15-min periods of time-lapse-photographed plume outlines, and used methods suggested by Venkatram (1978) to estimate w_*. As F_b was rather modest for these sources, 277 m^4 s^{-3} for the largest, F^* was only of the order of 0.01 and buoyancy effects on W_z were apparently negligible. Examples of these plots of W_z/h versus X are shown in Fig. 2.13. They were compared with Lamb's (1979) σ_z formulation by assuming that $W_z = 4.3\sigma_z$.

When $F^* > 0.1$ or so, buoyancy effects become very important. The most recent convective tank experiments have shown that buoyant plumes "loft" near the top of the mixing layer at such values of F^* (Willis and Hukari, 1984). At smaller values, plume rise is the

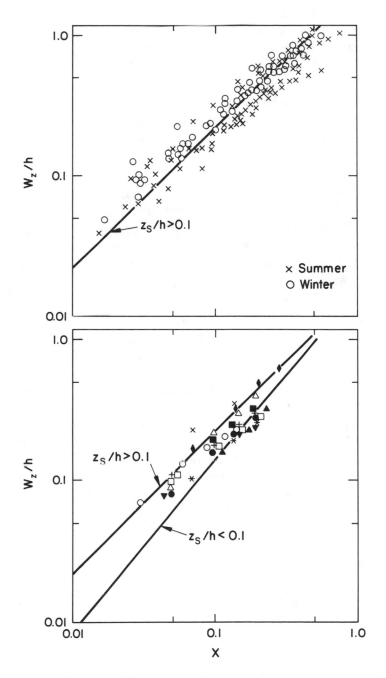

Figure 2.13. W_z/h versus X, using smelter-plume data from Carras and Williams (1983), compared with Lamb's (1979) formulas assuming that $W_z = 4.3\sigma_z$.

main buoyancy effect. Any plume rise model consistent with convective scaling can be put in the form $\Delta h = h f(F^*, z_s/h)$. The simplest one recommended by Briggs (1984) is $\Delta h = 3(F_b/U)^{3/5}H^{*-2/5} = 3hF^{*3/5}$. Note that for $F^* = 0.1$, it gives $\Delta h = 0.75h$ for the plume centerline; depending somewhat on z_s/h, this plume rise model does predict that the *top* of the plume will bump against $z = h$ at this value of F^* or a little less. What happens at larger F^* depends somewhat on the strength of the capping inversion above $z = h$. If it is weakly stable, the plume might penetrate into the stable air and remove itself from the convective boundary layer altogether. Significant partial penetration can happen if $F_b > 0.1U(h - z_s)^2 g\Delta\Theta_i/\Theta$ or so, where $\Delta\Theta_i$ is the change in potential temperature across an inversion capping the mixed layer. In such cases, more elaborate modeling efforts are required, as the diffusion cannot be described simply in terms of convective scaling (models for penetration of several types of elevated stable layers are given in Briggs, 1984). However, for typical capping strengths significant penetration does not occur until $F^* > 1$ or so. For a wide range of F^*, typically 0.1 to 1, buoyant plumes loft into the top of the mixed layer, yet are trapped by the capping layer (thus, they are buoyant relative to the mixed layer but denser than the air above).

This lofting of buoyant plumes appears to have two main effects. First, the residual buoyancy causes more rapid lateral spreading of the lofted plume, as with hot smoke on a ceiling. Second, the positive buoyancy of the plume material helps it resist downward mixing, until the plume sufficiently thins out. Such F^* effects were systematically explored in Briggs (1985), although this study was limited to data from only two power plants. For $F^* < 0.06$ cases, there was no apparent enhancement of σ_y measured at the *surface* due to buoyancy. As a practical approximation, the data in this range of F^* could be approximated by Lamb's (1979) equation for $X > 1$, increased by 50%: $\sigma_y = 0.5hX^{2/3}$. This was used as a baseline for looking for F^* enhancement of σ_y, by plotting $\sigma_y/(hX^{2/3})$ versus F^*. This plot showed much data scatter, but there was a definite upturn for $F^* > 0.1$ cases, in the "bumping" regime (however, very low wind speed cases, with $U < 1.2w_*$, did not conform). This upturn behaved roughly as $\sigma_y = 1.6F_b^{1/3}U^{-1}x^{2/3} = 1.6hF^{*1/3}X^{2/3}$, the "2/3 law" prediction that applies to the horizontal spread of a line of buoyant material, as well as to bent-over plume rise (Chen, 1980).

Briggs (1985) suggested that "mixdown" of lofted buoyant plumes ($F^* > 0.1$) would be delayed in proportion to F^*, until some $X \propto F^*$ (dimensionalization of this cancels U and h, with the result $x \propto F_b/w_*^3$). One simple way to get this result is to assume that thermals impinging on the underside of the lofted plume have a vertical velocity $\propto w_*$; then they can penetrate into the plume a vertical distance $\Delta z \propto w_*^2/(g\Delta\Theta_p/\Theta)$, where $\Delta\Theta_p/\Theta$ is the relative density difference of the plume material. Since the total buoyancy flux is conserved in a bent-over plume embedded in a well-mixed environment, $g(\Delta\Theta_p/\Theta)U\sigma_y\sigma_z \propto F_b$, the source buoyancy; assuming $\sigma_y \propto hF^{*1/3}X^{2/3}$, the "2/3 law" spread, substitutions above lead immediately to $\Delta z/\sigma_z \propto (X/F^*)^{2/3}$. Then at some critical value of X/F^*, the impinging thermal can deeply penetrate the lofted plume and begin mixing it substantially into the mixing layer by entrainment. At very large X/F^*, the plume becomes completely mixed down to the surface, with the result for uniform vertical mixing: $\int \chi dy = Q/(Uh)$.

Briggs (1985) tested this hypothesis by plotting $F^* > 0.1$ power plant observations in the form $C^y = Uh \int \chi dy/Q$ versus X/F^*, which is shown here as Fig. 2.14. Most of the data conform very well to

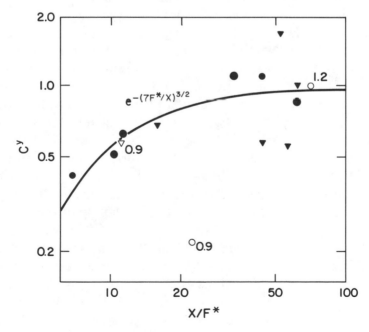

Figure 2.14. Convective scaling plot of C^y at surface versus $X/F^* = xw_*^3/F_b$ for two power plants, $F^* > 0.1$ cases. Open symbols, $U/w_* < 1.5$. Large filled circles, sampling duration times $(U/h) > 40$.

the hypothesis, especially when the sampling times in terms of h/U were > 40 (large symbols). At $X > 30F^*$, $C^y \simeq 1$, the uniform-mixing asymptote; at $X \simeq 10F^*$, only one-half of this concentration is observed at the surface. The experimental goals may have excluded measurements at smaller X/F^*, because of the absence of the plume material at the surface. More comprehensive testing of these ideas is needed. Recent additional analyses of buoyant plume data in terms of X and F^* have been presented by Hanna et al. (1986), Pierce (1986), Weil et al. (1986), and Willis (1986).

In summary, it appears that convective scaling provides a potent analysis tool for daytime diffusion measurements of both buoyant and passive plumes trapped within the mixed layer. The dimensionless buoyancy flux F^* is an extremely important additional parameter.

2.5. Statistical-Theory Analyses

Although Pasquill (1961) expressed a preference for diffusion modeling on the basis of wind fluctuations measured at the site, using some form of statistical theory, this approach has been taken up seriously only in the last decade. Perhaps steady improvements in instrumentation will make it increasingly attractive.

The first comprehensive effort of this type was undertaken by Draxler (1976). He looked at five experiments with surface sources and six experiments with elevated sources. The instrumentation was relatively crude in many of the early experiments, with sluggish or underdamped wind vane response; this undoubtedly is responsible for some of the large scatter in his results. Draxler started with equations such as Eq. (2.5), with $\sigma_y/(\sigma_a x)$ or $\sigma_z/(\sigma_e x)$ equal to functions f_1 or f_2 of t/T_L. As the Lagrangian time scale T_L can vary widely and was not known, Draxler used an entirely empirical strategy. First, for each trial he determined a time t_i at which f_1 or $f_2 = 0.5$ (he did this using a linear regression fit to log f versus log t). Then all data were plotted in the form f versus t/t_i; using these plots he suggested a best fit for most cases as $f_1 = f_2 = 1/(1 + 0.9(t/t_i)^{0.5})$. His plot for surface sources is shown here as Fig. 2.15. In the above cases the time scale T_L must then equal 0.62 t_i, to produce agreement with the large t/T_L asymptote of Taylor's (1921) theory: $f_{1,2} = (2T_L/t)^{1/2}$.

The values of t_i for individual runs varied tremendously, from about 10 to 10^4 s. Draxler plotted these separately for horizontal and vertical diffusion, and for surface and elevated sources, as a function of a bulk Richardson number. He discerned no trends with Ri for σ_y

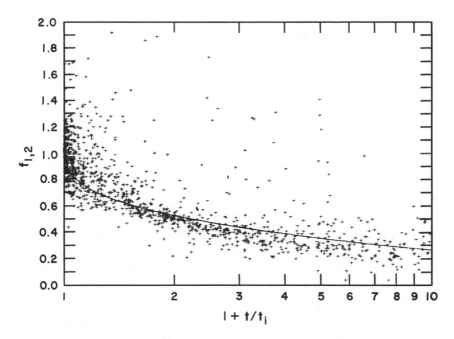

Figure 2.15. Draxler's (1976) $f_1 = \sigma_y/(\sigma_a x)$ and $f_2 = \sigma_z/(\sigma_e x)$ versus $(1 + t/t_i)$, where t_i is travel time for f_1 or $f_2 = 0.5$ (by regression analysis). Surface source, stable conditions. Line indicates $[1 + 0.9(t/t_i)^{0.5}]^{-1}$.

and only some broad difference between stable and unstable cases for vertical diffusion. He then recommended some rough-cut t_i values for various cases; these ranged from 50 s for surface source vertical diffusion in stable conditions to 1000 s for horizontal diffusion from elevated sources in all stabilities.

Draxler's σ_z values were not directly determined, but rather were derived from surface $\int \chi dy$ and the assumptions of Gaussian diffusion and no depletion, whereas considerable deposition did occur in some experiments. For σ_z from a ground source in unstable conditions, there was too much scatter in the total data set to determine f_2; assuming that the least amount of deposition occurred in the PGr experiment, and noting less scatter in that data set, he determined a best fit f_2 for the PGr data alone. This f_2 was radically different from those in the other cases: $f_2 = 1 - 1.5t/t_i + 1.88(t/t_i)^2$. He also noted that f_2 drops off more quickly than $t^{-1/2}$ for vertical diffusion from elevated sources in stable conditions; for this case, he suggested $f_2 = 1/[1 + 0.95(t/t_i)^{0.8}]$. In addition, he noted a trend toward larger f_1 for surface-source σ_y in stable conditions at $t > 550$ s; he attributed this to enhancement of σ_y caused by wind direction shear, and suggested still another empirical form of f_1 for this case.

Doran et al. (1978) considered statistical theory in more detail in order to explain differences in f_1 as a function of x. They did this by looking at forms of f_1 based on Lagrangian and Eulerian turbulent velocity spectra, and then computing f_1 versus x for some sample spectra based roughly on published measurements. They concluded that both the averaging time (due to numerical smoothing or instrument response) and the sampling time (total time of release or of sampling), as well as stability conditions, significantly affect $f_1(x)$. They also showed empirical $f_1(x)$ curves for four different surface release experiments. The comparison showed a clear trend towards more rapidly decreasing f_1 with shorter sampling times.

The most comprehensive and detailed published analysis of diffusion data in terms of wind fluctuation measurements was by Irwin (1983). He looked at ten different field programs for surface releases and seven programs for elevated releases. He also carefully considered the strengths and shortcomings of each experiment, including terrain, sampling time, and deposition effects. Because of deposition effects, and because σ_z from surface sources could be inferred only by assuming a Gaussian plume with no tracer loss, Irwin used only data with minimal suspected deposition for these cases—namely, the SO_2 observations from the PGr and RH experiments. For elevated sources, σ_z was inferred in the same manner for several experiments; however, surface deposition effects should be less than for surface-source plumes. Irwin compared this large array of data with the forms of f_1 or f_2 versus t or x published by Draxler (1976), Cramer (1976), and Pasquill (1976), as well as with two simplified versions of Draxler's recommendations. He also compared a restricted data set with Pasquill's σ_y and σ_z curves as classified by Turner's (1970) method, for all cases with data adequate for this stability classification system.

Irwin gave extensive figures and tables showing the relative performance of these models as a function of distance classes for both σ_y and σ_z, surface and elevated sources, and stable and unstable conditions. He chose as a performance measure the fractional error

$$E = 2(P - O)/(P + O), \qquad (2.6)$$

where P is a predicted value and O is an observed value; E can range from -2 to $+2$ and has the same magnitude, although with reversed signs, for either under- or overestimates by the same factor. For each case he showed the mean value and standard deviation of E, as well as the percentage of cases with P within a factor of 2 of O ($|E| < 2/3$). This presentation method highlighted changes in model performances with distance, weak areas of various models, and also many cases for which there were only small differences in model performance (in such cases, error is probably mostly due to measurement inadequacies and atmospheric variability). An example of these comparisons is given here in Fig. 2.16, for vertical diffusion. Irwin also showed the ratios of predicted and observed centerline surface concentrations; an example for the data subset that could be compared with Pasquill's methodology is shown here as Fig. 2.17. Here we see that the Pasquill/Turner approach worked about as well as the best statistical models for surface releases in stable conditions, but for the data taken in unstable conditions the statistical models tended to do far better.

A similar type of data analysis, but with inclusion of more "traditional" type models using stability classification schemes, was carried out for a diffusion experiment in Copenhagen by Gryning and Lyck (1984). The site, in a northern suburb, was chosen to be more representative of urban environments than classic tracer experiments, most of which were done at very flat and smooth sites; for this site, $z_0 \simeq 60$ cm. A conservative tracer, SF_6, was released at $z_s = 115$ m on a TV tower, and samplers were spaced about $2°$ on roughly circular arcs at $x \simeq 2, 4$, and 6 km. Most samplers were mounted on lamp posts along existing roads, so they measured essentially "surface" concentrations. Although the Gaussian plume assumption again had to be invoked to infer σ_z, at least in this experiment deposition was not a problem, and good quality turbulence measurements were obtained. A fractional error analysis, similar to that of Irwin (1983), showed that for σ_y the statistical theory models of Draxler (1976) and of Hay and Pasquill (1959) gave substantially better predictions than did other techniques (these included both the Pasquill/Turner and the Brookhaven methods). Pasquill's (1976) $f_1(x)$ function was shown to underestimate σ_y consistently, except at the lowest wind speeds. For $\int \chi dy$ at the surface as predicted by the Gaussian vertical plume distribution, Draxler's σ_z predictions worked best, followed closely by the Hay and Pasquill method and the Pasquill/Turner σ_z predictions (Turner, 1970). The authors also noted systematic variations in the fractional error of the Hay and Pasquill method with distance

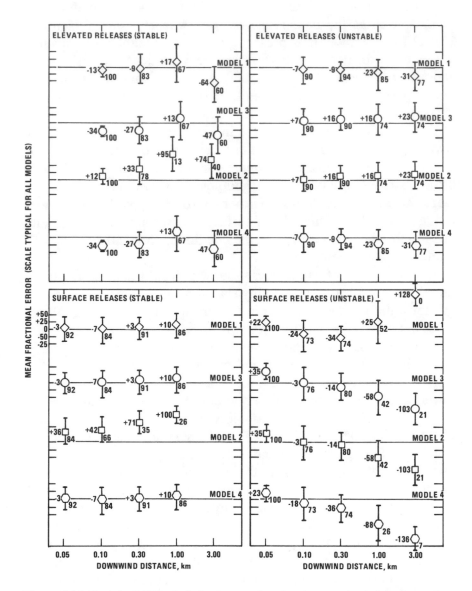

Figure 2.16. Irwin's (1983) analysis of mean fractional errors for inferred σ_z values compared with predictions of various models. Model 1 = Draxler (1976), Model 2 = Cramer (1976), Models 3 and 4 = simplified versions of Draxler (1976).

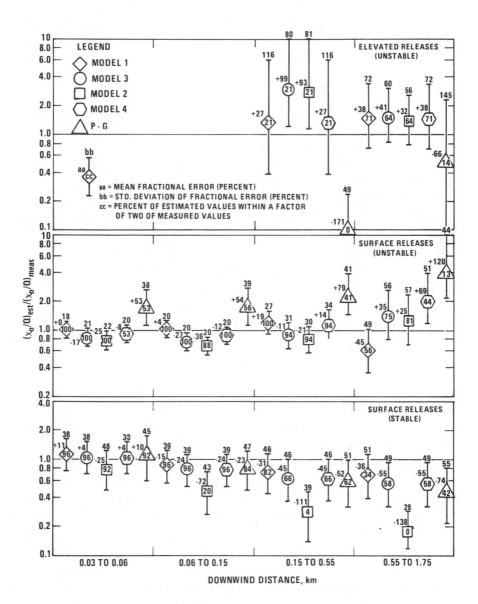

Figure 2.17. Irwin's (1983) analysis of ratios of predicted to observed χ_0 at surface for various models, as in Fig. 2.16, plus P-G = Pasquill/Gifford curves with Turner (1970) classification scheme.

and stability, suggesting some flaws in the basic assumptions of the statistical theory (e.g., similarity between Lagrangian and Eulerian velocity spectra).

An interesting new approach to σ_z analysis combining statistical and SLS theories was recently tried by Venkatram et al. (1984). The field data from Cinder Cone Butte, Idaho, were also unique; σ_z of elevated plumes in stable conditions was determined both from lidar cross sections and from successive photographs of oil fog illuminated by either a rotating carbon arc light or moonlight. A comparison of simultaneous lidar and photographic periods established the relationship $\sigma_z \simeq 0.27$ times the plume depth, nearly what it would be for a top-hat vertical distribution. They assumed that $f_2 = \sigma_z/(\sigma_w t)$ is a function of t/T_L, but made a *theoretical* estimate of T_L using SLS theory. It was assumed that the eddy diffusivity for passive diffusion equals that for heat, $K = K_h$; that $T_L = K/\sigma_w^2$, $\sigma_w = 1.3u_*$, and $k = 0.35$; and that K_h and $\partial\Theta/\partial z$ are given by the Businger et al. (1971) flux-profile relationship to derive neutral and stable asymptotes for T_L. For simplicity, it was assumed that the reciprocals of these asymptotes can be summed to get the result

$$1/T_L = 2.75\sigma_w/z_s + 3.66N, \tag{2.7}$$

where $N^2 = (g/\Theta)\partial\Theta/\partial z$. Thus, the only meteorological measurements needed to get a dimensionless plot in the form $\sigma_z/(\sigma_w t)$ versus t/T_L were σ_w, U, and $\partial\Theta/\partial z$ at the release height. The result is shown here as Fig. 2.18. On this scatter plot the authors plotted the interpolation formula $f_2 = (1 + t/2T_L)^{1/2}$, which gives the constant-K result $\sigma_z \simeq (2Kt)^{1/2}$ at large t. However, the plot shows some evidence of a somewhat more rapid drop of f_2 at large t, reminiscent of Draxler's (1976) empirical finding for the same case; at $t > 8T_L$, the Cinder Cone Butte data are represented better by $\sigma_z \propto \sigma_w t(t/T_L)^{-0.6}$. Venkatram et al. (1984) believed that the circled outliers reflect uncertainties and interpolation problems in the meteorological measurements. Overall, this attempt to characterize T_L for vertical diffusion in terms of measurable meteorology variables appears to order the σ_z observations well, with about a factor-of-6 drop in f_2 at the largest observed t/T_L values.

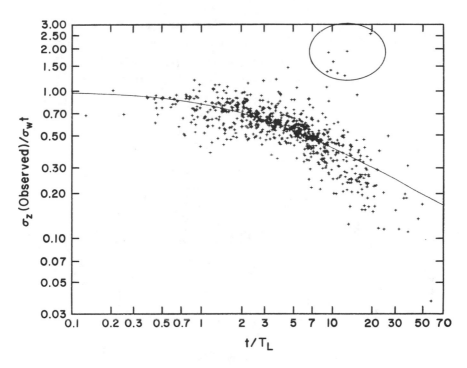

Figure 2.18. Venkatram et al. (1984) analysis of $f_2 = \sigma_z/(\sigma_w t)$ from Cinder Cone Butte lidar scans and photographs versus t/T_L, where it was assumed that $1/T_L = 2.75\sigma_w/z_s + 3.66N$.

2.6. Conclusions and Recommendations

We have reviewed three general techniques developed for analysis of diffusion data in terms of a few basic meteorological parameters that seem to characterize most aspects of atmospheric boundary layer turbulence. These parameters are mean wind speed U, the friction velocity u_*, the convective scaling velocity $w_* = (H^*h)^{1/3}$, the Obukhov length $L = -u_*^3/(kH^*)$, the mixed-layer depth h, and the standard deviations of lateral and vertical directions or speeds, σ_a and σ_e or σ_v and σ_w. In the surface layer, below about $0.1h$, there are well-known relationships for $(zu_*/H^*)(g/\Theta)\partial\Theta/\partial z$ and σ_w/u_* as functions of z/L and for U/u_* as a function of z/L and z/z_0, where z_0 is the surface roughness length; these relationships can be exploited to provide substitutes for the fast-response correlation measurements required to determine u_*, H^*, and L directly.

Each of the three methods—surface layer similarity, convective scaling, and statistical—has areas of strong and weak performance. Several taken together or hybrids of these methods, such as those in Venkatram et al. (1984), can handle diffusion within the mixed layer with considerably more skill than traditional methods using "stability classes" to characterize the whole diffusion process. Figure 2.19 is an attempt to map the areas of validity for the three analysis methods, according to stability, source height, and lateral or vertical diffusion. The strengths of each method are summarized below.

		UNSTABLE	NEUTRAL	STABLE
	Elevated	C	$S+h+u_{x*}/f$	$S+h$
		$(\sigma_e$ modified$)$	σ_e	σ_e
$\boxed{\sigma_z \text{ or} \atop \int\chi dy}$			$C+S$	
		$C,S,(\sigma_e)$		
	Surface	$S, (\sigma_e)$	$S, (\sigma_e)$	$S, (\sigma_e)$
	Elevated	C, σ_a	$\sigma_a , (S)$	σ_a
$\boxed{\sigma_y}$			$C+S$	
	Surface	C, σ_a	$\sigma_a , (S)$	$\sigma_a, (S^{short}_{term})$

Figure 2.19. Schematic view of stability and source-height regimes and the most effective analysis approaches. S = surface layer similarity scaling, C = convective scaling, σ_a or σ_e = statistical models. Commas may be read as "or". Parentheses indicate more limited success. Height of turbulent layer h, and u_*/f are additional considerations in some cases, as shown.

Surface layer similarity (SLS) methods using u_* and L as principal scales are based on well-understood principles of turbulence production near the surface, where the fluxes that generate the turbulence can be considered nearly equal to their surface values. Mechanical turbulence is produced by the action of wind shear on kinematic stress $\simeq u_*^2$ near the surface. Convective turbulence is produced by the release of potential energy due to vertical motions of density variations, $\simeq H^*$ near the surface. Together these form the Obukhov length L; below $z = |L|$, mechanical production of turbulence is stronger than buoyant production or destruction of turbulence. Hence, the dimensionless height z/L influences profiles of mean and turbulent quantities and diffusion rates.

SLS theory has been quite successful for vertical turbulence and diffusion near the surface, but not as much so for lateral components.

In unstable conditions, h is an important horizontal eddy scale even near the surface; at large $|h/L|$, σ_v and $d\sigma_y/dt$ are better characterized by the convective scale w_* than by u_*. Hybrid models have been proposed that use both scalings (Briggs, 1985). In near-neutral conditions, turbulence and diffusion velocities scale to u_* but with some site-to-site variations ranging more than about a factor of 2; horizontal turbulence eddies may be affected by terrain or roughness inhomogeneities, as wind shear near the ground plays a dominant role. In stable conditions, the large horizontal eddies are even more influenced by site conditions, even very small terrain slopes, so a more direct turbulence measure such as σ_a or σ_v is required to characterize σ_y for longer than 5-min sampling times. However, very-short-term σ_y is caused by smaller, more isotropic eddies, which are more controlled by u_*, z, and L and less responsive to terrain, so SLS is more effective here. Pasquill's stability classification system is loosely related to L, and his σ_y curves are limited to 3-min averages for similar reasons.

SLS scaling works well for vertical diffusion from surface sources up to $\sigma_z \simeq 0.2h$ in convective boundary layers. Then h becomes very influential, and vertical diffusion is better described in terms of convective scaling. The same is true for $z_s > 0.1h$. In stable and neutral boundary layers, too, vertical turbulence intensities drop with height, so z_s/h or σ_z/h is an additional consideration. In the stable boundary layers σ_w/u_* may be only a function of z/h, but in neutral boundary layers, the way it drops off also seems to be influenced by hf/u_* (Briggs and Binkowski, 1985).

Convective scaling using w_* and h can, of course, be used only in convective (unstable) conditions. Since this is the condition during which pollutants from elevated sources most often reach the ground, this scaling has great potential utility in diffusion modeling. According to laboratory measurements, numerical modeling, and the limited field observations that are analyzable using this scaling, it is quite effective for characterizing both lateral and vertical diffusion when $|h/L| > 10$ or so. At smaller $|h/L|$ on the unstable side of neutral, u_* has significant influence; here a convective-SLS hybrid approach can be effective. Also, below $z = |L|$ vertical turbulence and diffusion are best described in terms of SLS scaling. If $|h/L| > 10$, there is a region of overlap at $|L| < z < 0.1h$ called the "free convection" regime where convective and SLS scaling are equally valid for the vertical quantities.

It is difficult to obtain stable averages in convective conditions, so I use the word "effective" more loosely here. One important time scale is the large-eddy passage time, $1.5h/U$. It is desirable, but not always possible, to have a sample length at least $15h/U$ to get a representative "ensemble" of eddies acting on a plume. In addition, there are longer time scales of the order of 1 or 2 h affecting wind direction shifts and mean vertical velocity (Kaimal et al., 1976); perhaps these are loosely organized convective rolls nearly parallel to the wind direction, which take a long time to move over a point. This means that we must expect much more scatter in our diffusion measurements in convective field experiments than in the laboratory, in spite of our efforts to obtain long averages.

Statistical methods for characterizing diffusion using σ_a and σ_e (or $\sigma_v \simeq \sigma_a U$ and $\sigma_w \simeq \sigma_e U$) are quite versatile; yet, without refinements incorporating reasonable estimates of the Lagrangian time scale T_L, these methods can be expected to lose accuracy with increasing time or distance of travel. At very short distances, $\sigma_y/(\sigma_a x) = f_1$ and $\sigma_z/(\sigma_e x) = f_2$ are both nearly equal to unity. Statistical theory then states that f_1 and f_2 are functions of t/T_L, and determination of T_L is a problem. It can be inferred indirectly, but direct measurement is rarely possible. Estimates of it range from, essentially, $T_L \propto U^{-1}$ in Pasquill's (1976) f_1, to T_L depending on broad stability and release height categories in Draxler (1976), to T_L based on considerations of SLS theory in Venkatram et al. (1984).

Statistical methods work best for horizontal diffusion, because PBL turbulence is usually far more homogeneous in the horizontal than in the vertical. It was suggested in Sec. 2.5 that σ_a alone can give a good characterization of $\sigma_y(x)$ if dominant horizontal eddy sizes, D_e, correlate well with σ_a (i.e., $t/T_L \propto \sigma_a x/D_e$). In convective conditions, $D_e \sim h$, which corresponds to large σ_a values (larger than neutral), but h does not correlate well with σ_a within that stability category. In stable conditions, both D_e and σ_a are very site dependent, so generalization about their relationship is risky. Draxler's (1976) empirical approach, which assumes that T_L is constant in a given category of stable/unstable, surface/elevated source, vertical/lateral diffusion, has given good results, especially for σ_y. For instance, in Gryning and Lyck's (1984) experimental study it worked as well as the far more complex Hay and Pasquill (1959) statistical theory. To implement this approach for a given source, one needs only to plot f_1 or f_2 data divided into the above categories versus dimensional time of travel;

the optimum, constant value of T_L can be inferred from the average of many such trials. On the other hand, convective scaling experiments indicate that $T_L \propto h/w_*$ is a reasonable first-order physical model for unstable conditions; h/w_* typically ranges from about 500 s to 1500 s, so use of this time scale might reduce the scatter in f_1 (or f_2) compared with using the constant T_L assumption. In neutral conditions, f^{-1} and z_s/u_* may be effective time scales. Stable conditions remain an enigma for σ_y modeling over 15 min–1 h averaging times, because of large site variations in large horizontal eddy population; there does not seem to be enough information on the optimum T_L range at present.

For vertical diffusion, statistical approaches have given good short-range results, but these tend to degrade considerably after $x > 1$ km, roughly (e.g., see Fig. 2.16). The basic reason for this is that vertical turbulence varies strongly with z, so as a plume grows it advances into regions of substantially different turbulence statistics from those at $z = z_s$. In convective conditions there are further problems associated with the action of the largest eddies on the plume. To begin with stable conditions, recent experimental and theoretical information indicates that $\sigma_w/u_* \propto (1 - z/h)^{3/4}$ and that the vertical eddy scale is proportional to z near the surface and to L or σ_w/N higher up (Chs. 1, 5; Nieuwstadt, 1984; Venkatram et al., 1984). For a surface source, vertical diffusion will considerably slow down as σ_z approaches L, owing to the lid on eddy size; beyond this point, we are likely to get better results with SLS theory. For elevated sources, within the mixed layer ($z_s < h$), plume growth may be skewed downward because of more vigorous σ_w and slower u near the surface, where most turbulent shear production occurs. This depends strongly on z_s/h; unfortunately, h is difficult to measure accurately and at present can be only crudely estimated on the basis of near-surface in-situ measurements (Briggs and Binkowski, 1985). Acoustic soundings or bulk Ri profiles are the most accessible means to measure h at present. Truly "neutral" boundary layers, with turbulence time scales more limited by f^{-1} than by N^{-1} or h/w_*, are probably quite rare. However, when they exist, σ_w and eddy size $/z$ drop off with z much more slowly with height than in stable conditions; statistical theory is probably "safe" for this condition at ordinary heights of σ_e measurement.

For vertical diffusion in convective conditions, more caution with the statistical approach seems advised. Draxler (1976) represented indirect estimates of σ_z from surface releases with a *parabolic* form of f_2, first decreasing with t and then increasing with t^2, in strong contrast to the usual slowly dropping off behavior. This was probably due to the effect of "sweepout" of surface concentrations into rising thermals that Willis and Deardorff (1976a) observed occurring near $x \simeq 0.5hU/w_*$. One can force f_2 to give the proper result for surface $\int \chi dy$, but convective scaling provides a more direct and physically more complete approach here. For elevated sources, Irwin (1983) and Gryning and Lyck (1984) reported fairly good results with statistical approaches such as those used by Draxler (1976), Hay and Pasquill (1959), or Cramer (1976), with $f_2 = 1$. However, for most of these data $z_s/h < 0.1$, so the enhancement of surface concentrations due to downward skew of vertical velocities would not be large. Furthermore, the Copenhagen experiment, with its conservative tracer, does show some evidence of ground concentrations larger than the maximum predicted by straight Gaussian modeling ($C'_{ymax} = Uz_s(\int \chi dy)_{max}/Q = 0.484$) in spite of the fact that most measurements were considerably beyond the distance of maximum $\int \chi dy$ suggested by Briggs (1985), on the basis of laboratory and numerical experiments: $X' = (x/U)w_*/z_s \simeq 2$. For the seven cases in Gryning and Lyck (1984) with $2 < X'_{max} < 5.3$ and $U/w_* < 7$ (to exclude nearly neutral cases), the median and the mean C'_{ymax} were about 20% larger than 0.484. Thus, if $z_s > 0.1h$, an unmodified Gaussian-statistic approach is not advised. Alternatives are p.d.f. (probability density function) statistical models, which account for the downward skew of turbulent vertical velocities (Briggs, 1985; Ch. 4), or some modification using convective scaling, at least z_s/h.

Most research experiments use passive tracers, but the vast majority of diffusion applications involve buoyant sources (Ch. 3). For most cases, the traditional two-stage approach—plume rise followed by passive diffusion from an effective source height $z_e = z_s + \Delta h$—provides satisfactory modeling as long as Δh is measured or predicted with sufficient accuracy. This has not been much of a problem in stable conditions, as plumes are vertically well defined and clearly level off; Δh depends only weakly on U and $\partial \Theta/\partial z$ through the rise layer, and equations are available that treat these as either constants or variable with height (Briggs, 1984). It is important to know z_e relative to the depth of mixing, because above h there is usually no vertical diffusion except for plume thickness due to entrainment during plume rise (about $2\Delta h/3$). Neutral conditions are not usually problematic, because h is greater than during stable conditions, and U is larger and Δh is lower than in convective conditions. For most elevated sources, $\Delta h < z_s$, so Δh is not critical in neutral conditions. Formulas like Eq. (8.97) in Briggs (1984) have given good results for maximum ground concentration and have some direct observational support (limited in part by the frequency of the neutral case as a limit on Δh); however, there is a lack of data for testing of neutral Δh models in the case of surface sources, such as conflagrations. For unstable conditions, there are several problems with traditional approaches, but convective scaling offers possibilities of solutions. Particularly important is the dimensionless buoyancy flux, $F^* = F_b/(Uw_*^2h)$. When $F^* < 0.1$ the traditional two-stage approach can be used, but Δh predictions have not received solid validation yet; this is largely because of measurement problems in highly dispersive, looping conditions. There is also the question of whether effective z_e is reduced by the predominance of downdrafts, which is true of passive plumes. One way of taking care of this is to use a conservative equation for Δh, although this cannot be expected to work in all cases (for $\Delta h \ll h_s$, for instance). Briggs (1985) did show that the simple and supposedly conservative equation from Briggs (1984), $\Delta h = 3(F_b/U)^{3/5}H^{*-2/5}$, used in conjunction with a conventional Gaussian plume model gives good results compared with $\int \chi dy$ data from two power plants (restricted to $F^* < 0.1$ cases). When $F^* > 0.1$, plumes tend to "bump" the top of the mixing layer with some excess buoyancy that makes them behave very differently from passive plumes; hence, the traditional buoyant plume modeling approach is invalid. The excess buoyancy causes increased lateral spreading and a delay in downward mixing. This was discussed at length in Sec. 2.4.2, where the importance of F^* for modeling buoyant plumes in convective conditions was strongly emphasized (see Ch. 4 also).

Finally, I again caution against inferences of σ_z based on surface measurements of $\int \chi dy$ and no vertical χ profile data, especially when the tracer is suspected to be nonconservative. Such measurements are useful, but should be treated as and accepted for what they are, that is, surface $\int \chi dy$.

References

Agterberg, R., F. T. M. Nieuwstadt, H. van Duuren, A. J. Hasselton, and G. D. Krijt, 1983: *Dispersion Experiments with Sulphur Hexafluoride from the 213 m High Meteorological Mast at Cabauw in the Netherlands*. Royal Netherlands Meteorological Institute, DeBilt, The Netherlands, 129 pp.

Barad, M. L., and D. A. Haugen, 1959: A preliminary evaluation of Sutton's hypothesis for diffusion from a continuous point source. *J. Meteor.*, **16**, 12–20.

Barenblatt, G. I., 1979: *Similarity, Self-Similarity, and Intermediate Asymptotics*. Consultants Bureau, New York, NY, 218 pp.

Briggs, G. A., 1982a: Simple substitutes for the Obukhov length. *3rd Joint Conference on Applications of Air Pollution Meteorology*, Amer. Meteor. Soc., Boston, 68–71.

Briggs, G. A., 1982b: Similarity forms for ground-source surface-layer diffusion. *Bound.-Layer Meteor.*, **23**, 489–502.

Briggs, G. A., 1984: Plume rise and buoyancy effects. *Atmospheric Science and Power Production*, D. Randerson, Ed., U. S. Dept. of Energy DOE/TIC–27601, 327–366. (Available from NTIS as DE84005177.)

Briggs, G. A., 1985: Analytical parameterizations of diffusion: the convective boundary layer. *J. Climate Appl. Meteor.*, **24**, 1167–1186.

Briggs, G. A., and F. S. Binkowski, 1985: Research on diffusion in atmospheric boundary layers: a position paper on status and needs. U. S. Environmental Protection Agency, EPA/600/S3–85–072. (Available from NTIS as PB86–122587/AS.)

Briggs, G. A., and K. R. McDonald, 1978: Prairie Grass revisited. Optimum indicators of vertical spread. *Proceedings of the 9th International Technical Meeting on Air Pollution Modeling and Its Application*, **No. 103**, NATO/CCMS, 209–220.

Briggs, G. A., W. L. Eberhard, J. E. Gaynor, W. R. Moninger, and T. Uttal, 1986: Convective diffusion field measurements compared with laboratory and numerical experiments. *5th Joint Conference on Applications of Air Pollution Meteorology*, Amer. Meteor. Soc., Boston, 340–343.

Brost, R. A., J. C. Wyngaard, and D. H. Lenschow, 1982: Marine stratocumulus layers. Part II: Turbulence budgets. *J. Atmos. Sci.*, **39**, 818–836.

Businger, J. A., J. C. Wyngaard, Y. Izumi, and E. F. Bradley, 1971: Flux-profile relationships in the atmospheric surface layer. *J. Atmos. Sci.*, **28**, 181–189.

Calder, K. L., 1949: Eddy diffusion and evaporation in flow over aerodynamically smooth and rough surfaces: a treatment based on laboratory laws of turbulent flow with special reference to conditions in the lower atmosphere. *Quart. J. Mech. Appl. Math.*, **II**, 153.

Carras, J. N., and D. J. Williams, 1983: Observations of vertical plume dispersion in the convective boundary layer. *6th Symposium on Turbulence and Diffusion*, Amer. Meteor. Soc., Boston, 249–252.

Chaudhry, F. H., and R. N. Meroney, 1973: Similarity theory of diffusion and the observed vertical spread in the diabatic surface layer. *Bound.-Layer Meteor.*, **3**, 405–415.

Chen, J. C., 1980: Studies on Gravitational Spreading Currents. Rep. No. KH–R–40, W. M. Keck Laboratory of Hydraulics and Water Resources, California Institute of Technology, Pasadena, 436 pp.

Cramer, H. E., 1957: A practical method of estimating the dispersal of atmospheric contaminants. *Proceedings of the Conference on Applied Meteorology*, Amer. Meteor. Soc., Boston, 33–55.

Cramer, H. E., 1976: Improved techniques for modeling the dispersion of tall stack plumes. *Proceedings of the 7th International Technical Meeting on Air Pollution Modeling and Its Application*, No. 51–NATO–CCMS, 731–780.

Cramer, H. E., and F. A. Record, 1957: Field studies of atmospheric diffusion and the structure of turbulence. *Amer. Indust. Hygiene Assoc. Quart.*, **18**(2), 126–131.

Deardorff, J. W., and G. E. Willis, 1975: A parameterization of diffusion into the mixed layer. *J. Appl. Meteor.*, **14**, 1451–1458.

Doran, J. C., T. W. Horst, and P. W. Nickola, 1978: Variations in measured values of lateral diffusion parameters. *J. Appl. Meteor.*, **17**, 825–831.

Draxler, R. R., 1976: Determination of atmospheric diffusion parameters. *Atmos. Environ.*, **10**, 99–105.

Eberhard, W. L., W. R. Moninger, T. Uttal, S. W. Troxel, J. E. Gaynor, and G. A. Briggs, 1985: Field measurements in three dimensions of plume dispersion in the highly convective boundary layer. *7th Symposium on Turbulence and Diffusion*, Amer. Meteor. Soc., Boston, 115–118.

Fuquay, J. J., C. L. Simpson, and W. T. Hinds, 1964: Prediction of environmental exposures from sources near the ground based on Hanford experimental data. *J. Appl. Meteor.*, **3**, 761–770.

Garratt, J. R., 1982: Observations in the nocturnal boundary layer. *Bound.-Layer Meteor.*, **22**, 21–48.

Golder, D., 1972: Relations among stability parameters in the surface layer. *Bound.-Layer Meteor.*, **3**, 47–58.

Gryning, S. W., and E. Lyck, 1984: Atmospheric dispersion from elevated sources in an urban area: comparison between tracer experiments and model calculations. *J. Climate Appl. Meteor.*, **23**, 651–660.

Gryning, S. E., P. Van Ulden, and S. E. Larson, 1983: Dispersion from a continuous ground-level source investigated by a K-model. *Quart. J. Roy. Meteor. Soc.*, **109**, 355–364.

Hanna, S. R., 1981: Turbulent energy and Lagrangian time scales in the planetary boundary layer. *5th Symposium on Turbulence, Diffusion, and Air Pollution*, Amer. Meteor. Soc., Boston, 61–62.

Hanna, S. R., G. A. Briggs, J. Deardorff, B. A. Egan, F. A. Gifford, and F. Pasquill, 1977: Meeting Review: AMS Workshop on Stability Classification Schemes and Sigma Curves—Summary of Recommendations. *Bull. Amer. Meteor. Soc.*, **58**, 1305–1309.

Hanna, S. R., J. C. Weil, and R. J. Paine, 1986: Evaluation of a tall stack plume model. *5th Conference on Applications of Air Pollution Meteorology*, Amer. Meteor. Soc., Boston, 198–201.

Hay, J. S., and F. Pasquill, 1959: Diffusion from a continuous source in relation to the spectrum and scale of turbulence. *Atmospheric Diffusion and Air Pollution*, F. N. Frenkiel and P. A. Sheppard, Eds., *Advances in Geophysics*, **6**, Academic Press, New York, 345–365.

Hicks, B. B., 1981: An examination of turbulence statistics in the surface boundary layer. *Bound.-Layer Meteor.*, **21**, 389–402.

Hicks, B. B., 1985: Behavior of turbulent statistics in the convective boundary layer. *J. Climate Appl. Meteor.*, **24**, 607–614.

Horst, T. W., 1979: Lagrangian similarity modeling of vertical diffusion from a ground-level source. *J. Appl. Meteor.*, **18**, 733–740.

Irwin, J. S., 1979: Estimating plume dispersion—a recommended generalized scheme. *4th Symposium on Turbulence, Diffusion, and Air Pollution*, Amer. Meteor. Soc., Boston, 62–69.

Irwin, J. S., 1983: Estimating plume dispersion—a comparison of several sigma schemes. *J. Climate Appl. Meteor.*, **22**, 92–114.

Islitzer, N. F., and R. K. Dumbauld, 1963: Atmospheric diffusion-deposition studies over flat terrain. *Int. J. Air Water Pollut.*, **7**, 999–1022.

Kaimal, J. C., J. C. Wyngaard, Y. Izumi, and O. R. Coté, 1972: Spectral characteristics of surface-layer turbulence. *Quart. J. Roy. Meteor. Soc.*, **98**, 563–586.

Kaimal, J. C., J. C. Wyngaard, D. A. Haugen, O. R. Coté, Y. Izumi, S. J. Caughey, and C. J. Readings, 1976: Turbulence structure in the convective boundary layer. *J. Atmos. Sci.*, **33**, 2152–2169.

Kaimal, J. C., W. L. Eberhard, W. R. Moninger, J. E. Gaynor, S. W. Troxel, T. Uttal, G. A. Briggs, and G. E. Start, 1986: Project CONDORS: Convective Diffusion Observed by Remote Sensors. NOAA/ERL/Boulder Atmospheric Observatory, Report No. 7, Boulder, CO 80303, 305 pp.

Klug, W., 1968: Diffusion in the atmospheric surface layer: a comparison of similarity theory with observations. *Quart. J. Roy. Meteor. Soc.*, **94**, 555–562.

Lamb, R. G., 1978: A numerical simulation of dispersion from an elevated point source in the convective planetary boundary layer. *Atmos. Environ.*, **12**, 1297–1304.

Lamb, R. G., 1979: The effects of release height on material dispersion in the convective planetary boundary layer. *4th Symposium on Turbulence, Diffusion, and Air Pollution*, Amer. Meteor. Soc., Boston, 27–33.

Lowry, P. H., D. A. Mazzarella, and M. E. Smith, 1951: Ground-level measurements of oil-fog emitted from a hundred-meter chimney. *Meteor. Monogr.*, **1**(4), 30–35.

Lumley, J. L., and H. A. Panofsky, 1964: *The Structure of Atmospheric Turbulence.* Wiley Interscience, New York, 239 pp.

Monin, A. S., and A. M. Obukhov, 1954: Basic laws of turbulent mixing in the atmosphere near the ground. *Tr. Akad. Nauk SSSR Geofiz. Inst.*, No. **24 (151)**, 163–187.

Moninger, W. R., W. L. Eberhard, G. A. Briggs, R. A. Kropfli, and J. C. Kaimal, 1983: Simultaneous radar and lidar observations of plumes from continuous point sources. *21st Radar Meteorology Conference*, Amer. Meteor. Soc., Boston, MA, 246–250.

Nieuwstadt, F. T. M., 1980: Application of mixed-layer similarity to the observed dispersion from a ground-level source. *J. Appl. Meteor.*, **19**, 157–162.

Nieuwstadt, F. T. M., 1984: Some aspects of the turbulent stable boundary layer. *Bound.-Layer Meteor.*, **30**, 31–55.

Panofsky, H. A., H. Tennekes, D. H. Lenschow, and J. C. Wyngaard, 1977: The characteristics of turbulent velocity components in the surface layer under convective conditions. *Bound.-Layer Meteor.*, **11**, 355–361.

Pasquill, F., 1961: The estimation of the dispersion of windborne material. *Meteor. Mag.*, **90**, 33.

Pasquill, F., 1962: *Atmospheric Diffusion.* D. Van Nostrand, London, 297 pp.

Pasquill, F., 1966: Lagrangian similarity and vertical diffusion from a source at ground level. *Quart. J. Roy. Meteor. Soc.*, **92**, 185–195.

Pasquill, F., 1976: Atmospheric Dispersion Parameters in Gaussian Plume Modeling—Part II: Possible Requirements for Change in the Turner Workbook Values. EPA–600/4–76–030B, U. S. Environmental Protection Agency, Research Triangle Park, NC, 44 pp.

Pierce, T. E., 1986: Estimating surface concentrations from an elevated, buoyant plume in a limited-mixed convective boundary layer. *5th Joint Conference on Applications of Air Pollution Meteorology*, Amer. Meteor. Soc., Boston, 331–334.

Singer, I. A., and M. E. Smith, 1966: Atmospheric dispersion at Brookhaven National Laboratory. *Int. J. Air Water Pollut.*, **10**, 125–135.

Sivertsen, B., S. E. Gryning, A. A. M. Holtslag, and J. S. Irwin, 1987: Atmospheric dispersion modeling based upon boundary layer parameterization. *Air Pollution Modeling and Its Application V*, C. De Wispelaere, F. A. Schiermeier, and N. V. Gillani, Eds., Plenum, New York, 177–192.

Sutton, O. G., 1947: The theoretical distribution of airborne pollution from factory chimneys. *Quart. J. Roy. Meteor. Soc.*, **73**, 426–436.

Taylor, G. I., 1921: Diffusion by continuous movements. *Proc. London Math. Soc.*, Ser. 2, **20**, 196.

Turner, D. B. 1970: *Workbook of Atmospheric Dispersion Estimates*. U. S. Environmental Protection Agency, Atmospheric Sciences Research Laboratory, Research Triangle Park, NC, 84 pp.

Venkatram, A., 1978: Estimating the convective velocity scale for diffusion applications. *Bound.-Layer Meteor.*, **15**, 447–452.

Venkatram, A., 1980: Dispersion from an elevated source in a convective boundary layer. *Atmos. Environ.*, **14**, 1–10.

Venkatram, A., 1982: A semi-empirical method to compute concentrations associated with surface releases in the stable boundary layer. *Atmos. Environ.*, **16**, 245–248.

Venkatram, A., D. Strimaitis, and D. Dicristofaro, 1984: A semi-empirical model to estimate vertical dispersion of elevated releases in the stable boundary layer. *Atmos. Environ.*, **18**, 923–928.

Weber, A. H., K. R. McDonald, and G. A. Briggs, 1977: Turbulence classification schemes for stable and unstable conditions. *Joint Conference on Applications of Air Pollution Meteorology*, Amer. Meteor. Soc., Boston, 96–102.

Weil, J. C., 1979: Assessment of Plume Rise and Dispersion Models Using Lidar Data. PPSP–MP–24. Maryland Power Plant Siting Program, Department of Natural Resources, Annapolis, MD.

Weil, J. C., 1983: Application of advances in planetary boundary layer understanding to diffusion modeling. *6th Symposium on Turbulence and Diffusion*, Amer. Meteor. Soc., Boston, 42–46.

Weil, J. C., L. A. Corio, and R. P. Brower, 1986: Dispersion of buoyant plumes in the convective boundary layer. *5th Joint Conference on Applications of Air Pollution Meteorology*, Amer. Meteor. Soc., Boston, 335–338.

Willis, G. E., 1986: Dispersion of buoyant stack plumes in a convective boundary layer. *5th Joint Conference on Applications of Air Pollution Meteorology*, Amer. Meteor. Soc., Boston, 327–330.

Willis, G. E., and J. W. Deardorff, 1976a: A laboratory model of diffusion into the convective planetary boundary layer. *Quart. J. Roy. Meteor. Soc.*, **102**, 427–445.

Willis, G. E., and J. W. Deardorff, 1976b: Visual observations of horizontal platforms of penetrative convection. *3rd Symposium on Atmospheric Turbulence, Diffusion, and Air Quality*, Amer. Meteor. Soc., Boston, 9–12.

Willis, G. E., and J. W. Deardorff, 1978: A laboratory study of dispersion from an elevated source within a modeled convective boundary layer. *Atmos. Environ.*, **12**, 1305–1311.

Willis, G. E., and J. W. Deardorff, 1981: A laboratory study of dispersion from a source in the middle of the convectively mixed layer. *Atmos. Environ.*, **15**, 109–117.

Willis, G. E., and N. Hukari, 1984: Laboratory modeling of buoyant stack emissions in the convective boundary layer. *4th Joint Conference on Applications of Air Pollution Meteorology*, Amer. Meteor. Soc., Boston, 24–26.

Yokoyama, O., M. Gamo, and S. Yamamoto, 1977: On the turbulence quantities in the neutral atmospheric boundary layer. *J. Meteor. Soc. Japan*, **55**, 312–318.

CHAPTER 3

Plume Rise

Jeffrey C. Weil

3.1. Introduction

Most industrial pollution sources are stacks with discharges of momentum and heat as well as pollutants. The resulting plume rise can be considerable—hundreds of meters—and can substantially aid the dilution of plume constituents before they reach ground level. Thus, plume rise is an important factor to consider in diffusion modeling. For power plants and other moderate-to-large industrial sources, the major contribution to the rise is from the heat flux. For example, a modern power plant typically discharges ~ 100 MW of heat from its stack. Source momentum can be important for smaller sources, such as those typically found in light manufacturing. Although we will address the plume rise due to source momentum, we will give most attention to the effects of source buoyancy.

Plume rise varies not only with source conditions, but also with the local meteorological conditions—the wind speed, ambient stratification (i.e., potential temperature gradient), and ambient turbulence—and is a strong function of distance from the source. The distance dependence was ignored in much of the early empirical work (e.g., Holland, 1953), in which the rise was taken simply as the most distant observation as determined visually or by remote sensing. The diverse behavior of plumes dictates that one have a sound theoretical model along with good observations to assess the effects of the controlling variables. Furthermore, a simple model is generally more useful than a complex one, not only for diffusion applications but also in the interpretation of experimental data.

This chapter focuses on an integral model, in which one writes down the differential equations governing the total fluxes of mass, momentum, and energy through a plume cross section (e.g., Morton et al., 1956; Hoult et al., 1969; Briggs, 1975). The equations are closed using the entrainment assumption (Morton et al., 1956), which specifies that the rate of ingestion of ambient air into a plume is proportional to an "inflow" or "entrainment" velocity at the plume edge; this velocity is generally assumed to be equal to the local rise velocity times a dimensionless entrainment parameter. For practical purposes, the entrainment parameter can be considered a constant for a particular flow geometry, e.g., a bent-over plume in a cross wind.

This approach has been successful in predicting the rise and growth of plumes close to the source and the "leveled off" plume height in stable air. The success stems largely from the dominance of buoyancy effects over those of the ambient atmospheric turbulence in these problems; i.e., buoyancy-generated turbulence due to the plume's upward motion is the principal cause of mixing with ambient air. We discuss the governing equations for these problems as well as established results in Sec. 3.2.

There are two other important diffusion applications in which plume behavior is not totally resolved. One is the penetration of an elevated inversion by a buoyant plume (Sec. 3.3). The second is predicting the effect of ambient turbulence on plumes in a near neutral or a convective boundary layer (CBL), when the turbulence is sufficiently strong to bring the plume, or parts of it, to the surface (Sec. 3.4). This problem has benefitted from the improved understanding of the CBL over the past decade (see Ch. 1) and from laboratory simulations of dispersion (see Deardorff, 1985; Willis and Deardorff, 1983; Willis and Hukari, 1984).

Our discussion of plume rise addresses fundamental aspects and major problems, but it is not exhaustive in that some problems such as rise from multiple stacks, plume downwash, moisture effects, etc., are not covered. For these and other details the reader is referred to the review by Briggs (1984).

3.2. Rise in Neutral and Stable Environments: Established Results

The integral model presented below describes the ensemble-average properties of a turbulent, buoyant plume—in particular, its trajectory, spread, and mean density or species concentration. Clearly, in a single realization the instantaneous plume deviates from the smoothed outline of the averaged one as illustrated in Fig. 3.1.

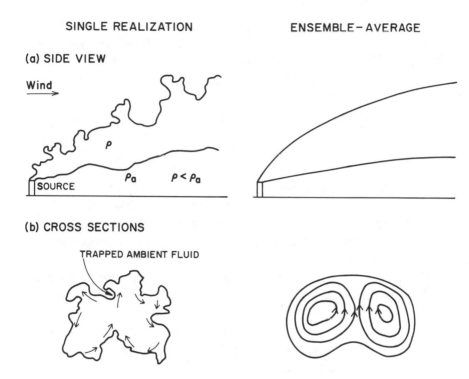

Figure 3.1. Illustration of instantaneous and average plume structure.

The upper surface of the instantaneous plume has irregular protu-
berances or bulges caused by the large eddies inside the plume. Such
eddies are comparable in size with the local radius and arise from
the relative motion between the upward moving plume and the ambient fluid; i.e., they are buoyancy generated. For example, one sees
such bulges and eddies in laboratory plumes produced in laminar
cross flows (Fan, 1967; Hoult and Weil, 1972) as well as in full-scale
plumes in the atmospheric boundary layer. In addition, the concentration profile in a single realization exhibits a well-mixed structure in
the plume center with relatively sharp gradients at the edges, whereas
the ensemble-average profile is more akin to a Gaussian distribution
(see Fan, 1967, for bent-over plumes and Rouse et al., 1952, for
vertical plumes).

The double-vortex structure in the average cross section (Fig. 3.1)
is inferred from the laboratory observations of Fan (1967), Richards
(1963), and Tsang (1971). Richards and Tsang conducted experiments with line thermals, which are cylindrical columns of buoyant

fluid having a horizontal axis; thermals can be produced by releasing a column of dense, salty water into a tank of fresh water. As the data of the above authors show, the flow structure and concentration isopleths in thermals are quite similar to those in the cross section of a bent-over plume.

Most of the entrainment in a bent-over plume or thermal occurs at the "front" or advancing surface; in turn, much of this is due to ambient fluid captured in the spaces between the bulges on the surface (see Fig. 3.1). For example, in spherical thermals, Woodward (1959) estimated that about 60% of the entrainment took place at the front with the balance at the bottom center. This estimate was based on the tracking of ambient particles into and around laboratory thermals (see her Fig. 3). A further review of thermal observations is given by Scorer (1978).

Even though entrainment in the model to be described occurs across a smooth (average) surface with a smooth radial inflow, in a given realization it actually occurs on the convoluted surface described above. The success of the model stems from the similarity in the problem: the turbulence length and velocity scales are proportional to the plume radius and rise velocity, respectively, in each cross section.

3.2.1. Governing Equations

The following equations apply to a plume in a steady, horizontal wind of constant direction but of speed U_a that can vary with height z above the surface. The plume axis lies in the x, z plane (Fig. 3.2) where s and U_{sc} are the distance and velocity along the centerline, and ϕ is the angle between the horizontal and the centerline. The equations were derived (Appendix A) for arbitrary distributions of properties inside the plume, but are presented here for "averaged" or uniformly distributed properties, i.e., for "top hat" profiles. The average cross section is assumed to be circular with radius r; thus, plume properties are constant within r and zero outside it.

Conservation of mass, horizontal momentum, and vertical momentum are given respectively by

$$\frac{d}{ds}\left(U_{sc}r^2\right) = E \tag{3.1}$$

$$\frac{d}{ds}\left(U_{sc}r^2\Delta u\right) = -r^2 w\frac{dU_a}{dz} \tag{3.2}$$

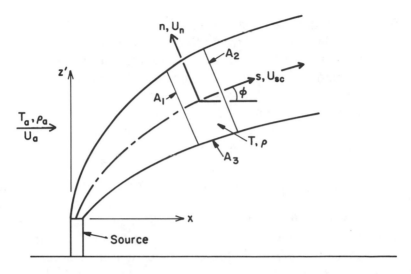

Figure 3.2. Schematic and nomenclature for plume in a crosswind.

$$\frac{d}{ds}\left(U_{sc}r^2w\right) = gr^2\frac{\Delta\rho}{\rho} , \tag{3.3}$$

where E is the rate of entrainment of ambient air by the plume; ρ is the plume density; g is the gravitational acceleration; and $\Delta u, w,$ and $\Delta\rho$ are the differences in the horizontal velocity, vertical velocity, and density between the plume and the ambient environment. The differences represent plume minus ambient variables except for $\Delta\rho$, which is defined to give a positive buoyancy force, $\Delta\rho = \rho_a - \rho$, for warm plumes in the atmosphere; here, a subscript a denotes ambient.

The entrainment assumption takes the same form in the two extremes of plume trajectories—a vertical plume ($\phi = 90°$) and a bent-over plume ($\phi \rightarrow 0$) or line thermal:

$$E = 2r\beta w , \tag{3.4}$$

where β is the dimensionless entrainment parameter and is empirically determined. The β is different in the above two extremes due to the differences in geometry and secondary flows: $\beta = 0.11$ and 0.6 for vertical and bent-over plumes, respectively (see Hoult and Weil, 1972).

For the bending-over stage, several entrainment assumptions are available to bridge the gap between the above β extremes (see Briggs, 1975, for a summary). Many of them (e.g., Hoult et al., 1969), consider two entrainment mechanisms—one due to velocity differences parallel to the plume axis and a second due to differences normal to the axis. The two mechanisms are assumed to be additive and the entrainment rate for each is taken to be the product of an entrainment parameter and the plume perimeter times the corresponding velocity difference.

In deriving Eqs. (3.2) and (3.3), it is assumed that the plume pressure is the same as in the local environment (i.e., it is in hydrostatic equilibrium) and that the density differences are sufficiently small that the Boussinesq approximation can be made. The latter stipulates that for $\Delta\rho/\rho \ll 1$, the only place density differences enter is in the buoyancy force term—the right-hand side of Eq. (3.3) (see Phillips, 1980). For a bent-over plume or line thermal, the left-hand side of Eq. (3.3) should be multiplied by $1 + k_v$, where k_v is an "added" or "virtual" mass coefficient (see Escudier and Maxworthy, 1973; Briggs, 1975) and has a value of 1 for a circular cylinder. The k_v accounts for the momentum of the ambient fluid displaced by the plume as the latter rises. Most models do not include this detail, but as discussed by Briggs (1975), incorporation of the added mass helps to explain the differences in entrainment parameters determined from plume growth, rise in neutral air, and rise in stable air.

For hot plumes in which the density difference is caused by a temperature difference ΔT, we have the energy equation:

$$\frac{d}{ds}\left(U_{sc}r^2\Delta T\right) = \frac{d\Theta_a}{dz}wr^2 , \qquad (3.5)$$

where Θ_a is the ambient potential temperature. Its gradient is given by $d\Theta_a/dz = dT/dz + g/c_p$, where c_p is the specific heat at constant pressure, and g/c_p is the adiabatic lapse rate (~ 0.01 K m^{-1}). Assuming the plume to be a perfect gas allows the density difference to be found from the equation of state:

$$p = \rho\mathcal{R}T , \qquad (3.6)$$

where p and $\mathcal{R}$ are the pressure and gas constant, respectively; thus, $\Delta\rho/\rho = \Delta T/T_a$.

In general, the density difference leading to the buoyancy force on a plume could be due to dissolved salts, as in a water-tank experiment, or to molecular weight differences, as in a wind-tunnel simulation. Since buoyancy is the principal driver of the vertical motion, it has been customary to derive a general equation for buoyancy conservation (see Morton et al., 1956; Turner, 1973; Briggs, 1984). This is simply another way of stating conservation of mass—salt or molecular species—or of energy, as the case may be. If we multiply Eq. (3.5) by g, replace ΔT by $T_a \Delta \rho / \rho$, and note that $\Theta_a \simeq T_a$, we have an equation for buoyancy conservation:

$$\frac{dF}{ds} = N^2 w r^2 \, , \tag{3.7}$$

where F is the buoyancy flux defined by

$$F = U_{sc} r^2 g \frac{\Delta \rho}{\rho} \, , \tag{3.8a}$$

$$N^2 = \frac{g}{\Theta_a} \frac{d\Theta_a}{dz} = -\frac{g}{\rho_a^*} \frac{d\rho_a^*}{dz} \, , \tag{3.8b}$$

and ρ_a^* is the ambient potential density. In stable conditions ($N^2 > 0$), N is the Brunt-Väisälä frequency, i.e., the natural frequency of oscillation of a fluid particle if perturbed from its equilibrium position (e.g., see Phillips, 1980). For plumes, N^{-1} is the time scale for the depletion of the buoyancy flux and for the maximum rise in a stable environment; in the atmosphere, a typical value of N^{-1} is 1 min.

Equation (3.7) can be used for inert pollutants or tracers. However, in some problems, there may be sources of heat and/or suspended material resulting from physical and chemical transformations. One should then be more careful in analyzing the mass, energy, and species conservation equations and should not simply use Eq. (3.7). An example is the equation set for condensing vapor plumes (Weil, 1974) where the latent heat is an appreciable energy source.

To solve for the plume trajectory—rise z' above the stack versus downwind distance x—we must specify the mass, momentum, and energy (or buoyancy) fluxes at the source and add the kinematic conditions:

$$\frac{dz'}{ds} = \frac{w}{U_{sc}}, \quad \frac{dx}{ds} = \frac{U_a + \Delta u}{U_{sc}} \, . \tag{3.9}$$

3.2.2. Rise Near the Source: Neutral Stratification

The governing equations can be solved to yield useful analytical expressions for the plume trajectory in a number of simple situations. Here, we discuss rise in a neutral environment ($N^2 = 0$) and a uniform wind (no shear), ignoring ambient atmospheric turbulence. As discussed later (Sec. 3.2.3), these results are also valid in stable conditions close to the source, i.e., before the stratification has significantly affected the buoyancy flux.

Near the source, the plume trajectory can follow three different scaling laws, each valid in a different distance regime. The actual trajectory depends on the wind speed and the source momentum (F_m) and buoyancy (F_b) fluxes, which are given by

$$F_m = \frac{\rho_0}{\rho_a} w_0^2 r_0^2 \tag{3.10}$$

$$F_b = w_0 r_0^2 g \frac{\Delta \rho_0}{\rho_a} , \tag{3.11}$$

where subscript 0 denotes conditions at the source. In a neutral environment, the buoyancy flux is conserved [see Eq. (3.7)] as is the vertical momentum flux of a nonbuoyant plume [$\Delta\rho = 0$; see Eq. (3.3)].

Very near the source, the plume is dominated by its initial mass and momentum fluxes and its axis departs only slightly from the vertical ($\phi \sim 90°$). Hoult et al., (1969) found that the trajectory is then given approximately by

$$\frac{z'}{\ell_m} = \left(\frac{R}{\alpha R + \beta} \right)^{1/2} \left(\frac{x}{\ell_m} \right)^{1/2} , \tag{3.12}$$

where α and β are the entrainment parameters corresponding to velocity differences parallel and normal to the plume centerline, respectively ($\alpha = 0.11$, $\beta = 0.6$), R is the speed ratio

$$R = \frac{w_0}{U_a} , \tag{3.13}$$

and ℓ_m is the momentum length scale:

$$\ell_m = \left(\frac{\rho_0}{\rho_a} \right)^{1/2} \frac{w_0}{U_a} r_0 . \tag{3.14}$$

In the original Hoult et al., (1969) definition of ℓ_m, the density ratio ρ_0/ρ_a was assumed to be unity and omitted, but it should be included in general (e.g., see Poreh and Kacherginsky, 1981).

R is an important parameter determining when plume downwash can occur in the lee of a stack (e.g., Briggs, 1984; Snyder, 1981; Overcamp and Hoult, 1971). For most stacks (excluding cooling towers), downwash can be avoided provided that $R > 1.5$ (Briggs, 1984).

For a bent-over plume or jet (no buoyancy) in a crosswind, the radius is predicted to grow as

$$r = \beta'z + r_0 ,\qquad (3.15)$$

where $\beta' \simeq 0.4$ (Briggs, 1975; Manins, 1979). Far from the source and for most stacks, $r_0 \ll \beta'z$. When the plume equations are being solved to derive the analytical trajectories below [Eqs. (3.16), (3.17), and (3.19)], the above inequality is assumed (see Appendix B); this results in the "point source" solution.

In the following, the entrainment parameter β', governing the growth of a bent-over jet or plume, is distinguished from the parameter β (unprimed), corresponding to the rise. The β' relates to the actual mass of the plume whereas β includes the "added" mass discussed earlier; the parameters are related by $\beta = \beta'(1 + k_v)^{1/2}$.

For a pure jet ($F_b = 0$), the trajectory is controlled by the initial momentum flux and the entrainment rate. Far from the source in the bent-over stage (small ϕ), it is

$$\frac{z'}{\ell_m} = \left(\frac{3}{\beta^2}\right)^{1/3} \left(\frac{x}{\ell_m}\right)^{1/3} ,\qquad (3.16)$$

where β is assumed to be constant, 0.6 (Hoult and Weil, 1972). Briggs (1975, 1984) gave the same result except that he empirically found β to vary with R according to $\beta = 0.4 + 1.2/R$.

For a highly buoyant plume, the trajectory is governed by the buoyancy flux and the entrainment rate. In the bent-over stage, it is given by the familiar "two-thirds law" (Slawson and Csanady, 1967; Briggs, 1975):

$$\frac{z'}{\ell_b} = \left(\frac{3}{2\beta^2}\right)^{1/3} \left(\frac{x}{\ell_b}\right)^{2/3} ,\qquad (3.17)$$

where ℓ_b is the buoyancy length scale:

$$\ell_b = \frac{F_b}{U_a^3} . \tag{3.18}$$

For a source with significant momentum and buoyancy fluxes, the trajectory in the bent-over region is

$$\frac{z'}{\ell_b} = \left[\frac{3}{\beta^2} \left(\frac{\ell_m}{\ell_b} \right)^2 \frac{x}{\ell_b} + \frac{3}{2\beta^2} \left(\frac{x}{\ell_b} \right)^2 \right]^{1/3} \tag{3.19}$$

(see Briggs, 1975, and Appendix B). This equation shows that the plume will be dominated by buoyancy if $x \gg \ell_m^2/\ell_b$ and by momentum if the inequality is reversed.

Hoult and Weil (1972) gave an approximate method for determining the region of applicability of the three simple trajectories—Eqs. (3.12), (3.16), and (3.17)—by equating the rise from any two to define a "critical" or transition distance x_{ci}. Three distances were found for the intersection of the three trajectories:

$$\frac{x_{c1}}{\ell_b} = \left(\frac{3}{2\beta} \right)^2 \left(\frac{\alpha R + \beta}{R} \right)^3 \frac{\ell_m}{\ell_b} , \qquad \text{(Eqs. 3.12 and 3.16)} \tag{3.20}$$

$$\frac{x_{c2}}{\ell_b} = \left(\frac{R}{\alpha R + \beta} \right)^3 \left(\frac{2\beta^2}{3} \right)^2 \left(\frac{\ell_m}{\ell_b} \right)^3 , \qquad \text{(Eqs. 3.12 and 3.17)} \tag{3.21}$$

and

$$\frac{x_{c3}}{\ell_b} = 2 \left(\frac{\ell_m}{\ell_b} \right)^2 , \qquad \text{(Eqs. 3.16 and 3.17),} \tag{3.22}$$

where it was assumed that the entrainment parameter, β, was the same for bent-over jets and plumes. (If the values of β are different, the β appropriate for a jet or a plume should be used in the above; e.g., see Overcamp and Ku, 1986.) In the general case where a plume can follow the three trajectories, Eq. (3.12) applies for $x < x_{c1}$, Eq. (3.16) for $x_{c1} < x < x_{c3}$, and Eq. (3.17) for $x > x_{c3}$. For highly buoyant plumes ($\ell_m \ll \ell_b$), the $x^{1/3}$ regime disappears.

In evaluating the analytical expressions, Hoult and Weil (1972) found that Eq. (3.12) described the "near-field" rise of jets in cross-flows from several laboratory experiments provided that $R > 4$. For smaller R values, Eq. (3.12) overestimated the observed rise, although the rise may have been diminished by enhanced mixing of the jet with the stack wake or wind tunnel boundary layer. The region of validity of Eq. (3.12) is generally so small that it has limited use in applications.

The most useful results in applications are Eqs. (3.16) and (3.17). Hoult and Weil (1972) found reasonable agreement between Eq. (3.16), with $\beta = 0.6$, and a number of experimental trajectories subject to the constraint $x_{c1} < x < x_{c3}$ (see Fig. 3.3). Briggs (1975, 1984) obtained similar results for jets ($F_b = 0$), but with the entrainment parameter given by $\beta = 0.4 + 1.2/R$. In a more recent study, Overcamp and Ku (1986) found that the "one-third" law (Eq. 3.16) described experimental trajectories of momentum-dominated plumes quite well with $\beta = 0.6$, provided that a correction was made for the non-negligible source size. With this correction, they found little or no dependence of β on R over the range $3 < R < 20$ although there may have been a weak dependence over a broader (R) range.

One remaining issue about momentum-dominated rise is whether to include the buoyancy effect as in Eq. (3.19) or to ignore it

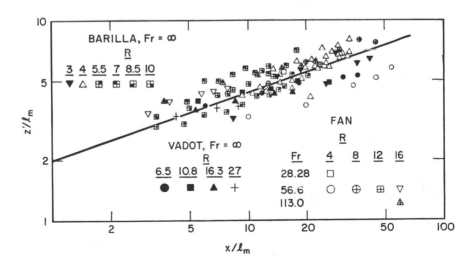

Figure 3.3. Comparison of laboratory plume trajectories in the momentum-dominated region ($x_{c1} < x < x_{c3}$) with the "one-third" law, Eq. (3.16). From Hoult and Weil (1972).

(Eq. 3.16). On physical grounds, incorporation of the buoyancy term is correct. However, as a practical matter, the difference between Eqs. (3.16) and (3.19) is quite small over the range $x_{c1} < x < x_{c3}$ as shown in Fig. 3.4 for example; oddly, Eq. (3.16) appears to be a slightly better approximation to the numerical solution to the plume equations in Fig. 3.4 than is Eq. (3.19). In view of these small differences, the results of Hoult and Weil (1972), and those of Overcamp and Ku (1986), we believe that Eq. (3.16) with $\beta = 0.6$ should give useful engineering estimates of momentum-dominated rise (i.e., for $x < x_{c3}$).

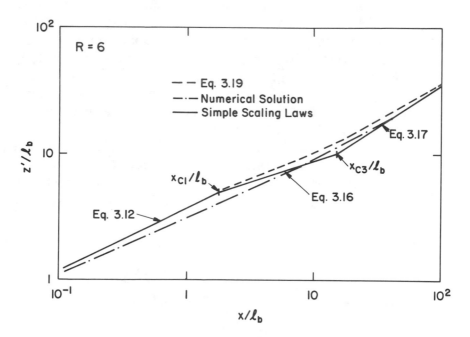

Figure 3.4. Comparison of approximate trajectory expressions for momentum-dominated plumes, Eqs. (3.16) and (3.19), with the numerical solution to the plume equations (Eqs. 3.1–3.4 and 3.7–3.8).

Of all the simple rise predictions, probably the best-documented is the two-thirds law, Eq. (3.17), having been shown to be in agreement with numerous field and laboratory data (see Briggs, 1975, for a summary). Figure 3.5 shows such agreement for a variety of observations compiled by Weil (1982); Eq. (3.17) with $\beta = 0.6$ fits the various data sets equally well regardless of the size of the buoyancy (heat) source. The same value of β found for field and laboratory experiments demonstrates that ambient turbulence is not an important

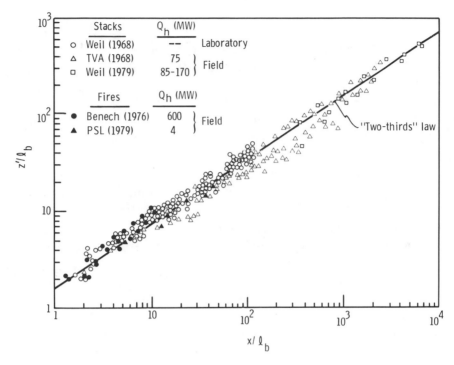

Figure 3.5. Observed trajectories of buoyancy-dominated plumes compared with the "two-thirds" law, Eq. (3.17). From Weil (1982).

factor in predicting the mean rise close to the source since it was not simulated in the laboratory studies; they were conducted in a laminar crossflow. However, in strong turbulence as in the CBL, the scatter in the observed rise and the mean plume outline are significantly broadened, as discussed later.

The above results apply to plume rise in a uniform wind. Using the same integral approach as described in Sec. 3.2.1, Djurfors and Netterville (1978) took into account the effect of wind shear on the rise of a buoyant plume, adopting a power law velocity profile as a model for the actual profile. By considering the rise above a "virtual point source," they obtained an analytical expression for the rise. However, in an evaluation of their model with field observations, they were unable to find any improved performance over the uniform wind model—the two-thirds law—because of the scatter in the observations.

3.2.3. Rise in Stable Air

During the night and early morning, a plume generally rises into stably stratified air ($N^2 > 0$) and levels off when its density is in equilibrium with the surrounding air, i.e., $\Delta\rho \to 0$. Plumes can then be observed to remain compact and distinct for up to tens of kilometers. Dispersion of such plumes and the resulting ground-level concentrations then depend on whether the "final" height is above or below the stable boundary layer height.

In this section, we consider plume rise in a uniformly stratified environment, N^2 = a positive constant, and either a zero or a uniform wind. Rise in stable air is the only stratification regime in which the issue of "final rise" is really settled.

In zero or very light winds, plumes rise vertically, reach some maximum height, and then settle back to an equilibrium height. The initial overshoot of the equilibrium height is due to the vertical momentum of the plume. Morton et al. (1956) were the first to solve the governing equations for the maximum height as a function of the buoyancy flux and the stratification; they also conducted laboratory experiments to determine the final height and the appropriate entrainment parameter. They found the maximum rise, Δh_{max}, to be (see Briggs, 1969)

$$\Delta h_{max} = 5\, F_b^{1/4} N^{-3/4} . \tag{3.23}$$

Briggs (1969) and Weil (1982) found good correlation between Eq. (3.23) and field observations covering a broad range of heat sources.

In a uniform wind, the plume trajectory in the bent-over region is given by Briggs (1975) as

$$\frac{z'}{(F_b/U_a N^2)^{1/3}} = \left(\frac{3}{\beta'^2}\right)^{1/3} \left(N' \frac{F_m}{F_b} \sin\frac{N'x}{U_a} + 1 - \cos\frac{N'x}{U_a}\right)^{1/3} \tag{3.24}$$

$$\text{for}\quad \frac{N'x}{U_a} \le \pi ,$$

where $N' = N/(1 + k_v)^{1/2}$; Eq. (3.24) is derived in Appendix B. This expression, valid for a point source, includes the source momentum as well as the buoyancy, and it accounts for the added mass. For small momentum fluxes ($N'F_m/F_b \ll 1$), Eq. (3.24) reduces to the "two-thirds" law for $x/U_a \ll N'^{-1}$; this can be shown by approximating $\cos(N'x/U_a)$ in Eq. (3.24) as $1 - (N'x/U_a)^2$. Also, with $F_m = 0$, the maximum rise occurs at $N'x/U_a = \pi$, and the predicted final rise very far downstream is 84% of the maximum (see Fay et al., 1970).

Slawson and Csanady (1967) and Fay et al. (1970) solved the plume equations ignoring the "added" mass; in their solutions N' and β' are replaced by N and β, respectively. To demonstrate the effect of the added mass, we compare calculated plume trajectories with the Hewett et al. (1971) wind tunnel data, which exhibit very little scatter. Figure 3.6 shows these data together with the Hewett et al. calculated trajectory for $F_m = 0$ and $k_v = 0$ (dashed curve). The solid curve is Eq. (3.24) with $F_m = 0$ and $k_v = 1$ as assumed for a line thermal (Escudier and Maxworthy, 1973). Note that z' is

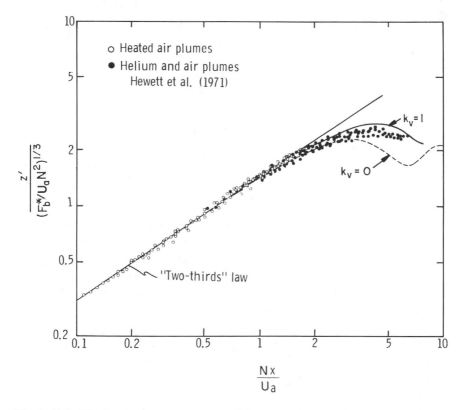

Figure 3.6. Wind-tunnel measurements of buoyant plume trajectories in a stable environment compared with Eq. (3.24) (solid curve) for $F_m = 0$ and $k_v = 1$ and with F_b replaced by F_b^* (Eq. 3.25); the dashed curve is the numerically calculated trajectory of Hewett et al. (1971) for $k_v = 0$. Adapted from Hewett et al. (1971).

nondimensionalized with a buoyancy flux F_b^* (Eq. 3.25) rather than the F_b in Eq. (3.24); this is discussed following Eq. (3.25).

Figure 3.6 shows that the Hewett et al. calculated trajectory underestimates the maximum plume rise as well as the downstream distance required to achieve that rise. Equation (3.24) with $k_v = 1$ is in better overall agreement with the data—peaking at the correct distance ($Nx/U_a \simeq 3.4$) although slightly overestimating the maximum rise. This agreement is attributed to the modeled buoyancy flux being distributed only over the "actual" plume mass, which is smaller than the "apparent" mass—actual plus "added." In the model with $k_v = 0$, the buoyancy flux is distributed over the apparent mass. Therefore, the temperature excess and the maximum plume rise are smaller, and the latter is attained sooner (at smaller x) than with $k_v = 1$.

Hewett et al. defined the buoyancy flux as

$$F_b^* = F_b T_0/T_a ,$$ (3.25)

where $T_0/T_a \simeq 1.4$ in their heated plume experiments. This definition led to a larger value of β, 0.71, than the 0.6 commonly reported. We can reconcile this by examining the two-thirds law; since $z' \propto F_b^{1/3}\beta^{-2/3}$, a 40% increase in F_b would result in an 18% larger β (as found) to yield the same plume rise.

Figure 3.7 compares the wind tunnel results of Hewett et al. with the field observations of Fay et al. (1970). Considering observations for $Nx/U_a > 1.55$ to be in the "leveled off" region, Fay et al. found the geometric mean rise to be $2.27 \left(F_b^*/U_a N^2\right)^{1/3}$, as shown in Fig. 3.7. The wind tunnel data agree with this result and simulate the mean of the field observations rather well. Once again, the agreement suggests that ambient turbulence can be neglected in simulating the mean trajectory since the experiments were conducted in a laminar airstream. However, ambient turbulence does partially account for the broad scatter in the field data. Other potential causes of such scatter are stack and building downwash, local circulations induced by nearby water bodies, and flow over hilly terrain.

On the basis of an extensive survey of field and laboratory observations, Briggs (1975, 1984) concluded that the "final" plume rise in stable air was given by

$$\Delta h = 2.6 \left(\frac{F_b}{U_a N^2}\right)^{1/3} ;$$ (3.26)

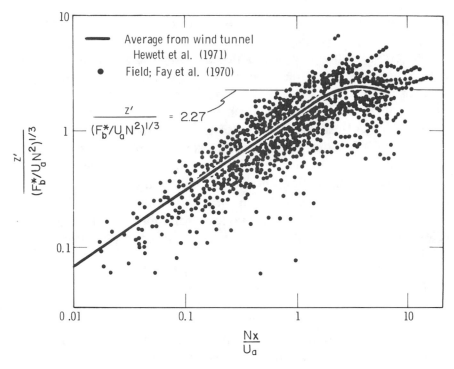

Figure 3.7. Comparison between wind tunnel data of Hewett et al. (1971) and field observations of Fay et al. (1970) for plume rise in a stable environment. From Hewett et al. (1971).

this formula is recommended for applications. The difference between the coefficient of 2.6 and the value 2.27 shown in Fig. 3.7 lies in the definition of the buoyancy flux. Replacing F_b^* by F_b using the typical value of $T_0/T_a = 1.4$, we find that the Fay et al. coefficient becomes 2.54, which is close to 2.6.

3.3. Penetration of Elevated Inversions

During the daytime, plume rise usually takes place in a convective boundary layer capped by stable air. The latter can be in the form of a thin inversion, characterized by a potential temperature jump, $\Delta\Theta_i$, or it could be a thick layer characterized by the potential temperature gradient, $d\Theta_i/dz$. Depending on the height and strength $(\Delta\Theta_i$ or $d\Theta_i/dz)$ of the inversion and the buoyancy flux, the plume may partially or completely penetrate the inversion or it may be fully trapped below it. In the second situation, the plume can mix down to the surface, leading to potentially high ground-level concentrations. Thus, it is important to know the fraction of the plume that is trapped.

At present, the general practice in applied dispersion modeling is to assume complete penetration if the "effective stack height" h_e exceeds the inversion height h, or no penetration if $h_e < h$. The $h_e = z_s + \Delta h$, where z_s is the physical stack height and Δh is the final rise, calculated on the basis of the wind speed and turbulence within the CBL. Thus, the inversion strength, which is the main impediment to penetration, is not a factor. This unphysical criterion can be improved with existing knowledge, although one must recognize that the modeling of penetration is still not completely settled.

In the following, we review penetration models that contain the relevant physical variables, first for a thin inversion and then for a thick one.

3.3.1. Thin Inversion

Existing models

For a zero wind and a vertical plume, Briggs (1969, 1975) argued that a plume should completely penetrate an elevated inversion if its mean temperature excess, $\Delta \Theta_T$, when it arrives at the inversion, exceeds the jump, $\Delta \Theta_i$. Here, subscript T denotes a top hat profile; a Gaussian profile will be considered below. The Morton et al. (1956) model predicts that $\Delta \Theta_T$ varies as

$$\Delta \Theta_T / \Theta_a = \frac{5}{6 \alpha g} \left(\frac{9}{10} \alpha \right)^{-1/3} \frac{F_b^{2/3}}{h'^{5/3}} , \qquad (3.27)$$

where

$$h' = h - z_s . \qquad (3.28)$$

Choosing $\alpha = 0.08$, Briggs (1975) found that complete penetration would occur for h' satisfying

$$h' < 6.9 \frac{F_b^{2/5}}{b_i^{3/5}} , \qquad (3.29)$$

where

$$b_i = g \frac{\Delta \Theta_i}{\Theta_a} . \qquad (3.30)$$

He also showed that Eq. (3.29) agrees well with laboratory data for buoyant plumes. However, plumes with significant vertical momentum were not able to penetrate as high an inversion because of their greater entrainment and dilution below the inversion.

As mentioned earlier, the actual temperature profile in an ensemble-averaged plume is not a top hat and may be more akin to a Gaussian distribution, where the maximum temperature excess, $\Delta\Theta_m$, is $2\Delta\Theta_T$ and the standard deviation, σ_r, is $r/\sqrt{2}$ (Manins, 1979). Thus, even when $\Delta\Theta_T < \Delta\Theta_i$, the center of the plume may be buoyant relative to air above the inversion and could penetrate it; i.e., partial penetration could occur. However, no experimental data exist on partial penetration of vertical plumes.

In applications, a more relevant problem is the penetration of an elevated inversion by a bent-over plume. This problem is more complicated than the zero-wind situation because the plume has a finite cross section nearly perpendicular to the stable interface. Two models have been proposed for this problem: one by Briggs (1975) based on buoyancy depletion of the plume as it crosses the inversion, and a second by Manins (1979) based on the plume density or temperature distribution when its centerline reaches the inversion. In both, the inversion is idealized as a jump of zero thickness.

Briggs (1975) assumed that a plume rising into an elevated inversion would attain an equilibrium height, z'_{eq}, when its buoyancy flux was completely depleted, i.e., $F = 0$. He modified the buoyancy conservation equation (Eq. 3.7) to account for the change in F for that portion of the plume crossing the inversion at any one time. This follows from Eq. (3.7) by 1) replacing dF/ds by $(dF/dz')(dz'/ds) = (w/U_{sc})(dF/dz')$, and 2) including $d\Theta_i/dz$ on the right-hand side of that equation inside an area integral over the plume cross section; i.e.,

$$\frac{w}{U_{sc}}\frac{dF}{dz'} = -\frac{1}{\pi}\int\int \frac{g}{\Theta_a}\frac{d\Theta_i}{dz}w\, dy\, dz \ . \qquad (3.31)$$

With $\int (g/\Theta_a)(d\Theta_i/dz)dz = b_i$, w assumed to be uniform over the plume, and $U_{sc} \simeq U_a$, Eq. (3.31) becomes

$$\frac{dF}{dz'} = -\frac{U_a b_i Y}{\pi} \ , \qquad (3.32)$$

where Y is the plume width at h'. As a further simplification, Briggs chose the plume cross section to be rectangular with a depth equal to the rise z', and a width $Y = 0.5z'$; the plume area, $0.5z'^2$, is the same as that for a round plume, $\pi r^2 = \pi \beta'^2 z'^2$, with $\beta' \simeq 0.4$.

Replacing Y by $0.5z'$, Eq. (3.32) can be integrated from $z' = 2h'/3$ where $F = F_b$, to $z' = z'_{eq}$ where $F = 0$, with this result:

$$\frac{z'_{eq}}{h'} = \frac{2}{3}(1 + 9\pi P)^{1/2}. \tag{3.33}$$

The dimensionless buoyancy flux P is defined by

$$P = \frac{F_b}{U_a b_i h'^2}. \tag{3.34}$$

From the assumed geometry, the fraction f of the plume trapped by the inversion is

$$f = h'/z'_{eq} - 0.5 \tag{3.35}$$

if $2/3 < z'_{eq}/h' < 2$. No penetration occurs if $z'_{eq}/h' < 2/3$, and complete penetration occurs if $z'_{eq}/h' > 2$.

Manins (1979) adopted a Gaussian distribution as a reasonable approximation to the temperature profile in a bent-over plume and assumed that penetration would commence when $\Delta\Theta_m = \Delta\Theta_i$, $\Delta\Theta_m$ being the maximum excess temperature. For a Gaussian distribution, $F = U_a \sigma_r^2 g \Delta\Theta_m / \Theta_a$, and $\sigma_r^2 = r^2/2$, he found that $\Delta\Theta_m = \Delta\Theta_i$ when

$$P = \frac{\beta'^2}{2} = 0.08; \tag{3.36}$$

this criterion established incipient penetration.

For $P > 0.08$, he argued that partial penetration would take place for that portion of the plume with $\Delta\Theta > \Delta\Theta_i$. For a Gaussian distribution, the above condition leads to a simple expression for the fraction of the plume trapped, $f = 0.08/P$. However, for $P > 0.12$, this expression overestimated f as determined from laboratory experiments. As a result, Manins modified it to account for the initial overshoot of a plume above h and for re-entrainment of plume material within the inversion; his modified expression has the form

$$f = \frac{0.08}{P} - (P - 0.08). \tag{3.37}$$

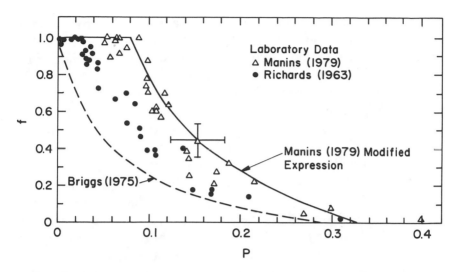

Figure 3.8. Models and laboratory measurements of the fraction of a plume trapped by an elevated inversion as a function of the dimensionless buoyancy flux P; models idealize inversion as a jump of zero thickness. Adapted from Manins (1979).

Manins (1979) and Richards (1963) conducted laboratory experiments of inversion penetration by buoyant plumes and line thermals, respectively. Figure 3.8 compares Briggs' (1975) and Manins' (1979) models for f with these experimental data. As expected, Manins' expression (Eq. 3.37) is in good agreement with his data, but for the most part it overestimates the f obtained in Richards' experiments. On the other hand, Briggs' model [Eqs. (3.33) and (3.35)] underestimates the fraction trapped in both experiments.

The differences in the experimental results as well as in the two models cannot be fully resolved at present, but there are several possible causes. Considering the experiments first, one potential source of the f difference is Manins' experimental technique, which may have overestimated f because it relied on concentration measurements only along the plume centerplane ($y = 0$) and not over the entire cross section (see Briggs, 1984). Richards' technique avoided this problem. A second possibility is the finite thickness Δh_i of the inversion and differences in $\Delta h_i / h'$ between the two experiments. Manins (private communication) reported $\Delta h_i / h'$ values ranging from 0.1 to 0.5 with a mean of 0.3 and $\sim 90\%$ of the values less than 0.4. Richards (1963) did not report Δh_i, but similar experiments conducted on axisymmetric thermals earlier (Richards, 1961) suggest that

$\Delta h_i/h' \sim 0.1$. A finite Δh_i would lead to a larger effective inversion height (i.e., top), and hence, a larger P required for the onset of penetration. A third possibility is the geometrical difference in the source configuration in the two experiments and in the ratio of the effective initial radius r_0 to h'; $r_0/h' \simeq 0.03$ in Manins' experiments and ~ 0.15 in Richards'.

On physical grounds, it seems that the onset of penetration must be determined by the density distribution when the plume reaches the inversion as in Manins' model. The buoyancy depletion model (Briggs) predicts that some penetration occurs for vanishingly small P even if $\Delta\Theta_m < \Delta\Theta_i$; this is not plausible since plume material above the inversion would be negatively buoyant and sink back to h. Thus, in the limit as $f \to 1$, Manins' argument is the more appropriate. However, for sufficiently large P, when a significant fraction of the plume has penetrated, further penetration could be governed by buoyancy depletion. This possibility is suggested by calculations and comparisons with data discussed below.

Further model considerations

In the following, we consider the effect of Δh_i on the plume's penetration capability and an alternative temperature distribution for determining the onset of penetration.

The buoyancy depletion model is easily modified to account for an inversion of finite thickness. Equation (3.31) can be rewritten as:

$$\frac{dF}{dz'} = -\frac{g}{\Theta_a}\frac{\Delta\Theta_i}{\Delta h_i}U_a \int\int_{A_i} dy dz , \tag{3.38}$$

where $\Delta\Theta_i$ is the total change in Θ_a over the inversion layer, w is assumed to be uniform over the plume, $U_{sc} \simeq U_a$, and the integral is over the area, A_i, of the plume within the inversion layer. Integrating Eq. (3.38) from z'_1 where the upper edge of the plume is at the inversion base and $F = F_b$ to z'_{eq} where $F = 0$, we have

$$F_b = \frac{U_a b_i}{\Delta h_i} \int_{z'_1}^{z'_{eq}} A_i dz' \tag{3.39}$$

or

$$P\delta = \int_{\eta_1}^{\eta_{eq}} \frac{A_i}{h'^2} d\eta , \tag{3.40}$$

where

$$\eta = \frac{z'}{h'} , \quad \delta = \frac{\Delta h_i}{h'} \tag{3.41}$$

and subscripts 1 and *eq* refer to z'_1 and z'_{eq}, respectively. For a circular plume, $z'_1 = h'/(1 + \beta')$.

We have calculated the area integral (Eq. 3.40) to find z'_{eq} as a function of P and δ, assuming the plume cross section to be a circle with $r = \beta'z'$. From the given geometry, the fraction of the plume below the inversion top, $h + \Delta h_i$, is

$$f = 1 - \frac{1}{\pi}\left[\cos^{-1}\lambda - \lambda\left(1 - \lambda^2\right)^{1/2}\right] , \qquad (3.42)$$

where

$$\lambda = \frac{1 + \delta - \eta_{eq}}{\beta'\eta_{eq}} . \qquad (3.43)$$

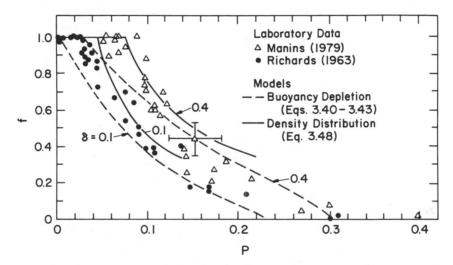

Figure 3.9a. Models and laboratory measurements of the fraction of a plume trapped by an elevated inversion as a function of the dimensionless buoyancy flux P; models include a finite inversion thickness, where δ is the ratio of the thickness to the inversion height above the stack.

Figure 3.9a compares Eq. (3.42) for $\delta = 0.1$ and 0.4 (dashed lines) with the data of Manins (1979) and Richards (1963). The dashed lines bracket almost all the data for $P > 0.12$ or $f < 0.5$, suggesting that the buoyancy depletion concept may apply when a significant fraction of the plume has penetrated the inversion. The bracketing of the Richards data for smaller P is fortuitous because 1) δ is not expected to be as large as 0.4 in his experiments, and 2) the initial departure of f from 1 should be governed by the plume density or temperature distribution at h' rather than by buoyancy depletion.

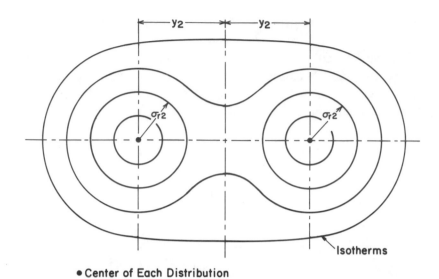

● Center of Each Distribution

Figure 3.9b. Schematic of model plume cross section containing two Gaussian distributions of buoyancy.

Richards' (1963) measurements and others show that a thermal cross section contains two concentrated cores of material rather than a single axisymmetric distribution. To gain insight on the effect of a non-axisymmetric distribution on incipient penetration, say $1 > f > 0.5$, we have calculated f for a cross section consisting of two Gaussian distributions, each offset a distance y_2 from the plume centerline and axisymmetric about their individual centers (Fig. 3.9b). The buoyancy flux in each distribution is $F_b/2$ and the maximum temperature excess, $\Delta\Theta_{m2}$, is therefore

$$\Delta\Theta_{m2} = \frac{\pi F_b\Theta_a/2}{\pi U_a g \sigma_{r2}^2} = \frac{F_b\Theta_a}{2U_a g \sigma_{r2}^2}, \qquad (3.44)$$

where σ_{r2} is the standard deviation of each. The total variance σ_r^2 about the plume centroid is

$$\sigma_r^2 = \sigma_{r2}^2 + y_2^2 ; \qquad (3.45)$$

for simplicity, σ_r^2 is assumed to be equal to that of an axisymmetric plume, $\sigma_r^2 = \beta'^2 z'^2/2$.

Assuming that the ratio $y_2/\sigma_{r2} = a$ is known, we can find $\sigma_{r2}, y_2, \Delta\Theta_{m2}$, and f. For $a > 1$ and $\Delta\Theta$ near $\Delta\Theta_{m2}$, the isotherms are circles about each buoyant core, and f is the same as that found by Manins (1979) for a single Gaussian distribution, $f = \Delta\Theta_i/\Delta\Theta_{m2}$. Combining Eqs. (3.44) and (3.45) with a known, we have

$$\Delta\Theta_{m2} = \frac{1 + a^2}{2} \frac{F_b}{U_a \sigma_r^2 g/\Theta_a} \qquad (3.46)$$

and

$$f = \frac{\Delta\Theta_i}{\Delta\Theta_{m2}} = \frac{2}{1 + a^2} \frac{\beta'^2/2}{P} . \qquad (3.47)$$

Thus, for $a = 1$, the double Gaussian distribution gives the same P for incipient penetration ($f = 1$) as the single, axisymmetric Gaussian distribution, i.e., $P = \beta'^2/2 = 0.08$. For $a > 1$, the required P (for $f = 1$) is less, $P = \beta'^2/(1 + a^2)$, and for $a < 1$, it is probably about the same as for $a = 1$, owing to overlap of the isotherms from the two distributions. Richards' (1963) data suggest $a = 1.8$.

A simple correction to Eq. (3.47) can be made for the nonzero inversion thickness by assuming that the temperature jump $\Delta\Theta_i$ occurs at $(1 + \delta)h'$ rather than at h'. This means that the plume radius is larger by the factor $1 + \delta$, and the expression for f is then

$$f = \frac{(1 + \delta)^2}{1 + a^2} \frac{\beta'^2}{P} . \qquad (3.48)$$

Equation (3.48) is plotted in Fig. 3.9a for $\delta = 0.1$ and 0.4 with $a = 1.8$. The lines bracket most of Manins' data for $P < 0.14$, and the line for $\delta = 0.1$ is in reasonable agreement with Richards' data for $0.05 < P < 0.14$. These comparisons suggest that the departure of f from 1 is sensitive to the assumed temperature distribution and the inversion thickness. Moreover, they show that the expected range of δ in the above experiments can partially account for the scatter in the data and the differences in the two data sets, especially for $P < 0.14$.

3.3.2. Thick Inversions

In the following we discuss penetration criteria only for bent-over plumes. The inversion is considered to be thick if the entire plume cross section lies within it when complete penetration occurs ($f = 0$); i.e., $2r \le \Delta h_i$, where $2r$ is the plume depth. For a round plume and

$f = 0$, $z'_{eq} = h'/(1 - \beta')$ and $2r = 2\beta'h'/(1 - \beta')$; thus, the criterion for a thick inversion is $\Delta h_i/h' \gtrsim 2\beta'/(1 - \beta') \simeq 1.3$. In this section, penetration and trapping are measured with respect to the height of the inversion base h'.

Briggs (1984) proposed a simple, conservative approach for estimating f by assuming that the stable lapse rate extended from the inversion layer down to the stack top. The plume equilibrium height was then found from Eq. (3.26) using the given $d\Theta_i/dz$, and the fraction of the plume trapped was based on the height of the lower plume edge relative to h'. To estimate f, Briggs used Eq. (3.35), which is based on a rectangular plume geometry. With the above assumptions

$$f = 0 \quad \text{if} \quad h' < 0.5z'_{eq}$$

$$f = 1 \quad \text{if} \quad h' > 1.5z'_{eq}$$

$$f = \frac{h'}{z'_{eq}} - 0.5 \quad \text{if} \quad 0.5z'_{eq} < h' < 1.5z'_{eq} \, ,$$

$$(3.49a)$$

where

$$z'_{eq} = 2.6 \left(\frac{F_b}{N_i^2 U_a} \right)^{1/3} \quad \text{and} \quad N_i^2 = \frac{g}{\Theta_a} \frac{d\Theta_i}{dz} \, . \qquad (3.49b)$$

This model is conservative in that the plume initially rises in unstratified air and thus should not lose any buoyancy. To account for this initial phase, Berkowicz et al. (1986) assumed that buoyancy depletion would not begin until the upper edge of the plume reached the inversion base or $z' = 2h'/3$. Their expression for z'_{eq} is

$$\frac{z'_{eq}}{h'} = \left[2.6^3 P_s + (2/3)^3 \right]^{1/3} \, , \qquad (3.50)$$

where

$$P_s = \frac{F_b}{U_a N_i^2 h'^3} \, . \qquad (3.51)$$

Equation (3.50) can be derived from the vertical momentum equation using the customary assumptions for a bent-over plume (Appendix B) but neglecting any vertical momentum acquired by the plume during its rise to the height $2h'/3$.

Figure 3.10 compares the predicted f from the above models with a small sample of field data. In general, the model trends agree with

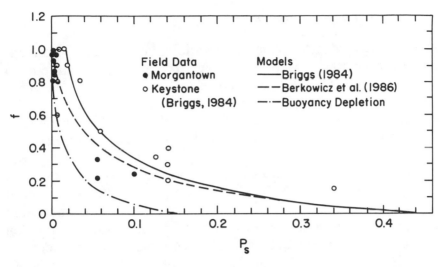

Figure 3.10. Models and measurements of the fraction of plume trapped by an elevated stable layer of uniform stratification as a function of the dimensionless buoyancy flux P_s.

the data, differing most and with each other for $P_s < 0.06$. Briggs' model works better for $P_s > 0.06$ or $z'_{eq}/h' > 1$ than for $P_s < 0.06$, as should be expected, since that is when the assumed temperature distribution is most applicable; the Berkowicz et al. model performs similarly in this P_s regime. For $P_s < 0.06$, the two models show P_s tending to 0.02 (Briggs) and 0 (Berkowicz et al.) as $f \to 1$. The second behavior makes the most sense, since for a small amount of buoyancy some portion of a plume should be able to penetrate the base of an inversion having a finite $d\Theta_i/dz$.

The Morgantown data support the Berkowicz et al. model as $f \to 1$. We attach more significance to this data because f was numerically determined from horizontal integrals of the plume aerosol distribution measured via lidar and their variation with height. For the Keystone data, f was estimated with more difficulty from isopleths of aerosol content. In the small P_s regime, the buoyancy depletion model [Eqs. (3.40)–(3.43)] also agrees well with the Morgantown data, and it provides a lower bound to the observations. [In applying this model to thick inversions, we replace $P\delta$ by P_s; see Eqs. (3.30), (3.34), (3.41), and (3.51) for this equivalence.] On the basis of this discussion, the Berkowicz et al. model appears to be most consistent with the observations and the expected behavior of the f versus P_s curve in the limits of $f = 1$ and 0. However, further work is necessary to understand the plume penetration process fully.

3.3.3. Summary and Recommendations

Existing models for plume penetration of elevated inversions apply either for "thin" inversions characterized by a temperature jump $\Delta\Theta_i$ or for "thick" ones characterized by the gradient $d\Theta_i/dz$ in the elevated layer. For thin inversions, incipient penetration—the departure of f from 1—is governed by the plume density distribution at h'; once a significant fraction of the plume has penetrated (say $f < 0.5$), the variation of f with P may be governed by buoyancy depletion at the inversion, provided that one accounts for the inversion thickness Δh_i (see Sec. 3.3.1). For thick inversions, the Berkowicz et al. model appears to be the best simple description of existing data; however, for sufficiently large $d\Theta_i/dz$, the plume density distribution at h could be important and should be considered in the model. To provide for a more systematic variation of f with P, Δh_i or its dimensionless counterpart δ needs to be incorporated in the models and alternative density distributions (to that used by Manins, 1979) should be considered (Sec. 3.3.1).

Recognizing our earlier caveats, we can make some tentative recommendations on model applications. First, if observations of the temperature structure are available to define $\Delta\Theta_i$ and Δh_i, the Manins (1979) model can be used as a conservative approach for estimating f if $\delta < 0.4$ and the Berkowicz et al. (1986) model can be used for larger δ. A δ of 0.4 corresponds to the upper end of the δ range in Manins' experiments. It is recognized that there is an inconsistency in the treatment of penetration and in the variation of f with P in the two models; this needs to be resolved with further modeling and laboratory experiments. Second, if only an early morning temperature profile is available, the Berkowicz et al. model can be applied using a $d\Theta_i/dz$ determined from the profile just above the calculated or observed h and assuming that the stratification aloft remains unchanged from its morning value. An alternative approach is simply to assume a typical value for $d\Theta_i/dz$, e.g., 0.01 K m^{-1} (Weil and Brower, 1984).

The analysis in Sec. 3.3.1, which includes δ as an additional parameter, should be continued to resolve the inconsistency in the models for "thin" and "thick" inversions. In addition, further laboratory experiments should be conducted over a wide range of conditions—δ, P, and $d\Theta_i/dz$, etc.—to give more information on the penetration process. To avoid ambiguities, f should be determined from area-integrated concentrations over the plume cross section; furthermore, it would be beneficial to conduct these experiments both with and without ambient convection below the inversion.

3.4. Ambient Turbulence Effects on Plumes

Dispersion modeling, and especially the Gaussian approach, has influenced the way we think about turbulence effects on plumes. In the Gaussian model, the dispersion process is assumed to be divided into two stages: 1) a plume rise phase where ambient turbulence can be neglected, and 2) a dispersion phase where such turbulence dominates. Thus, it has been the practice to think in terms of a "final rise," Δh, and an "effective stack height," $h_e = z_s + \Delta h$, and to develop models to parameterize Δh in terms of source conditions, mean wind speed, and the local turbulence. However, nearly all plume-rise measurements in a near-neutral or convective boundary layer show that a plume is either still rising at the maximum observational distance (if it has not encountered an elevated inversion) or it is "looping."

There are two main ways in which ambient turbulence can affect plumes. First, if the small-scale turbulence (with eddy size $\lesssim r$) is sufficiently intense, it can increase the plume growth rate beyond that given by buoyancy-induced turbulence and lead to an asymptotic rise. Second, if the large-scale turbulence is strong, it can carry entire sections of the plume up and down, thereby dispersing the plume by meandering.

Whether one or both of the above processes occur is determined by the turbulence structure of the boundary layer. In the CBL, measurements (Kaimal et al., 1976; Willis and Deardorff, 1974) and numerical modeling (Deardorff, 1972) show that the turbulence is dominated by large-scale elements consisting of thermal plumes (updrafts) and downdrafts. As discussed in Ch. 1, these elements extend from the surface to the CBL top, are long-lived, and have length and velocity scales proportional to h and w_*, respectively; w_* is the convective velocity scale. This suggests that large-scale turbulence should be considered at least as seriously as the small-scale turbulence, and in fact may dominate. Indeed, field observations as well as laboratory data (see Fig. 3.11) show the large-scale meandering or "looping" due to the large eddies in the CBL.

In the following, we review plume rise models based on both the small- and the large-scale turbulence.

(a) LABORATORY

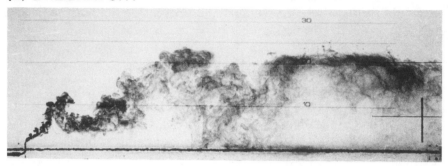

(b) FIELD

Figure 3.11. Photographs of laboratory and full-scale looping plumes in a convective boundary layer. (a), from Willis and Deardorff (1983); (b), from Schiermeier and Niemeyer (1970).

3.4.1. Small-Scale Turbulence

The entrainment model discussed in Sec. 3.2.1 has been extended in three ways to include ambient turbulence and predict an asymptotic rise. These include 1) a change in the entrainment velocity, 2) incorporation of momentum and buoyancy losses due to "detrainment," and 3) plume "breakup" due to intense small-scale turbulence. Of the three, only the third has been linked to current boundary layer understanding.

Following the first approach, Briggs (1971) and Slawson and Csanady (1967) assumed that ambient turbulence in the inertial subrange would eventually dominate the entrainment process, and that the most important eddies would be those comparable in size with the plume radius. The velocity v_e of these eddies depends on the turbulence dissipation rate in the environment ϵ_a and the plume radius through $v_e = (\epsilon_a r)^{1/3}$; v_e is assumed to replace βw in the entrainment expression when $v_e > \beta w$. Using an empirical form for ϵ_a, Briggs calculated an asymptotic rise, but this occurred at distances much beyond those typical of maximum ground-level concentrations. With additional assumptions, he predicted that an "effective" final rise would occur at shorter range and arrived at this simple formula:

$$\Delta h = 1.6 \frac{F_b^{1/3}}{U_a} (3.5x^*)^{2/3} , \qquad (3.52)$$

where x^* is the distance at which the entrainment expression changes. This formula has been widely used in applications, but neither it nor the model prediction that r grows as $x^{3/2}$ has been confirmed by observations.

Priestley (1956), Netterville (1985), and Turner (1963) proposed different models to describe the momentum and buoyancy losses of a plume due to "detrainment" or erosion of the plume by the ambient turbulence. For example, Turner superposed the usual entrainment velocity βw, and a constant "outflow" velocity u_0, which was assumed to be proportional to the ambient turbulence (the root-mean-square component). The resulting entrainment rate (E) was $2r(\beta w - u_0)$ and could lead to an r that decreased to zero; r defined the region within which the turbulence was buoyancy- or self-generated. Turner found that the maximum rise predicted for round thermals agreed with laboratory observations in decaying homogeneous turbulence. However, this model and in particular the idea of a shrinking r have not been evaluated in the field.

According to the breakup model (Briggs, 1975), a plume's organized internal structure would be suddenly destroyed by intense small-scale ambient turbulence entrained by the plume. Since this turbulence is dependent on the dissipation rate, Briggs argued that breakup would occur when the ambient rate, ϵ_a, was equal to or exceeded that in the plume, ϵ. Using dimensional arguments, he assumed that ϵ was proportional to w^3/r or w^3/z', whereas ϵ_a was based on the neutral surface layer expression for weak convection or near-neutral conditions and the CBL result for strong convection.

In the neutral surface layer, the dissipation rate is given by

$$\epsilon_a = \frac{u_*^3}{kz} , \tag{3.53}$$

where u_* is the friction velocity and k is the von Kármán constant ($k \sim 0.4$). Equating ϵ_a and ϵ, Briggs (1975) found the final rise for a buoyant plume to be

$$\Delta h = 1.2 \left(\frac{F_b}{U_a u_*^2} \right)^{3/5} (z_s + \Delta h)^{2/5} , \tag{3.54}$$

which must be solved iteratively. However, Briggs (1984, private communication) suggested a simple approximation that satisfied the small and large $\Delta h / z_s$ limits of Eq. (3.54); the approximate formula is

$$\Delta h = 1.2 F_{*n}^{3/5} (1 + 1.2 F_{*n})^{2/5} z_s \tag{3.55}$$

where $F_{*n} = F_b / U_a u_*^2 z_s$. For a jet, Briggs (1975) suggested that

$$\Delta h = \frac{0.9}{\beta} \left(\frac{F_m}{U_a u_*} \right)^{1/2} . \tag{3.56}$$

The formulas for buoyant rise have received some testing as discussed below, whereas that for the jet has not.

In the CBL, Briggs argued that the plume segments contributing most to the ground-level concentrations are those in downdrafts, and that the final rise should be based on the ϵ_a within a downdraft. This ϵ_a is less than the average value, $0.5 w_*^3 / h$, over the CBL. Assuming ϵ_a to be $0.25 w_*^3 / h$, Briggs (1984) found the final rise to be

$$\Delta h = 3 F_*^{3/5} h , \tag{3.57}$$

where

$$F_* = \frac{F_b}{U_a w_*^2 h} \tag{3.58}$$

is a dimensionless buoyancy flux. For jets, the predicted rise was

$$\Delta h = \frac{1.3}{\beta^{6/7}} \left(\frac{F_m}{U_a w_*} \right)^{3/7} h^{1/7} . \tag{3.59}$$

We note that in earlier work, Briggs (1975) reported the coefficient in Eq. (3.57) to be 4.3 based on an assumed $\epsilon_a = 0.1w_*^3/h$ within downdrafts; thus, there is some uncertainty in the coefficient and a suitable value depends in part on the application. If one uses Eq. (3.57) in a diffusion model for the purpose of predicting ground-level concentrations, it is recommended that one try both coefficients (3.0 and 4.3) and choose the one that produces the best agreement with observations; for example, Weil and Brower (1984) found that the coefficient 4.3 worked well in their Gaussian plume model predictions of concentrations downwind of power plants. In practice, the simplest approach for selecting the appropriate final rise expression—that for near-neutral or strongly convective conditions—is to choose the one yielding the lowest rise (e.g., see Briggs, 1984); for buoyant plumes, the choices are Eqs. (3.54) or (3.55) and (3.57).

Verification of final rise formulas is difficult because observations must be made at large distances (up to about 5 km) where the plume can be detected only by remote sensing, e.g., by lidar. Even then, it is difficult to find situations where the plume levels because of ambient turbulence and not because of stable stratification, i.e., an elevated inversion.

Weil (1979) compared the breakup model to lidar measurements of plume rise out to about 5 km from a power plant. Out of 11 periods, typically of a 1-hour duration, there were 3 in which the rise was reported to have been turbulence limited. However, further analysis suggested that only one was so limited, and the observed rise was in excellent agreement with Eq. (3.54). In the remaining 10 cases, the maximum rise was either the most distant observation or it was limited by an elevated inversion. For the four periods in which the neutral breakup model was applicable (i.e., it gave a lower rise than the convective model), Eq. (3.54) overestimated the rise by 6% on average; in the remaining seven periods when Eq. (3.57) was appropriate, it underestimated the rise by 13% on average.

Another approach for assessing "final-rise" models is to evaluate the ground-level concentrations based on the plume rise prediction. This, of course, assumes that the diffusion model is correct. Briggs (1984) followed this approach in testing the breakup model, using measurements of maximum ground-level concentrations around two British power plants. Although a number of assumptions had to be made concerning the surface heat flux and F_b, the breakup model resulted in reasonably good estimates of surface concentrations. This suggests that the model is useful in parameterizing ground-level concentrations, but it does not mean that a "final rise" due to turbulence actually occurs.

3.4.2. Large-Scale Turbulence

Models of increasing detail have been developed to predict the effect of large-scale ambient turbulence on plumes in a convective boundary layer. These range from a simple breakup criterion to the tracking of plume segments in a field of updrafts and downdrafts in the CBL.

In early work, Weil and Hoult (1973) proposed that rise termination should occur when the plume rise velocity w was equal to the vertical turbulence component σ_w in the environment. Beyond this point, they assumed that plume segments would be convected equally well by updrafts and downdrafts and that buoyant rise would be negligible. They predicted the final rise to be

$$\Delta h = \frac{2}{3\beta^2} \frac{F_b}{U_a \sigma_w^2} , \tag{3.60}$$

where σ_w was assumed to be $0.49w_*$ and characteristic of the large-scale eddies. With $\beta = 0.6$, Eq. (3.60) leads to $\Delta h/h = 7.7F_*$.

Csanady (1973) proposed a similar model that assumed more generally that final rise occurred when $w = \mu\sigma_w$, where μ was an empirically chosen coefficient; σ_w was to be either assumed or given. Both Csanady and Weil and Hoult showed the utility of their models in estimating ground-level concentrations downwind of tall stacks.

Briggs (1975) carried this idea one step further by assuming that the plume rise took place relative to the updrafts and downdrafts. Plume segments caught in downdrafts would eventually be carried to the surface when the rise velocity, which was proportional to $x^{-1/3}$, was less than the downdraft speed. For a plume rising relative to the mean downdraft speed $\overline{w}_d$, the distance x_d where the lower plume edge touches the ground satisfies

$$z_s + 0.5z_b' - \frac{\overline{w}_d}{U_a}x_d = 0 , \tag{3.61}$$

where z_b' is the rise from the two-thirds law (Eq. 3.17). Briggs chose the final rise to be that given by the two-thirds law at $x = x_d$, or

$$\Delta h = \frac{F_b}{U_a \overline{w}_d^2} \left(1 + \frac{2z_s}{\Delta h}\right)^2 , \qquad (3.62)$$

which must be solved iteratively; $\overline{w}_d$ was assumed to be $0.4w_*$.

This model has received little direct testing with plume rise observations. However, it was a key component of Venkatram's (1980) dispersion model, which performed reasonably well in predicting SO_2 concentrations around tall stacks.

Weil et al. (1986) extended the above ideas in a simple dispersion model for the CBL (see Ch. 4) under the assumption of an infinitely long time scale for the ambient turbulence. They superimposed the plume rise and the ambient convection (w_a) velocities, but in their model w_a is a random variable specified by a probability density function (p.d.f.), $p_w(w_a)$. The plume centerline height z_c is also a random variable given by

$$z_c = z_s + z_b' + \frac{w_a x}{U_a} . \qquad (3.63)$$

The ensemble-averaged concentration is computed from the p.d.f. of z_c, p_z, which in turn is found from p_w by

$$p_z = p_w[U_a(z_c - z_s - z_b')/x]\left|\frac{dw_a}{dz_c}\right| . \qquad (3.64)$$

Here the argument w_a of p_w has been rewritten in terms of z_c, z_s, z_b', x, and U_a using Eq. (3.63). An advantage of this model is that one does not have to specify a final rise; instead, the relative magnitudes of z_b' and $w_a x/U_a$ in Eq. (3.63) determine when plume segments are carried to the surface (i.e., $z_c = 0$) to yield high concentrations.

This model can also be used to calculate the mean plume trajectory and vertical dispersion. Such calculations have been done using an early numerical version of the model containing the following assumptions: 1) p_w is divided into an updraft and a downdraft portion, with that in the downdraft part given by

$$p_w(w_a) = (\pi/2)(w_a/\overline{w}_d^2) \ \exp\left[-\pi(w_a/2\overline{w}_d)^2\right] ;$$

a similar form is used for the updraft velocities but with $\overline{w}_d$ replaced by $\overline{w}_u$, the mean updraft velocity. 2) $\overline{w}_d$ and $\overline{w}_u$ are assumed to be $0.4w_*$ and $0.6w_*$, respectively, in accord with Lamb's (1978) work. 3) Plume segments initially in updrafts reverse their direction at h, and the acquired downdraft velocity w_d is related to the updraft velocity w_u by $w_d = -A_u w_u / A_d$, where A_d and A_u are the average horizontal areas covered by downdrafts and updrafts, respectively; $A_u / A_d = 2/3$ as determined from continuity, $A_u \overline{w}_u + A_d \overline{w}_d = 0$, and the values of $\overline{w}_u$ and $\overline{w}_d$. A similar assumption governs the reversal of plume segments at the surface.

Figure 3.12 compares the mean centerline and spread (σ_z) from the model with a laboratory simulation for a highly buoyant plume, $F_* = 0.11$, as observed by Willis and Deardorff (1983). The dimensionless distance X is the ratio of the travel time (x/U_a) to the convective time scale (h/w_*). The modeled trajectory and outline follow the data quite well out to $X \sim 0.7$. This supports the idea of superimposing the plume rise and the vertical plume displacements (meandering) caused by the convection. The difference in the boundaries beyond $X \sim 0.7$ is attributed to the difference in the quantities that they represent: the modeled boundary is $\overline{z}_c \pm \sigma_z$ whereas the observed boundary is the mean plume envelope.

3.4.3. Dispersion Applications

I believe that the p.d.f. model is a better representation of dispersion in the CBL and the action of the large-scale turbulence on plumes than the Gaussian model. In addition, the p.d.f. approach does not require a "final rise," which is a difficult quantity to predict with confidence and to observe. The model simulates the average, near-source behavior of buoyant plumes in a laboratory convection tank as shown in Fig. 3.12. It also predicts crosswind-integrated concentrations near the surface that are in good agreement with the laboratory data and ground-level concentrations that are in reasonable agreement with field observations downwind of tall stacks (Weil et al., 1986).

Compared with the p.d.f. approach, the Gaussian model performs about the same in predicting concentrations downwind of tall stacks if one uses Briggs breakup model for final rise and convective scaling for estimating the dispersion parameters σ_y and σ_z (see Weil et al., 1986). I recommend the above approaches for plume rise and dispersion estimation if one adopts the Gaussian model. In the comparisons cited here, the good performance of the p.d.f. and Gaussian models is restricted to weakly-to-moderately buoyant plumes, i.e., $F_* < 0.1$.

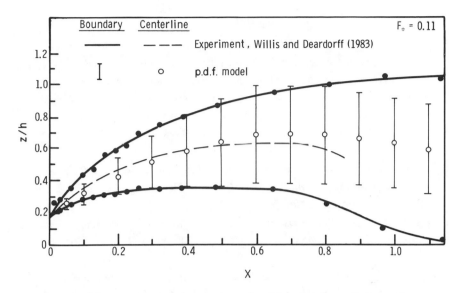

Figure 3.12. Laboratory measurements and p.d.f. model predictions of average centerline and spread of a buoyant plume in the convective boundary layer.

For highly buoyant plumes ($F_* > 0.1$) that do not penetrate an elevated inversion, a different dispersion model is necessary. As discussed in Ch. 4, it is then more useful to think in terms of energy arguments. Plume segments can be brought to the surface when the large eddies have sufficient kinetic energy, i.e., $\rho w_a^2/2$, to overcome the potential energy difference between the plume and the environment, $\Delta\rho gh$.

References

Berkowicz, R., H. R. Olesen, and U. Torp, 1986: The Danish Gaussian air pollution model (OML): Description, test and sensitivity analysis in view of regulatory applications. *Air Pollution Modeling and Its Application V*, C. De Wispelaere, F. A. Schiermeier, and N. V. Gillani, Eds., Plenum, New York, 453–481.

Briggs, G. A., 1969: *Plume Rise*. USAEC Critical Review Series, TID–25075, NTIS, 81 pp.

Briggs, G. A., 1971: Some recent analyses of plume rise observations. *Proceedings of the Second International Clean Air Congress*, H. M. Englund and W. T. Beery, Eds., Academic Press, New York, 1029–1032.

Briggs, G. A., 1975: Plume rise predictions. *Lectures on Air Pollution and Environmental Impact Analyses*, D. A. Haugen, Ed., Amer. Meteor. Soc., Boston, 59–111.

Briggs, G. A., 1984: Plume rise and buoyancy effects. *Atmospheric Science and Power Production*, D. Randerson, Ed., U.S. Dept. of Energy DOE/TIC–27601, available from NTIS as DE84005177, 327–366.

Csanady, G. T., 1973: Effect of plume rise on ground level pollution. *Atmos. Environ.*, **7**, 1–16.

Deardorff, J. W., 1972: Numerical investigation of neutral and unstable planetary boundary layers. *J. Atmos. Sci.*, **29**, 91–115.

Deardorff, J. W., 1985: Laboratory experiments on diffusion: The use of mixed-layer scaling. *J. Climate Appl. Meteor.*, **24**, 1143–1151.

Djurfors, S., and D. Netterville, 1978: Buoyant plume rise in nonuniform wind conditions. *J. Air Pollut. Control Assoc.*, **28**, 780–784.

Escudier, M. P., and T. Maxworthy, 1973: On the motion of turbulent thermals. *J. Fluid Mech.*, **61**, 541–552.

Fan, L., 1967: Turbulent buoyant jets into stratified or flowing ambient fluids. California Institute of Technology, Pasadena, CA, Report KH–R–15, 195 pp.

Fay, J. A., M. P. Escudier, and D. P. Hoult, 1970: A correlation of field observations of plume rise. *J. Air Pollut. Control Assoc.*, **20**, 391–397.

Hewett, T. A., J. A. Fay, and D. P. Hoult, 1971: Laboratory experiments of smokestack plumes in a stable atmosphere. *Atmos. Environ.*, **5**, 767–789.

Hildebrand, F. B., 1976: *Advanced Calculus for Applications* (second ed.) Prentice-Hall, 733 pp.

Holland, J. Z., 1953: A meteorological survey of the Oak Ridge area: Final report covering the period 1948–1952. U. S. Weather Bureau, USAEC Report ORO–99, 554–559.

Hoult, D. P., and J. C. Weil, 1972: A turbulent plume in a laminar crossflow. *Atmos. Environ.*, **6**, 513–531.

Hoult, D. P., J. A. Fay, and L. J. Forney, 1969: A theory of plume rise compared with field observations. *J. Air Pollut. Control Assoc.*, **19**, 585–590.

Kaimal, J. C., J. C. Wyngaard, D. A. Haugen, O. R. Coté, Y. Izumi, S. J. Caughey, and C. J. Readings, 1976: Turbulence structure in the convective boundary layer. *J. Atmos. Sci.*, **33**, 2152–2169.

Lamb, R. G., 1978: A numerical simulation of dispersion from an elevated point source in the convective boundary layer. *Atmos. Environ.*, **12**, 1297–1304.

Manins, P. C., 1979: Partial penetration of an elevated inversion layer by chimney plumes. *Atmos. Environ.*, **13**, 733–741.

Morton, B. R., G. I. Taylor, and J. S. Turner, 1956: Turbulent gravitational convection from maintained and instantaneous sources. *Proc. Roy. Soc. London*, **A234**, 1–23.

Netterville, D. D. J., 1985: Plume rise in turbulent winds. *7th Symposium on Turbulence and Diffusion*, Amer. Meteor. Soc., Boston, 23–26.

Overcamp, T. J., and D. P. Hoult, 1971: Precipitation in the wake of cooling towers. *Atmos. Environ.*, **5**, 751–765.

Overcamp, T. J., and T. Ku, 1986: Effect of a virtual origin correction on entrainment coefficients as determined from observations of plume rise. *Atmos. Environ.*, **20**, 293–300.

Phillips, O. M., 1980: *The Dynamics of the Upper Ocean.* Cambridge University Press, Cambridge, 336 pp.

Poreh, M., and A. Kacherginsky, 1981: Simulation of plume rise using small wind-tunnel models. *J. Wind Eng. Ind. Aerodyn.*, **7**, 1–14.

Priestley, C. H. B., 1956: A working theory of the bent-over plume of hot gas. *Quart. J. Roy. Meteor. Soc.*, **82**, 165–176.

Richards, J. M., 1961: Experiments on the penetration of an interface by buoyant thermals. *J. Fluid Mech.*, **11**, 369–384.

Richards, J. M., 1963: The penetration of interfaces by cylindrical thermals. *Quart. J. Roy. Meteor. Soc.*, **89**, 254–264.

Rouse, H., C.-S. Yih, and H. W. Humphreys, 1952: Gravitational convection from a boundary source. *Tellus*, **4**, 201–210.

Schiermeier, F. A., and L. E. Niemeyer, 1970: *Large Power Plant Effluent Study (LAPPES). Vol. 1: Instrumentation, Procedures, and Data Tabulations (1968).* Department of Health, Education, and Welfare, Public Health Service, Environmental Health Service, National Air Pollution Control Administration, Raleigh, NC, Publication APTD 70-2.

Scorer, R. S., 1978: *Environmental Aerodynamics.* Halsted Press, New York, 488 pp.

Slawson, P. R., and G. T. Csanady, 1967: On the mean path of buoyant, bent-over chimney plumes. *J. Fluid Mech.*, **28**, 311–322.

Snyder, W. H., 1981: Guideline for fluid modeling of atmospheric diffusion. Environmental Sciences Research Laboratory, U. S. Environmental Protection Agency, Research Triangle Park, NC, Report EPA–600/8–81–009, 185 pp.

Spiegel, E. A., and G. Veronis, 1960: On the Boussinesq approximation for a compressible fluid. *Astrophys. J.*, **131**, 442-447.

Tsang, G., 1971: Laboratory study of line thermals. *Atmos. Environ.*, **5**, 445–471.

Turner, J. S., 1963: The motion of buoyant elements in turbulent surroundings. *J. Fluid Mech.*, **16**, 1–16.

Turner, J. S., 1973: *Buoyancy Effects in Fluids*. Cambridge University Press, 367 pp.

Venkatram, A., 1980: Dispersion from an elevated source in a convective boundary layer. *Atmos. Environ.*, **14**, 1–10.

Weil, J. C., 1974: The rise of moist, buoyant plumes. *J. Appl. Meteor.*, **13**, 435–443.

Weil, J. C., 1979: Assessment of plume rise and dispersion models using lidar data. Martin Marietta Environmental Center, Baltimore, Report No. PPSP–MP–24, 73 pp.

Weil, J. C., 1982: Source buoyancy effects in boundary layer diffusion. *Workshop on the Parameterization of Mixed Layer Diffusion*, Physical Sciences Laboratory, New Mexico State University, Las Cruces, 235–246.

Weil, J. C., and D. P. Hoult, 1973: A correlation of ground-level concentrations of sulfur dioxide downwind of the Keystone stacks. *Atmos. Environ.* **7**, 707–721.

Weil, J. C., and R. P. Brower, 1984: An updated Gaussian plume model for tall stacks. *J. Air Pollut. Control Assoc.*, **34**, 818-827.

Weil, J. C., L. A. Corio, and R. P. Brower, 1986: Dispersion of buoyant plumes in the convective boundary layer. *5th Joint Conference on Applications of Air Pollution Meteorology*, Amer. Meteor. Soc., Boston, 335–338.

Willis, G. E., and J. W. Deardorff, 1974: A laboratory model of the unstable planetary boundary layer. *J. Atmos. Sci.*, **31**, 1297–1307.

Willis, G. E., and J. W. Deardorff, 1983: On plume rise within the convective boundary layer. *Atmos. Environ.*, **17**, 2435–2447.

Willis, G. E., and N. Hukari, 1984: Laboratory modeling of buoyant stack emissions in the convective boundary layer. *4th Joint Conference on Applications of Air Pollution Meteorology*, Amer. Meteor. Soc., Boston, 24–26.

Woodward, B., 1959: The motion in and around isolated thermals. *Quart. J. Roy. Meteor. Soc.*, **85**, 144–151.

Appendix A
Integral Equations for a Buoyant Plume in a Crosswind

The derivation of the integral equations for a dry plume in a dry atmosphere follows that given in Weil (1974) for moist plumes except that here the water vapor and liquid water terms are ignored and there is assumed to be no internal heat generation (i.e., from condensing water vapor). It is assumed that 1) the flow is steady and incompressible, 2) the density differences between the plume and the ambient air are sufficiently small to allow the Boussinesq approximation to be made, and 3) the pressure in the plume is the hydrostatic value in the undisturbed atmosphere. With these assumptions, the differential equations governing the conservation of mass, horizontal momentum, vertical momentum, and energy are given respectively by (see Spiegel and Veronis, 1960)

$$\nabla \cdot \vec{V} = 0 \qquad (A1)$$

$$\nabla \cdot (\vec{V} U_x) = \nu \nabla^2 U_x \qquad (A2)$$

$$\nabla \cdot (\vec{V} U_z) = \frac{\rho_a - \rho}{\rho} g + \nu \nabla^2 U_z \qquad (A3)$$

$$\nabla \cdot (\vec{V} T) + U_z \frac{g}{c_p} = \frac{K}{\rho c_p} \nabla^2 T , \qquad (A4)$$

where $\vec{V}$ is the plume velocity vector, ν is the kinematic viscosity, K is the thermal conductivity, and U_x, U_z are the horizontal and vertical velocity components of the plume.

Equations (A1) to (A4) are all of the divergence form

$$\nabla \cdot (\vec{V} P) = G + \nabla^2 H , \qquad (A5)$$

where P denotes some plume variable. This equation can be integrated over a control volume bounded by the plume edge, A_3, and the two normal sections, A_1 and A_2, which are a distance ds apart; see Fig. 3.2 of the main text. If the stresses and heat conduction are ignored along the plume axis (i.e., in the s direction) and it is recognized that these same quantities disappear at the plume edge (i.e., at A_3), the integral of Eq. (A5) over the control volume can be written as:

$$\frac{d}{ds} \int_A U_s P \, dA = - \int_C U_n P \, dC + \int_A G \, dA , \qquad (A6)$$

where the divergence theorem has been used to integrate the left hand side of Eq. (A5); see Sec. 6.13 of Hildebrand (1976).

In Eq. (A6), the area (A) integrals are over the plume cross section, and the C integral is around the perimeter of A. Velocities U_s and U_n are in the s direction and normal to the plume edge, respectively.

The variables P_a in the atmosphere are assumed to vary only with z so that we have the identity

$$\nabla \cdot (\vec{V}P_a) = U_z \frac{dP_a}{dz} . \tag{A7}$$

By subtracting the integral of Eq. (A7) over the plume control volume from Eq. (A6), we have

$$\frac{d}{ds}\int_A U_s(P-P_a)dA = -\int_C U_n(P-P_a)dC + \int_A GdA - \int_A U_z \frac{dPa}{dz}dA . \tag{A8}$$

If it is assumed that all plume variables are equal to their local ambient value at the plume boundary, then the C integral in Eq. (A8) disappears.

Applying Eq. (A6) to Eq. (A1) and Eq. (A8) to Eqs. (A2) to (A4), we have

$$\frac{d}{ds}\int_A U_s dA = -\int U_n dC \tag{A9}$$

$$\frac{d}{ds}\int_A U_s(U_x - U_{xa})dA = -\frac{dU_a}{dz}\int U_z dA \tag{A10}$$

$$\frac{d}{ds}\int_A U_s U_z dA = \int_A g\left(\frac{\rho_a - \rho}{\rho}\right)dA \tag{A11}$$

$$\frac{d}{ds}\int_A U_s(T - T_a)dA = -\left(\frac{dT_a}{dz} + \frac{g}{c_p}\right)\int_A U_z dA . \tag{A12}$$

In the above, it has been assumed that the ambient wind is horizontal and that the vertical gradients in ambient properties (dU_a/dz, dT_a/dz) do not vary significantly across the plume cross section; therefore, they can be taken outside the area integrals [Eqs. (A10) and (A12)].

The following average plume properties can now be defined with the effective plume radius r undefined for the moment:

$$w = \frac{1}{\pi r^2 U_{sc}} \int_A U_s U_z dA \qquad (A13)$$

$$\Delta u = \frac{1}{\pi r^2 U_{sc}} \int_A U_s (U_x - U_{xa}) dA \qquad (A14)$$

$$\Delta T = \frac{1}{\pi r^2 U_{sc}} \int_A U_s (T - T_a) dA \; , \qquad (A15)$$

where U_{sc} is the velocity in the s direction along the plume centerline.

Although r should perhaps be defined in terms of the region where the velocity U_s differs from its free stream counterpart, $U_a \cos \phi$, we instead use Eq. (A9) and the entrainment assumption. Equation (A9) states that the rate of change of material in the plume equals the entrainment E which is specified in terms of r and the velocity differences between the plume and the ambient wind. Thus, we can write

$$\frac{d}{ds}(r^2 U_{sc}) = E, \qquad (A16)$$

which serves as the definition of r.

Equations (3.2), (3.3), and (3.5) of the main text can now be obtained by substituting Eqs. (A13) to (A15) into Eqs. (A10) to (A12).

Appendix B
Trajectories of Bent-Over Plumes: Analytical Results

B1. Neutral Atmosphere

The assumption of a bent-over plume means that the plume centerline is inclined at a small angle ϕ with respect to the horizontal and that the following approximations can be made: $w \ll U_a$, $U_{sc} \sim U_a$, and $ds \sim dx$. In a neutral atmosphere ($N^2 = 0$), the solution to the plume equations is simplified because the buoyancy flux F is conserved as shown by Eq. (3.7) of the main text; therefore, $F = F_b$, the source buoyancy flux. In the following derivation, we include the "added" mass, which is appropriate for bent-over plumes as discussed in Sec. 3.2.1; thus, the equation for conservation of vertical momentum is

$$(1 + k_v)\frac{d}{ds}(U_{sc}r^2w) = gr^2\frac{\Delta\rho}{\rho} . \tag{B1}$$

If we multiply Eq. (B1) by U_{sc}, we have

$$(1 + k_v)U_{sc}\frac{d}{ds}(U_{sc}r^2w) = U_{sc}r^2g\frac{\Delta\rho}{\rho} = F_b , \tag{B2}$$

which is readily integrated to yield

$$U_{sc}r^2w = \frac{F_b}{1 + k_v}\int \frac{ds}{U_{sc}} + \frac{F_m}{1 + k_v} . \tag{B3}$$

Using the bent-over approximations and assuming no wind shear, we can approximate the integral $\int ds/U_{sc}$ as x/U_a, and Eq. (B3) as

$$r^2w = \frac{F_b}{(1 + k_v)}\frac{x}{U_a^2} + \frac{F_m}{(1 + k_v)U_a} . \tag{B4}$$

With the same assumptions, the horizontal momentum equation [Eq. (3.2)] can be integrated to yield

$$U_{sc}r^2\Delta u = w_or_o^2\Delta u_o \tag{B5}$$

or

$$\frac{\Delta u}{U_a} \simeq -\frac{w_o}{U_{sc}}\frac{r_o^2}{r^2} , \tag{B6}$$

since $\Delta u_o = -U_a$ and $U_{sc} \sim U_a$. This result shows that $\Delta u / U_a$ is indeed small once the plume has grown to several times the initial radius r_o, consistent with our assumption that $U_{sc} \sim U_a$ and $w \ll U_a$, U_{sc}.

The mass conservation equation can be written as

$$\frac{d}{dx}(U_a r^2) = 2r\beta'w \tag{B7}$$

or

$$\frac{dr}{dx} = \beta' \frac{w}{U_a} , \tag{B8}$$

where we have again used the bent-over approximations and replaced β by β' in Eq. (3.4); recall that β' is the entrainment parameter relating to the actual plume radius and mass. The final equation needed to solve for the plume trajectory is the kinematic relationship:

$$\frac{dz'}{dx} = \frac{w}{U_a} . \tag{B9}$$

Combining Eqs. (B8) and (B9), we have

$$\frac{dr}{dz'} = \beta' \quad \text{or} \quad r = r_o + \beta' z' . \tag{B10}$$

The plume trajectory is now found for a "point source" where it is assumed that $r_o \ll \beta' z'$ or

$$r \simeq \beta' z' . \tag{B11}$$

Equations (B9) and (B11) can be substituted into Eq. (B4) to give

$$(\beta' z')^2 U_a \frac{dz'}{dx} = \frac{F_b}{(1 + k_v)} \frac{x}{U_a^2} + \frac{F_m}{(1 + k_v) U_a} , \tag{B12}$$

which can be integrated to yield

$$z' = \left[\frac{3 F_m x}{\beta'^2 (1 + k_v) U_a^2} + \frac{3 F_b x^2}{2 \beta'^2 (1 + k_v) U_a^3} \right]^{1/3} . \tag{B13}$$

Introducing the buoyancy (ℓ_b) and momentum (ℓ_m) length scales from Eqs. (3.14) and (3.18), respectively, and noting that $\beta^2 = \beta'^2(1 + k_v)$, we can cast Eq. (B13) into the dimensionless form of Eq. (3.19):

$$\frac{z'}{\ell_b} = \left[\frac{3}{\beta^2} \left(\frac{\ell_m}{\ell_b} \right)^2 \frac{x}{\ell_b} + \frac{3}{2\beta^2} \left(\frac{x}{\ell_b} \right)^2 \right]^{1/3} . \tag{B14}$$

Equations (3.16) and (3.17) of the main text result from assuming that $F_b = 0$ and $F_m = 0$, respectively.

B2. Stable Atmosphere with Constant N^2

The solution for the plume trajectory in a stable atmosphere is greatly simplified when N^2 is constant, and that solution is derived here using the "bent-over" approximations given in the previous section; again, we include the "added" mass in the vertical momentum equation.

If we multiply Eqs. (3.3) and (3.7) of the main text by $(1 + k_v)U_{sc}$ and U_{sc}, respectively, we have the following equations for the conservation of vertical momentum and buoyancy:

$$(1 + k_v)U_{sc}\frac{d}{ds}(U_{sc}wr^2) = U_{sc}r^2g\frac{\Delta\rho}{\rho} = F \tag{B15}$$

$$U_{sc}\frac{dF}{ds} = N^2U_{sc}wr^2. \tag{B16}$$

We now differentiate Eq. (B15) with respect to s and multiply the result by U_{sc} to get

$$(1 + k_v)U_{sc}\frac{d}{ds}\left[U_{sc}\frac{d}{ds}(U_{sc}wr^2) \right] = U_{sc}\frac{dF}{ds}. \tag{B17}$$

Equation (B16) can be substituted into Eq. (B17) to get the following second-order differential equation for the vertical momentum flux:

$$(1 + k_v)U_{sc}\frac{d}{ds}\left[U_{sc}\frac{d}{ds}(U_{sc}wr^2) \right] - N^2U_{sc}wr^2 = 0 . \tag{B18}$$

By defining the variable ζ as

$$\zeta = \int_0^s \frac{N'd\eta}{U_{sc}} , \tag{B19}$$

where $N' = N/(1 + k_v)^{1/2}$, we can rewrite Eq. (B18) as

$$\frac{d^2}{d\zeta^2}(U_{sc}wr^2) - U_{sc}wr^2 = 0 . \tag{B20}$$

This equation has the general solution

$$U_{sc}wr^2 = A \cos \zeta + B \sin \zeta , \tag{B21}$$

where A and B are constants to be determined from the initial conditions, i.e., at $\zeta = 0$. The A is found from the initial momentum flux F_m, but in the bent-over region, F_m acts on the apparent mass—actual plus added; therefore, we have $A = F_m/(1 + k_v)$. We evaluate B from the derivative of the vertical momentum flux at the source,

$$\frac{dM}{ds} = \frac{F_b}{(1 + k_v)U_{sc}} \tag{B22}$$

or

$$\frac{dM}{d\zeta} = \frac{dM}{ds} \cdot \frac{ds}{d\zeta} = \frac{F}{(1 + k_v)N'} , \tag{B23}$$

having obtained $ds/d\zeta$ from Eq. (B19). With the above results for A and B, the momentum flux is given by

$$U_{sc}wr^2 = \frac{F_m}{1 + k_v} \cos \zeta + \frac{F_b}{(1 + k_v)N'} \sin \zeta . \tag{B24}$$

Using the bent-over approximations, we can rewrite the variable ζ as

$$\zeta = \frac{N' x}{U_a} , \tag{B25}$$

and we obtain the following expression for wr^2:

$$wr^2 = \frac{F_m}{(1 + k_v)U_a} \cos \frac{N'x}{U_a} + \frac{F_b}{(1 + k_v)N'U_a} \sin \frac{N'x}{U_a} . \tag{B26}$$

In the above, the r can be replaced by $\beta'z'$ (Eq. B11) and w by $U_a dz'/dx$ (Eq. B9) to get the following differential equation for z':

$$(\beta'z')^2\frac{dz'}{dx} = \frac{F_m}{(1 + k_v)U_a} \cos \frac{N'x}{U_a} + \frac{F_b}{(1 + k_v)N'U_a} \sin \frac{N'x}{U_a} . \quad (B27)$$

With $z' = 0$ at $x = 0$, this equation can be integrated to yield

$$\frac{z'}{(F_b/N^2U_a)^{1/3}} = \left(\frac{3}{\beta'^2}\right)^{1/3} \left(N'\frac{F_m}{F_b} \sin \frac{N'x}{U_a} + 1 - \cos \frac{N'x}{U_a}\right)^{1/3},$$
$$(B28)$$

which is Eq. (3.24) of the main text and is valid for $N'x/U_a < \pi$.

For distances greater than $\pi U_a/N'$, z' oscillates about its final equilibrium position and for $n < N'x/\pi U_a < n + 1$, where n is odd, dz'/dx is negative. In their model, Fay et al. (1970) assumed that entrainment takes place whenever there is relative motion, positive or negative, and therefore, they took the absolute value of w in Eq. (3.4). This assumption leads to a final rise that is less than the maximum value found from Eq. (B28). When $F_m = 0$, the final rise is 84% of the maximum value z'_{max}, which is

$$z'_{max} = \left(\frac{6}{\beta'^2}\right)^{1/3} \left(\frac{F_b}{N^2U_a}\right)^{1/3} ; \quad (B29)$$

the z'_{max} occurs at $N'x/U_a = \pi$.

CHAPTER 4

Dispersion in the Convective Boundary Layer

Jeffrey C. Weil

4.1. Introduction

Dispersion in the planetary boundary layer (PBL) is controlled by the mean and turbulent structure of that layer. In the past decade, significant progress has occurred in our understanding of PBL structure and diffusion, especially for the daytime convective boundary layer (CBL). As discussed in Ch. 1, this has resulted from the collective efforts of numerical modeling, laboratory simulations, and field observations. In this chapter I discuss how this better understanding has improved the theoretical basis and performance of dispersion models for routine applications. I shall focus on point sources of nonbuoyant (i.e., passive) or buoyant material at arbitrary height in the CBL. The CBL is especially important in the case of tall stacks because they generally contribute their maximum ground-level concentrations (GLC) during convection (see Venkatram, 1980; Weil and Brower, 1984).

The high GLCs for an elevated release occur as a result of the vigorous mixing between the surface and the base of an elevated inversion capping the CBL. As we saw in Ch. 1, this mixing is driven principally by upward heat flux that arises from solar heating of the ground. The heating gives rise to large-scale convective motions in the form of updrafts and downdrafts that are responsible for the "looping" character of plumes in unstable conditions. These convective elements have dimensions that scale with h, the CBL depth, and velocities that scale with w_*, the convective velocity scale, which we introduced in Ch. 1. At midday over land, typical values of h and w_* are 1 to 2 km and 2 m s^{-1}, respectively.

The current regulatory practice in dispersion modeling adopts the Gaussian plume equation and assumes lateral (σ_y) and vertical (σ_z) dispersion parameters given by the Pasquill-Gifford (PG) curves (e.g., Gifford, 1961; Turner, 1971). The curves are given for six "classes" of atmospheric stability ranging from extremely unstable (A) to very stable (F). The stability is estimated from Turner's (1964) criteria, which are based on near-surface winds, cloud cover, ceiling height, etc. The limitations of this combined approach—the Pasquill-Gifford-Turner (PGT) technique—are well-known. First, the PG curves are based on passive tracer releases from a ground-level source and on surface concentrations measured out to only ~ 800 m from the source. They do not apply to an elevated plume, which has dispersion characteristics quite different from those for a surface release (Lamb, 1982; Hunt, 1982). Second, the Turner method is an empirical approach that does not account for the PBL's vertical structure, mean or turbulent. In addition, it predicts that neutral conditions occur most frequently during daytime when in fact convective conditions usually exist (see Weil and Brower, 1984).

In applications, the PGT method is used not only for near-surface sources but also for tall stacks, including those with large heat releases and plume rise (e.g., the CRSTER model; EPA, 1977). This practice has been criticized by several workshops (Hanna et al., 1977; Weil, 1985) and in peer reviews of models conducted by the American Meteorological Society for the U. S. Environmental Protection Agency (EPA) (Smith, 1984). Undoubtedly, the use of the PGT method for tall stacks is one of the prime reasons for the poor model performance summarized in Smith's report.

The improved knowledge of PBL structure has motivated the development of better dispersion models. This has been guided by two constraints: 1) the models should not be too complex numerically because they must be used repeatedly, and with fast turnaround, in applications; and 2) they must capture the essential physics of the PBL and dispersion. These competing constraints require that both the developers and users of models have a good appreciation of the important features of the CBL and of dispersing plumes.

At the risk of being redundant of other chapters, we first discuss the key aspects of CBL structure and methods for estimating it (Sec. 4.2). Next we review some important theoretical concepts for dispersion and characteristics of plumes in the CBL (Sec. 4.3). With this background, we then discuss applied dispersion models: first for passive releases (Sec. 4.4) and then for buoyant plumes (Sec. 4.5).

4.2. CBL Structure and Controlling Variables

4.2.1. Structure

We assume that the CBL is horizontally homogeneous; i.e., its statistical properties do not vary in the horizontal (x, y) plane but only with height z. We also assume that the CBL is quasi-steady, or that its structure is unaffected by its changing boundary conditions (e.g., its depth, surface heat flux) during the day. As we saw in Ch. 1, these assumptions are rarely strictly true but can be a good approximation in practice. Moreover, they lead to a very simple description of the turbulence structure—turbulence velocities when nondimensionalized by w_* can be represented as a function of z/h over most of the CBL. This idealization is frequently a good representation in the daytime over land and therefore serves as a useful description for many dispersion applications.

As discussed in Ch. 1, turbulence in the CBL is generated mostly by heat flux and to a lesser extent by surface friction or shear stresses, which are characterized by the friction velocity u_*. Heat flux effects extend to h whereas those of shear are confined mostly to $-z < L$, where L is the Monin-Obukhov length (Ch. 1). This prompted Deardorff (1972) to suggest $-h/L$ as a stability parameter. Since $-L$ is typically 1 to 100 m, $-h/L$ is usually large (> 10) indicating that the bulk of the CBL is dominated by convection. The stability can also be characterized by w_*/u_*, which is related to $-h/L$ by $w_*/u_* = (-h/kL)^{1/3}$ where k is the von Kármán constant (~ 0.4; see Ch. 1 for the definitions of w_* and L). In some diffusion models, u_* is replaced by U, the mean wind speed, by assuming a typical "resistance" coefficient, $u_*/U(\sim 0.05)$; the stability parameter is then w_*/U. However, it is more correct to use u_* as a measure of friction effects because it accounts for the local roughness.

In strong convection ($-h/L > 10$), the CBL has an approximate triple-layer structure consisting of a surface layer, a mixed layer, and an interfacial layer (Ch. 1). In the surface layer ($z < 0.1h$), strong gradients appear in both the mean wind and potential temperature, and they are strongest for $-z < L$. The lateral (σ_v) and longitudinal (σ_u) turbulence components can be parameterized by either of the following expressions (Panofsky et al., 1977)

$$\frac{\sigma_u}{u_*} = \frac{\sigma_v}{u_*} = \left[4 + 0.6(-h/L)^{2/3}\right]^{1/2} \tag{4.1}$$

$$\frac{\sigma_u}{u_*} = \frac{\sigma_v}{u_*} = \left[12 + 0.5(-h/L)\right]^{1/3} . \tag{4.2}$$

Both appear to be acceptable fits to observations. The expressions effectively interpolate between the neutral stability limit ($\sigma_v \propto u_*$) and the strongly convective limit ($\sigma_v \propto w_*$).

For the vertical turbulence component (σ_w), Panofsky et al. (1977) offered the following alternative expressions for the surface layer:

$$\frac{\sigma_w}{u_*} = \left[1.6 + 2.9(-z/L)^{2/3}\right]^{1/2} \tag{4.3}$$

$$\frac{\sigma_w}{u_*} = 1.3[1 + 3(-z/L)]^{1/3} , \tag{4.4}$$

giving some preference to Eq. (4.4). These expressions are based on Monin-Obukhov similarity; they interpolate between the neutral limit, $\sigma_w \propto u_*$, and the free-convection limit, $\sigma_w = 1.34(g\overline{w\theta_0}z/T_0)^{1/3}$ (Wyngaard et al., 1971), which is valid for $-L < z < 0.1h$. Here, g is the gravitational acceleration, $\overline{w\theta_0}$ is the surface virtual temperature flux, and T_0 is the temperature. Figure 1.13 (Ch. 1) shows the observed variation of σ_w/u_* with z/L and its approach to the neutral and free-convection limits.

In the mixed layer ($0.1h < z < 0.8h$), the mean wind, potential temperature, and turbulence components exhibit little height variation (Ch. 1; see also Kaimal et al., 1976; Caughey, 1982). Thus, in diffusion modeling they are often assumed to be uniform with height with

$$\sigma_u, \sigma_v, \sigma_w \sim 0.6 w_* . \tag{4.5}$$

The σ_u, σ_v results are consistent with the strong convection limit of Eqs. (4.1) and (4.2), and the σ_w value agrees with the free-convection result at $z \sim 0.1h$. The laboratory simulations of Deardorff and Willis (1985) are in approximate agreement with the field data summarized by Caughey (1982); this lends credence to the use of a convection tank for turbulence and diffusion investigations.

In the interfacial layer ($0.8h < z < 1.2h$, roughly) the mean wind and potential temperature approach their "free atmosphere" values (above $1.2h$) and often exhibit large increases or jumps across the layer (Kaimal et al., 1976; Caughey, 1982). The turbulence components decrease gradually to their smaller values above h.

For $-h/L > 10$, the convection is well developed, with the convective elements consisting of randomly distributed thermal plumes surrounded by large downdraft regions. At $-h/L > 50$, the convection patterns reveal organized structures resembling hexagonal convection cells (Deardorff, 1982). The convective elements extend from the surface to the entrainment zone and are long-lived, having time scales (h/w_*) of about 10 min.

One of the quantities of much recent interest in diffusion modeling is the probability density function (p.d.f.) of the vertical velocity $p_w(w)$. This interest stems primarily from the non-Gaussian vertical dispersion patterns of passive plumes, first discovered in the laboratory (Willis and Deardorff, 1976a, 1978) and in numerical simulations (Lamb, 1978, 1979).

Lamb (1982) calculated p_w from Deardorff's (1974) velocity field, which was computed through large-eddy simulation (i.e., three-dimensional, time-dependent, fine-mesh numerical solution of the equations of motion) for a domain of 5 km × 5 km in the horizontal plane and 2 km high. Figure 4.1 shows these p.d.f.'s which are nondimensionalized using w_*. The key point illustrated is that the p.d.f.'s

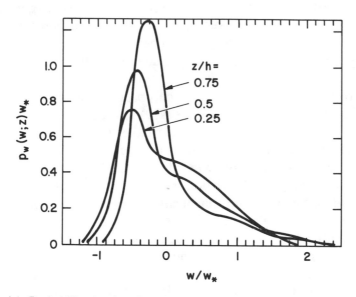

Figure 4.1. Probability density of vertical velocity at three heights in the convective boundary layer. From Lamb (1982).

are not symmetrical about $w = 0$, and, hence, are not Gaussian, but are positively skewed ($\overline{w^3} > 0$). The mode or most probable velocity is negative and approximately equal to $\overline{w}_d(z)$, the mean downdraft velocity at height z. Most (60%) of the area under the p_w curve is on the negative side of the w axis, indicating the higher probability of occurrence of downdrafts.

The field observations of Caughey et al. (1983) are in good agreement with Lamb's data at $z/h \cong 0.4$, but they tend to be nearly Gaussian at $z/h = 0.75$ and above, whereas Lamb's results still exhibit positive skewness. The field observations are supported by the laboratory data of Deardorff and Willis (1985) for $z/h > 0.5$, but the latter show somewhat smaller modes than either the field or Lamb data at lower heights. In contrast, observations from the Minnesota experiment shown in Fig. 1.13 indicate that there is positive skewness throughout the CBL, in agreement with Lamb's results. Some of the above differences may be explained by sampling errors (e.g., see Deardorff and Willis, 1985). However, overall it is clear that the features of these various p.d.f.'s are in relatively good agreement.

The preceding discussion applies for $-h/L > 10$ and perhaps best to very unstable conditions ($-h/L > 50$) in which the CBL structure is insensitive to $-h/L$ except very near the surface (Caughey, 1982). For $-h/L < 10$, the structure is less well documented, but it appears to depend systematically on $-h/L$, with effects of shear-generated turbulence extending to greater depths as $-h/L$ decreases. For example, the σ_w/u_* profile calculated numerically by Deardorff (1972) shows a progressive decrease in magnitude at middle-level regions of the CBL as $-h/L$ decreases from 4.5 to 0 (neutral stability). The results are broadly consistent with the observations over the ocean reported by Nicholls and Readings (1979) and Pennell and LeMone (1974). Another characteristic of this weaker stability regime is the prevalence of horizontal roll vortices (rather than thermals), which have axes aligned close to the wind direction (LeMone, 1973).

The turbulence characteristics, including p_w, of this weaker convective regime need greater clarification, not only for PBL understanding *per se* but also for dispersion modeling. In particular, medium to tall stacks can produce high GLCs under near-neutral stability with strong and persistent winds (Venkatram and Paine, 1985).

In summary, the horizontally homogeneous, quasi-steady model provides a useful representation of CBL turbulence structure for dispersion applications. Clearly, the morning and evening transitional periods present difficulty, as do the effects of spatial inhomogeneities. These problems require further attention.

4.2.2. Estimation of Key CBL Variables

As we have discussed, the key variables controlling the CBL turbulence structure are $\overline{w\theta_0}$, h, and u_* from which other important parameters, w_* and L, can be determined. Additionally, the mean wind speed $U(z)$ is needed for dispersion applications. For such applications, these variables are usually not measured directly and must be calculated by indirect methods. In the following, we discuss methods applicable when at least the following data are available: wind and temperature measurements from a 10-m tower, cloud-cover observations, and early-morning temperature profiles over the lowest ~ 3 km of the PBL.

Surface heat flux and friction velocity

The sensible heat flux Q_0 can be estimated from a simple energy balance at the surface, an approach often called the "energy budget" method. In steady state, this balance is expressed by

$$R_n = Q_0 + LE + G_s \, , \tag{4.6}$$

where R_n is the net radiation to the surface, LE is the surface-to-air latent heat flux, and G_s is the net heat flux to the soil (Sellers, 1965).

Holtslag and van Ulden (1983) have developed detailed equations for the terms in the above expression and have successfully tested them against field observations. For Q_0 they give

$$Q_0 = \frac{(1 - \beta_1) + \gamma/s}{1 + \gamma/s}(1 - c_G)R_n - \beta_2 \, , \tag{4.7}$$

where

β_1 and β_2 are dimensionless and dimensional variables, respectively, which characterize the surface moisture conditions; $\beta_2 = 20\beta_1$ W m^{-2} and β_1 ranges from 0 to 1,

$\gamma = c_p/\lambda$, where λ is the latent heat of water vaporization,

$s = \partial q_s/\partial T$, where q_s is the saturation specific humidity,

c_G is a dimensionless soil heat flux coefficient ($= 0.1$) relating G_s to R_n($G_s = c_G R_n$).

Equation (4.7) shows that Q_0 is strongly correlated with R_n as has been found by other authors (e.g., Yap and Oke, 1974).

The terms β_1, β_2, γ, and s enter Eq. (4.7) through the latent heat flux expression (see Holtslag and van Ulden, 1983), which when substituted into Eq. (4.6) along with the G_s expression yields Q_0 as a function of R_n. The β_1 accounts for that portion of LE that is correlated with R_n, and β_2 for the uncorrelated portion. Holtslag and van Ulden found that $\beta_1 = 1$ for a moist, grass-covered surface in The Netherlands and 0.45 for a much drier vegetated surface (Prairie Grass, USA); for bare, dry soil, $\beta_1 = 0$. The major uncertainty in applying this technique lies in characterizing the surface moisture condition. It would be beneficial to calibrate β_1 with heat flux measurements for a variety of surfaces and relate it to other more easily measured variables.

In Eq. (4.7), the R_n can be provided either by measurements or a simple model. Holtslag and van Ulden (1983) gave an expression for R_n in terms of the insolation, surface air temperature, cloud cover, and local albedo; they found good correlation between it and R_n measurements. During the day, the major component of R_n is the insolation. If the latter is not measured, it can be estimated from the solar constant, elevation angle (i.e., latitude, time of day, and month), and the cloud cover (see Holtslag and van Ulden, 1983).

Weil and Brower (1983) proposed simpler Q_0 and R_n expressions in which

$$Q_0 = \mu R_n \tag{4.8}$$

$$R_n = (1 - \alpha)Q_r - I_n , \tag{4.9}$$

where μ is a dimensionless coefficient assumed to be constant for a particular site and meteorological conditions,

α is the albedo,

Q_r is the insolation,

I_n is an assumed constant net long-wave radiation ($= 49$ W m^{-2}).

Weil and Brower chose $\mu = 0.47$ based on an annual average surface energy budget (Sellers, 1965), for the 40°–50°N latitude zone. With $\alpha = 0.2$, their expression resulted in $Q_0 = 0.4Q_r$, which was in good agreement with measurements from the Minnesota experiment

(Izumi and Caughey, 1976) and from a power plant site. The result, $Q_0 = 0.4Q_r$, has been used to estimate heat flux in diffusion models for vegetated, middle-latitude sites. For more general sites encompassing a broad range of surface cover, moisture, and temperature, the more complete expression of Holtslag and van Ulden (1983) is recommended.

In addition to the energy budget method, the heat flux can be estimated from the "profile" method, in which one uses the Monin-Obukhov (M-O) similarity profiles for $U(z)$ and mean potential temperature $\Theta(z)$ in the surface layer. The approach requires wind speed and temperature measurements from at least one and two heights, respectively.

The similarity profile for mean wind is (Lumley and Panofsky, 1964)

$$U(z) = \frac{u_*}{k} \left[\ln \frac{z}{z_0} - \psi_m(z/L) \right] , \qquad (4.10)$$

where z_0 is the roughness height and ψ_m is a stability function. For unstable conditions ($L < 0$), ψ_m is given by (Paulson, 1970)

$$\psi_m = 2\ln\left[(1 + \phi_m)/2\right] + \ln\left[(1 + \phi_m^2)/2\right] - 2\tan^{-1}\phi_m + \pi/2, \quad (4.11)$$

where ϕ_m is the M-O function for mean wind shear (Ch. 1). Paulson used

$$\phi_m = (1 - 15z/L)^{1/4} . \qquad (4.12)$$

With a wind speed measurement from one height, Q_0 provided by the budget method, and a z_0 estimate, Eq. (4.10) can be solved iteratively for u_* (Holtslag and van Ulden, 1983; Weil and Brower, 1983). The z_0 can be determined from visual inspection of the site and tabulated z_0 values for various terrain types (Wieringa, 1980) or it can be deduced from the wind profile if $U(z)$ measurements are available at multiple heights (Nieuwstadt, 1978).

To determine Q_0 by the profile method, one uses the similarity profile for the mean potential temperature difference $\Theta(z) - \Theta(z_0)$. The profile is similar to that for $U(z)$ except that u_* and ψ_m are replaced by a temperature scale θ_* and another similarity function, ψ_H (Holtslag and van Ulden, 1983). The M-O temperature scale θ_* is defined by

$$\theta_* = -\frac{Q_0}{\rho c_p u_*} \qquad (4.13)$$

and is the variable to be extracted from the temperature measurements; knowledge of θ_* and u_* determines Q_0. The simplest application of this method requires wind speed and temperature measurements at one and two heights, respectively. If measurements are available at more elevations, u_* and θ_* can be obtained by a least-squares fit between the observed and predicted profiles (Nieuwstadt, 1978; Berkowicz and Prahm, 1982). The method works best if applied within the friction layer, $-z < L$, where the wind and temperature gradients are largest. One advantage of the profile method is that it does not require knowledge of the surface moisture conditions.

CBL height

The CBL height h evolves with time in response to the continuous supply of heat at the surface. It can be predicted from one-dimensional (in time) "slab" models in which key assumptions are horizontal homogeneity and the neglect of radiation and latent heat effects (see Carson, 1973; Tennekes, 1973; Driedonks, 1982). For illustration, we consider the model of Carson who idealized the potential temperature distribution as uniform for $z < h$, to have a change step $\Delta\Theta_i$ at $z = h$, and to vary linearly with z for $z > h$ (see Fig. 4.2). The overshoot $\Delta\Theta_i$ is a measure of the degree of entrainment (at $z = h$) and is a function of time as are h, $\Delta\Theta_i$, and the mixed-layer temperature Θ_c. When there is no overshoot, $\Delta\Theta_i = 0$, and the CBL only "encroaches" on the elevated stable layer.

Under the model assumptions, the energy equation for the CBL reduces to

$$\frac{d\Theta_c}{dt} = -\frac{\partial \overline{w\theta}}{\partial z}, \qquad (4.14)$$

where $\overline{w\theta}$ is the vertical temperature flux. Since Θ_c is independent of z, $\partial\overline{w\theta}/\partial z$ is also, and

$$\frac{\partial \overline{w\theta}}{\partial z} = -\frac{\overline{w\theta}_0 - \overline{w\theta}_i}{h}, \qquad (4.15)$$

where $\overline{w\theta}_i$ is the temperature flux at the top of the CBL, i.e., that generated by entrainment of the stable air aloft. The linear variation of $\overline{w\theta}$ with z implied by Eq. (4.15) is supported by observation (Caughey, 1982). $\overline{w\theta}_i$ is found by integrating the energy equation across the interfacial layer (Lilly, 1968) with the approximate result

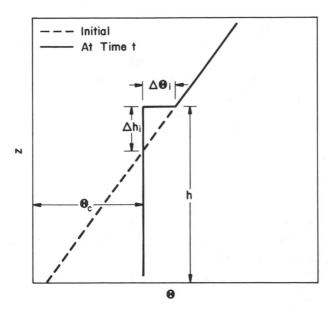

Figure 4.2. Schematic of potential temperature distribution in the Carson (1973) model.

$$\overline{w\theta}_i(t) = -\left[\frac{dh}{dt} - W(h)\right]\Delta\Theta_i , \qquad (4.16)$$

where $W(h)$ is the subsidence velocity of the synoptic flow. Using the temperature profile geometry (Fig. 4.2), one finds

$$\Delta\Theta_c = \gamma_i h - \Delta\Theta_i \quad \text{and} \quad \Delta\Theta_i = \gamma_i\Delta h_i , \qquad (4.17)$$

where γ_i is the potential temperature gradient above h.

From the above equations, Carson derived rate equations for h and Δh_i and solved them subject to the condition $\overline{w\theta}_i = -A_0\overline{w\theta}_0$, where A_0 is a constant ranging from 0 to 1. When $W(h) = 0$, the equation for h is

$$h = \left[\frac{2(1 + 2A_0)}{\gamma_i}\int_0^t \overline{w\theta}_0(\eta)d\eta\right]^{1/2} , \qquad (4.18)$$

which shows the importance of the integrated surface temperature flux and the stratification aloft (γ_i).

Weil and Brower (1983) extended Carson's model to an arbitrary initial temperature distribution with z and allowed for stress-induced mixing at h. The latter can be important when $\overline{w\theta}_0$ is small, e.g., in the early morning or on overcast days. An operational advantage of the arbitrary temperature distribution is the ease of adapting it to initial profiles that are very irregular, as sometimes found in early morning radiosondes. Weil and Brower found good agreement between predictions and observations of h and showed that use of the arbitrary distribution was superior to an assumed linear distribution (i.e., constant γ_i).

Tennekes' (1973) model is quite similar to Carson's except that h and $\Delta\Theta_i$ have nonzero initial values that must be determined from the initial temperature profile; the model requires a numerical solution to produce $h(t)$. (In Carson's model, h and $\Delta\Theta_i$ are zero initially.) Tennekes and van Ulden (1974) and Driedonks (1982) showed that the modeled h values agreed well with observations.

A simpler, often-used method for estimating h is to extrapolate an adiabat from the surface temperature, after the onset of surface heating, to the early morning temperature profile (Holzworth, 1972). The h is then taken as the height of intersection of the extrapolated and initial profiles. As pointed out in Panofsky and Dutton (1984), this approach is not correct after the time of maximum surface temperature (in the afternoon) because h remains constant or increases slightly even though the surface temperature decreases. This deficiency can be corrected by assuming h to remain constant after the maximum surface temperature, i.e., until evening transition.

Weil and Brower (1983) evaluated the modified Carson model, the Holzworth (1972) technique, and an adaptation of the latter used in EPA dispersion models. The observed h was determined from measured temperature profiles near the Keystone (Pennsylvania) power plant. They found that the modified Carson model was apparently slightly better than the Holzworth technique and that both of these approaches were superior to the EPA approach. Thus, while the Carson model or another slab approach is preferred theoretically, its operational superiority over the simpler Holzworth technique remains to be demonstrated. Figure 4.3 compares the modified Carson and EPA models with observations.

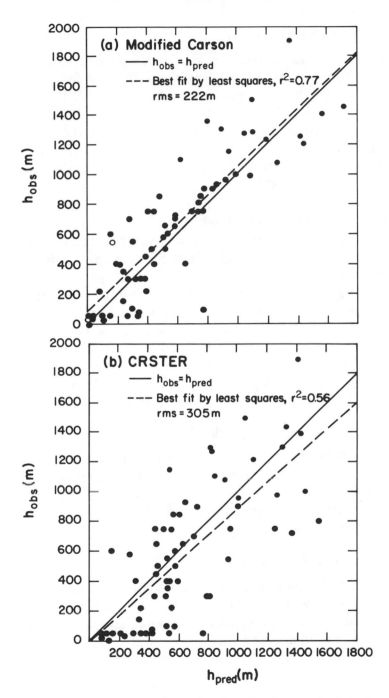

Figure 4.3. Comparison between observed and predicted convective boundary layer height for the (a) modified Carson (1973) model and (b) EPA approach as used in the CRSTER model (EPA, 1977). From Weil and Brower (1983).

Mean wind speed

As we have discussed earlier, the mean wind $U(z)$ exhibits strong gradients for $z < 0.1h$ but is approximately uniform for $z > h$. Thus, as explained in Ch. 1, a simple model of the entire CBL wind profile is Eq. (4.10) for $z < 0.1h$ and $U = $ constant in the mixed layer (see also Garratt et al., 1982; Weil and Brower, 1983); the constant is $U(0.1h)$ evaluated from Eq. (4.10). Figure 4.4 shows that this model agrees well with observed wind profiles from the Minnesota experiment (Izumi and Caughey, 1976).

A model for $U(z)$ very similar to the above was discussed by van Ulden and Holtslag (1985) who found that it also gave quite good performance. These authors gave an excellent review of methods for estimating PBL variables for both the convective and the stable boundary layers.

Under strongly baroclinic conditions substantial wind shear can exist throughout the CBL, as discussed in Ch. 1 (e.g., see Fig. 1.10). Our model may deviate considerably from observations under these conditions. Wyngaard (1985) discussed the baroclinic problem and offered a model for the wind variation across the CBL, but this requires knowledge of the geostrophic wind shear, which is not so readily available. Clearly, the problem merits further attention.

4.3. Characteristics of Dispersing Plumes

In the following, we first review important theoretical concepts useful in interpreting observations and in predicting diffusion characteristics. Then we briefly summarize key results from numerical modeling, laboratory experiments, and field observations of passive plume dispersion in the CBL.

4.3.1. Theoretical Background

We consider two idealized theories for the dispersion of passive plumes—Taylor's (1921) statistical theory and Yaglom's (1972) similarity model. Each is applicable under certain limiting but different conditions. In addition, we briefly discuss Lagrangian numerical modeling since it forms the basis of some predictions described in Section 4.3.2.

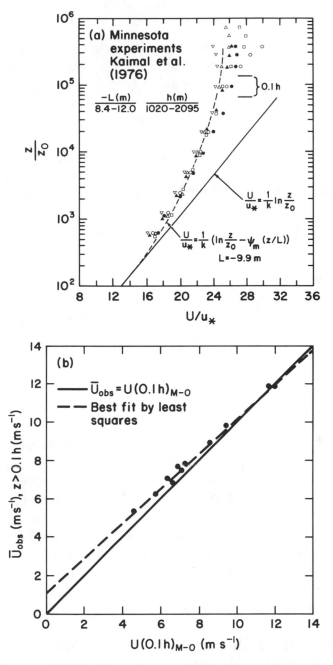

Figure 4.4. (a) Observed wind profiles approximating hourly averages compared with Monin-Obukhov (M-O) similarity theory (dashed curve) for heights $< 0.1h$; the logarithmic wind profile is shown for reference. (b) Observed vertically averaged wind over $0.1h < z < h$ compared with the M-O wind prediction at $z = 0.1h$. From Weil and Brower (1983).

Statistical theory

Taylor's (1921) theory applies to dispersion in a field of homogeneous and stationary turbulence, i.e., turbulence whose statistical properties are uniform in space and steady in time. An important variable characterizing the turbulent velocity of a fluid "particle" is its autocorrelation function. This defines the correlation between the particle velocity at one time $V'_i(t)$ and at some later time $V'_i(t + \tau)$ where τ is a time separation, and i indicates the velocity component. In general this function is different for each component. For the y direction, the dimensionless form of the function $\rho^y_L(\tau)$ is given by

$$\rho^y_L(\tau) = \frac{\overline{V'_y(t)V'_y(t + \tau)}}{\overline{V'^2_y}} ,$$ (4.19)

where the overbar denotes a time average.

As $\tau \to 0$, $\rho^y_L \to 1$ since the velocity is perfectly correlated with itself at zero lag, and as τ gets large, ρ^y_L approaches zero because the velocity becomes independent of its earlier value at time t. A measure of the time over which V'_y becomes independent of its value at t is the Lagrangian integral time scale:

$$T^y_L = \int_0^\infty \rho^y_L(\tau)d\tau .$$ (4.20)

Taylor's theory applies to the displacements of passive particles serially emitted from a point source in a turbulent flow. This problem is often labeled "single-particle dispersion" because it applies only to displacements of particles and not to clouds of material (or particles); the latter is termed "relative dispersion" (e.g., see Csanady, 1973; Pasquill and Smith, 1983). Since the particles are passive, they do not affect the flow and thus move with the local fluid velocity. Hence, the displacement Y in one realization is given by

$$Y = \int_0^t V'_y(t')dt' .$$ (4.21)

Following the analysis in Taylor (1921), Csanady (1973), or Pasquill and Smith (1983), the time rate of change of the mean square particle displacement, $\overline{dY^2}/dt = d\sigma^2_y/dt$, is found to be

$$\frac{d\sigma^2_y}{dt} = 2\sigma^2_v \int_0^t \rho^y_L(\tau)d\tau ,$$ (4.22)

where $\sigma_v^2 = \overline{V_y'^2}$. This equation can be integrated to yield

$$\sigma_y^2 = 2\sigma_v^2 \int_0^t \left(\int_0^{t'} \rho_L^y(\tau)d\tau \right) dt' . \tag{4.23}$$

Two limiting forms of Eq. (4.23) arise as a result of the $\rho_L^y(\tau)$ behavior and provide useful results for plume spread regardless of the precise variation of ρ_L^y with τ. For short times ($t \ll T_L^y$), $\rho_L^y \sim 1$ and

$$\sigma_y^2 = \sigma_v^2 t^2 ; \tag{4.24}$$

the plume growth is linear with time. For long times ($t \gg T_L^y$), the ρ_L^y approaches zero, but its integral remains finite; thus, Eq. (4.23) reduces to

$$\sigma_y^2 = 2\sigma_v^2 T_L^y t , \tag{4.25}$$

where we have identified the quantity in brackets in Eq. (4.23) as T_L^y. In this time limit, σ_y grows parabolically with t, which is a diffusive type of behavior.

Equations (4.24) and (4.25) show that the main quantities needed to apply this theory are σ_v and T_L^y, which must be specified for the particular flow of interest.

Expressions similar to Eqs. (4.24) and (4.25) can be derived for the mean square displacement in the other directions. Also note that by replacing t by x/U, where x is the downwind distance and U is the mean wind speed, σ_y is predicted to grow cither lincarly or parabolically with distance.

Now consider the application of statistical theory to the dispersion of passive material from an elevated point source in the CBL and under conditions of strong convection ($-h/L > 10$). By elevated, we mean that the source height z_s exceeds $0.1h$. In this height regime, we can idealize the turbulence structure as vertically homogeneous with $\sigma_v, \sigma_w \simeq 0.6w_*$ following our discussion in Sec. 4.2.1. Also, recall the Lagrangian time scale is of the order of h/w_*. Then, for short times, $t < h/w_*$ where $t = x/U$, Eq. (4.24) predicts that

$$\frac{\sigma_z}{h} \simeq \frac{\sigma_y}{h} \simeq 0.6X , \tag{4.26}$$

where the nondimensional distance X is

$$X = \frac{w_* x}{Uh}. \tag{4.27}$$

The X can also be thought of as a nondimensional time since it is the ratio of travel time (x/U) to the convective time scale (h/w_*).

As will be discussed later, Eq. (4.26) is consistent with observations in the CBL. The result for σ_y should apply over most of the surface layer because of the near invariance of σ_v and U with height in that regime; recall that the variation of U with z is greatest for $-z < L$. For long times $(t > h/w_*)$, the prediction of Eq. (4.25) is of somewhat limited use in the CBL as will be discussed in Sec. 4.3.2.

Similarity theory

Yaglom's (1972) theory governs vertical dispersion from a ground release $(z_s = 0)$ in the surface layer and is applicable to very unstable conditions $(-h/L > 50)$ when an extensive free-convection regime $(-L < z < 0.1h)$ exists. Yaglom argued that in this regime, the buoyant production rate of turbulence, $g\overline{w\theta_0}/T_0$, is the most relevant variable determining the mean vertical velocity $\overline{w}(t)$ of a plume or cloud. That is, surface friction effects and u_* are unimportant. Using dimensional arguments, he proposed that $\overline{w}(t)$ should depend on $g\overline{w\theta_0}/T_0$ and time according to

$$\overline{w}(t) = c_0 \left(\frac{g\overline{w\theta_0}t}{T_0}\right)^{1/2}, \tag{4.28}$$

where the bracketed quantity has dimensions of velocity squared, and c_0 is a dimensionless coefficient to be determined experimentally. One can make a similar argument about $\overline{w}$ as a function of z following the considerations of free-convection scaling (see Wyngaard et al., 1971; Pasquill and Smith, 1983); i.e.,

$$\overline{w}(z) \propto \left(\frac{g\overline{w\theta_0}z}{T_0}\right)^{1/3}. \tag{4.29}$$

Using either expression, the mean cloud height $\overline{Z}$ is found by integrating $d\overline{Z}/dt = \overline{w}(t)$ or $\overline{w}(z)$ with the result

$$\overline{Z}(t) = \frac{2c_0}{3} \left(\frac{g\overline{w\theta_0}}{T_0}\right)^{1/2} t^{3/2}. \tag{4.30}$$

Upon introducing $t = x/U$ and the nondimensional distance X, this equation can be recast in the nondimensional form

$$\frac{\overline{Z}}{h} = \frac{2c_0}{3}X^{3/2} .$$ (4.31)

A similar line of reasoning can be applied to the vertical spread of the cloud, σ_z, with the result

$$\frac{\sigma_z}{h} = \frac{2c_1}{3}X^{3/2},$$ (4.32)

where c_1 must be determined experimentally. The accelerated growth—i.e., the exponent of $3/2$ on X versus 1 for statistical theory—is tied to the increase of σ_w with height in the free-convection layer. One should recognize that this growth law can represent only the short time or distance behavior because the plume eventually will extend into the mixed layer where σ_w varies less rapidly with z. That is, the effects of a finite mixing region and a limiting turbulence length scale ($\propto h$) will enter. Also, recall that this theory applies only to surface sources. Despite these restrictions, the theory provides a useful limiting result on the dispersion rate since statistical theory cannot legitimately be applied in the surface layer because of the variation of σ_w with z.

Lagrangian numerical modeling

In the Lagrangian approach to dispersion modeling, one tracks or follows particles through a turbulent fluid using either a numerically calculated Eulerian velocity field or a stochastic model for the Lagrangian velocities. Lamb (1978, 1979, 1982) followed the first approach, and it is primarily his analysis that is briefly covered here; an excellent review of his work along with results and comparisons to experimental data are given in Lamb (1982). In principle, the Lagrangian approach is not limited by the restrictions of the previously discussed theories although most modeling has been done for stationary (steady) turbulent flows. Furthermore, one predicts the entire concentration field, not simply its second moments (σ_y, σ_z).

The ensemble-averaged concentration $C(x_i, t)$ of passive material from a point source is determined from

$$C(x_i, t) = Q \int_0^t p(x_i, t; x_{is}, t') dt' , \qquad (4.33)$$

where Q is the source strength, x_i is the position vector (x_1, x_2, x_3), and $p(x_i, t; x_{is}, t')$ is the probability density that material released at x_{is}, the source position, at time t' will be found at x_i at time t. Lamb obtained p from an ensemble of computer-generated particle trajectories using Deardorff's (1974) numerically calculated velocity field. The latter was generated from a large-eddy simulation of the CBL in which the grid dimensions were 125 m on a side in the horizontal plane and 50 m high. At each grid point, Deardorff's model provided the resolvable velocity, i.e., that due to eddies with length scales larger than the grid dimension, and the subgrid-scale energy—the total kinetic energy of eddies with dimensions smaller than the grid.

Lamb calculated the position, $X_i^n(t)$, of the nth particle from

$$\frac{d}{dt} X_i^n = \hat{u}_i [X_i^n(t), t] + u_i' [X_i^n(t), t] , \qquad (4.34)$$

where $\hat{u}_i$ is the resolvable velocity from Deardorff's model, and u_i' is a subgrid-scale velocity generated by a Monte Carlo scheme. The simulations used a release of 1225 particles.

Lamb's results have served as benchmark investigations of dispersion in the CBL, in part because of their tie to the numerical turbulence model, which has been favorably compared with observations; his results are discussed in Sec. 4.3.2.

These simulations have also stimulated and guided the construction of simpler Lagrangian models—those using a stochastic model for the Lagrangian velocities. Examples, with application to the CBL, are given by Baerentsen and Berkowicz (1984), van Dop et al. (1985), and de Baas et al. (1986). These models simulate many of the diffusion characteristics of the CBL and also require numerical solutions, but the latter are much simpler to generate than with Lamb's model. Even simpler models with analytical properties have been motivated by the Lagrangian method; they are discussed in Sec. 4.4 and 4.5.

4.3.2. Results of Experiments and Numerical Predictions

The salient features of passive plume dispersion in the CBL have been demonstrated through numerical predictions (Lamb, 1978, 1979, 1982), laboratory experiments (Willis and Deardorff, 1976a, 1978,

1981; Poreh and Cermak, 1984), and field observations (Moninger et al., 1983; Nieuwstadt, 1980). These studies show that the dispersion characteristics are source-height dependent and can be classified broadly into two regimes: surface layer sources where $z_s < 0.1h$, and elevated sources where $0.1h < z_s < h$. This division is consistent with the CBL turbulence structure as should be expected.

Overall diffusion patterns

Figure 4.5 presents a composite of results of the crosswind-integrated concentration (CWIC) patterns from the laboratory experiments of Willis and Deardorff (1976a, 1978, 1981). The isopleths of the nondimensional CWIC in a vertical (x-z) plane are shown for a near-surface release ($z_s/h = 0.067$) and two elevated sources ($z_s/h = 0.24, 0.49$). The CWIC, C^y, is defined by

$$C^y = \int_{-\infty}^{\infty} C(x,y,z)dy \qquad (4.35)$$

and is made dimensionless by dividing by Q/Uh, where Q is the source strength; the nondimensional vertical coordinate is $Z = z/h$.

For the surface source (Fig. 4.5a), the average plume centerline (dashed curve) as defined by the locus of maximum concentration ascends after a short distance, $x \sim Uh/w_*$. In contrast, the center-lines for the elevated sources (Fig. 4.5b,c) descend until they reach the ground. These behaviors can be understood in terms of the release of material into either updrafts or downdrafts. For a surface source, material emitted into the base of an updraft begins rising almost immediately, whereas that released into downdrafts remains near the ground and moves horizontally. The horizontal motion, *relative* to the mean wind, is caused by the divergent flow at the base of a downdraft; i.e., the descending flow, constrained by the surface, is converted to horizontal motion, which partly explains the nearly uniform σ_v with height in the CBL. Since downdrafts occupy most ($\sim 60\%$) of the horizontal area, more material remains near the ground initially as does the maximum concentration. However, after a significant amount of material is swept out of downdrafts and into neighboring updrafts, the centerline begins to rise or "lift off" the ground.

A simple estimate can be made of the time for lift-off to occur. It is of the order of the downdraft radius, $\sim 0.5h$, divided by a typical lateral velocity, $\sigma_v \simeq 0.6w_*$, or $t \sim h/w_*$. Equating this to the

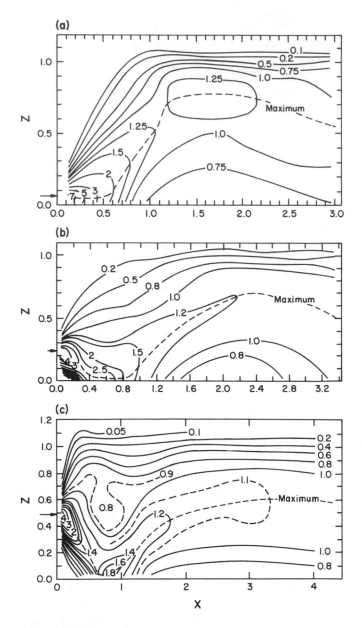

Figure 4.5. Laboratory convection tank results showing contours of nondimensional crosswind-integrated concentration (CWIC) as a function of dimensionless height Z and downwind distance X for sources at three release heights in a convective boundary layer. CWIC is nondimensionalized by Q/Uh; horizontal arrows denote the release height z_s. (a) $z_s/h = 0.067$, from Willis and Deardorff (1976a); (b) $z_s/h = 0.24$, from Willis and Deardorff (1978); (c) $z_s/h = 0.49$, from Willis and Deardorff (1981).

travel time x/U, we find that the distance at which centerline lift-off should occur is $x \sim Uh/w_*$. This estimate is in rough agreement with the observations (Fig. 4.5a), which show that lift-off occurs at $x \sim 0.5Uh/w_*$.

For elevated sources, the centerline descent is explained by the greater areal coverage of downdrafts, and thus, the higher probability of material being released into them. In addition, the downdrafts are long-lived so that material emitted into them tends to reach the surface. Figure 4.5 shows that the elevated centerlines reaching the surface remain there for some distance and then rise. The latter behavior is similar to that of the near-surface plume; it should not be too surprising because the downdrafts essentially create a distributed source of material at the surface. The similarity of the surface and elevated plumes is not unique to the CBL. In wind tunnel simulations of diffusion in a neutral boundary layer, Fackrell and Robins (1982) showed that the vertical concentration profile far downwind from an elevated source approaches that from a surface release. In contrast, near the source, the concentration profiles from the two release heights are quite different.

Aside from minor differences, the overall features of the diffusion patterns calculated numerically by Lamb (1978, 1979) are in good agreement with the laboratory simulations when suitably nondimensionalized. Figure 4.6 shows an example of this agreement for $z_s/h = 0.24$; see Lamb (1982) for others. Incidentally, in this example, the numerical predictions were made first and later confirmed by the laboratory data.

The field observations of Moninger et al. (1983) for $z_s/h = 0.5$ were quite similar to the numerical and laboratory simulations for the same release height (see Fig. 2.11). The consistency in the results from these widely differing investigations reinforces our understanding of the physics of dispersion in the CBL. It also demonstrates the utility of convective scaling, i.e., w_* and h as the relevant turbulence velocity and length scales.

The behavior of the plume centerlines for the surface and elevated sources is quite different than that given by the Gaussian plume model. The latter predicts that an elevated plume centerline remains elevated until a sufficient number of particle reflections occurs at the surface; the centerline then moves toward the surface (see Lamb, 1982). The essential cause of the difference between the "Gaussian" and observed behavior in the CBL is the positive skewness of w, as we have discussed.

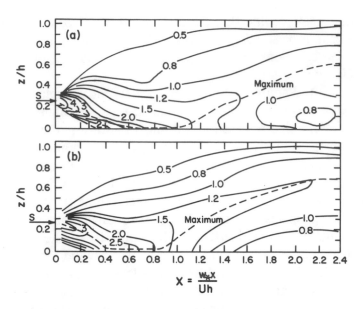

$$X = \frac{w_{*}X}{Uh}$$

Figure 4.6. Comparison between the crosswind-integrated concentration fields in a convective boundary layer as (a) predicted numerically by Lamb (1978); (b) observed in the laboratory tank experiments of Willis and Deardorff (1978). From Lamb (1982).

Dispersion parameters (σ_y, σ_z)

Statistical quantities of much interest in diffusion modeling are the lateral (σ_y) and vertical (σ_z) dispersion parameters, which are defined in terms of the second moments of particle displacements in the y and z directions. Lamb (1979) calculated these parameters from his numerical simulations, and the results are shown here (Fig. 4.7) in dimensionless form. As can be seen, the parameters segregate into the same two dispersion regimes—surface and elevated—that were discussed for the CWIC field. The σ_z/h data for the elevated releases $(z_s/h = 0.25, 0.50, 0.75;$ Fig. 4.7a) collapse to a nearly universal curve, $\sigma_z/h = 0.5X$, at short range, $X < 0.7$. This result is consistent with statistical theory and is attributed to the quasi-homogeneity of turbulence in the mixed layer. The numerical constant (0.5) is close to that (0.6) estimated earlier from statistical theory and σ_w observations (Eq. 4.26), thus establishing a link between numerical predictions, simple theory, and field data. The agreement between the numerical predictions and the laboratory simulations (Willis and Deardorff, 1978) is also very good as demonstrated in Fig. 4.8 for $z_s/h = 0.25$.

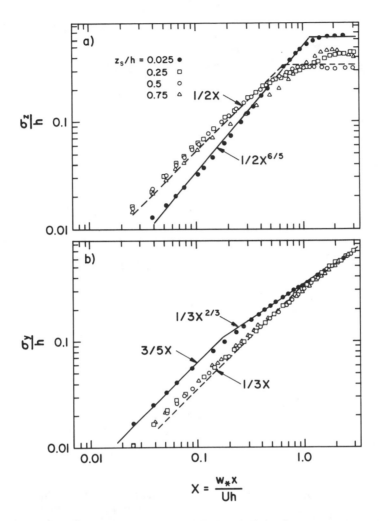

Figure 4.7. Lamb's (1979) numerically calculated dispersion parameters as a function of dimensionless distance for a point source in a convective boundary layer. (a) Vertical dispersion; (b) crosswind dispersion. Dashed and solid lines are approximations to the numerical results discussed in Lamb (1979, 1982). From Lamb (1982).

The σ_z/h predictions for the most elevated source ($z_s/h = 0.75$; Fig. 4.7a) exhibit a slightly smaller growth rate in the range $0.2 < X < 1$ than found for the releases at $z_s/h = 0.25$, 0.50. This is believed to be caused by the proximity of this source to the elevated inversion, which inhibits vertical motion (see Lamb, 1982).

For $X > 1$, the σ_z/h curves approach constants because of the trapping of material below $z = h$, and the homogenization of the

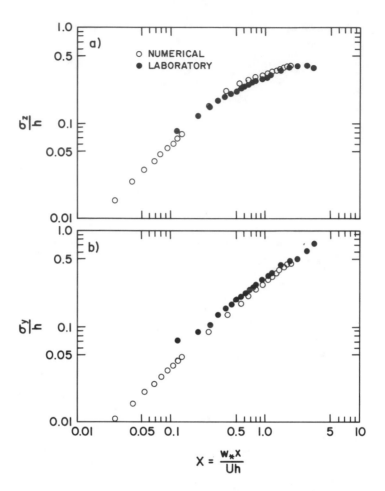

Figure 4.8. Comparison between numerical calculations (Lamb, 1978) and laboratory measurements (Willis and Deardorff, 1978) of dispersion parameters for a
source at height $0.25h$ in the convective boundary layer. (a) Vertical dispersion; (b)
crosswind dispersion. From Lamb (1982).

concentration field in the vertical direction. The latter is exhibited
by the approach of $C^y Uh/Q$ to 1 at $X \sim 2 - 3$ (see Fig. 4.5). The
limiting σ_z/h for a uniform concentration distribution in the vertical
is

$$\sigma_z/h = [1/3 - z_s/h + (z_s/h)^2]^{1/2}, \qquad (4.36)$$

which is source-height dependent. Lamb's (1979) σ_z/h results far
downstream are in approximate agreement with the above prediction.

The σ_z/h for the lowest source height ($z_s/h = 0.025$, Fig. 4.7a)
exhibits a growth rate that is greater than linear with X ($\sigma_z/h =$

$0.5X^{6/5}$) for small X, in qualitative agreement with Yaglom's (1972) theory (Eq. 4.32). Recall that the latter is for a true surface release, $z_s = 0$. An *elevated* release in the surface layer should initially grow linearly with t or X according to statistical theory because of the nonzero σ_w at the source height, i.e., $\sigma_w = 1.34(g\overline{w\theta_0}z_s/T_0)^{1/3}$. At greater times, the increase of σ_w with height should accelerate vertical spread such that $\sigma_z/h \propto X^{3/2}$. Thus, an approximate power law fit to the σ_z/h vs. X curve over both of these regimes would be expected to yield an exponent on X between 1 and 3/2 as found by Lamb (1979).

An indirect test of the Yaglom result is provided by Nieuwstadt's (1980) analysis of the Prairie Grass field data, which were obtained for $z_s \sim 0.5$ m. Adopting a Gaussian vertical distribution, which is a useful approximation for $X < 0.5$ (Deardorff and Willis, 1975), Nieuwstadt argued that $C^y Uh/Q \propto (\sigma_z/h)^{-1}$ if Yaglom's theory is correct. His plot of the dimensionless CWIC (Fig. 4.9a) indeed shows an approximate $X^{-3/2}$ dependence for $0.03 < X < 0.3$. The slower decline of the CWIC for $X < 0.03$ is probably due to the initial dominance of shear-driven turbulence with $\sigma_z \propto x$; the more rapid fall-off for $X > 0.3$ could be due to surface depletion.

The different σ_z behavior for surface and elevated sources discussed above is borne out at least qualitatively by some familiar σ_z results used in dispersion applications. These are the PG A σ_z curve for a ground source (Prairie Grass experiments) and the BNL B_2 σ_z curve for an elevated source at the Brookhaven National Laboratory (BNL; Singer and Smith, 1966); both curves apply to very unstable conditions. Figure 4.10 shows that the PG A σ_z is less than the BNL σ_z at small x but grows at a faster than linear rate with distance and eventually surpasses the BNL curve. In contrast, the BNL σ_z exhibits a slightly less than linear growth over all x. The qualitative similarity of these results to those in Fig. 4.7a lends further credibility to the numerical predictions.

Field, laboratory, and numerical data show that the crosswind concentration distribution is approximately Gaussian (e.g., Lamb, 1982). The Prairie Grass observations are in approximate agreement with $\sigma_y/h = 0.6X$ at short range ($X < 1$) as shown by the dashed line in Fig. 4.9b; this is consistent with the short-time limit of statistical theory (Eq. 4.26). There were few observations for $X > 1$ in that data set. However, analysis of other field data shows that σ_y/h increases less rapidly for $X > 1$. For example, Briggs (1985) recommended

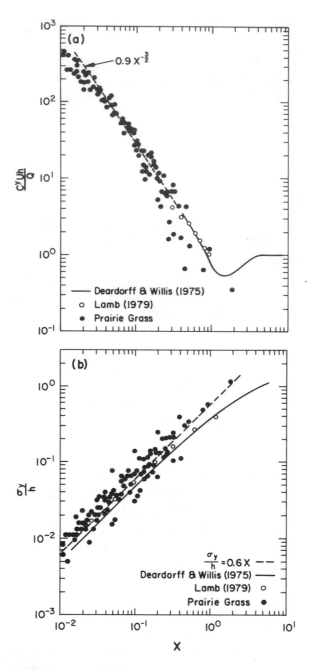

Figure 4.9. (a) Nondimensional crosswind-integrated concentration at ground level versus nondimensional downwind distance for a source near the surface. (b) Dimensionless crosswind dispersion parameter versus nondimensional downwind distance for a near-surface source. Adapted from Nieuwstadt (1980).

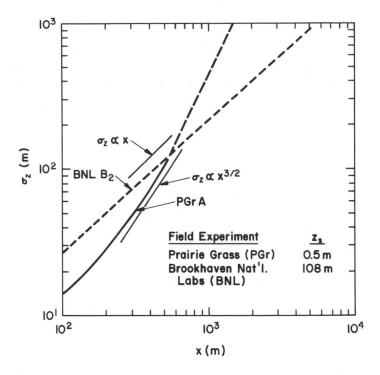

Figure 4.10. Comparison of vertical dispersion parameters for surface and elevated releases as determined from passive tracer concentrations in very unstable conditions.

$$\sigma_y/h = \frac{0.6X}{(1 + 2X)^{1/2}}, \tag{4.37}$$

which may provide a lower bound on σ_y at large X. The formulation of such an expression is guided by Taylor's (1921) large-time limit ($\sigma_y \propto t^{1/2}$); Eq. (4.37) reduces to $\sigma_y/h = 0.4X^{1/2}$ at large distances. This limit has not been observed in the CBL and is not expected to hold because of the presence of very large mesoscale eddies with correspondingly large time scales. More data are needed to confirm the large-time behavior.

Lamb's (1979) data and laboratory simulations both show that σ_y/h initially varies linearly with X in accord with statistical theory (e.g. Fig. 4.7b, 4.8b). However, these simulations show that σ_y is initially greater for a surface than an elevated release in contrast to what one would expect on the basis of field observations of σ_v; the latter are nearly independent of z (Caughey, 1982). The simulated σ_y behavior is attributed to the simulated σ_v, which is greater near the surface than in the middle of the CBL (see Caughey, 1982). This profile in turn is believed caused by the small aspect ratio of the tank—tank width/h (see Deardorff and Willis, 1985). The coefficient in the short-time expression for the elevated release, $\sigma_y/h \simeq (0.3 - 0.4)X$, is somewhat smaller than expected (0.6) on the basis of field observations of σ_y. We attribute this to the small aspect ratio.

4.4. Applied Dispersion Models: Passive Plumes

As shown in Sec. 4.3, the unique feature of passive plume dispersion in the CBL is the non-Gaussian nature of the vertical dispersion; this is linked to the skewness of w. The latter is most pronounced and well documented under strongly convective conditions ($-h/L > 10$) when surface friction is of little consequence in the mixed layer. In weaker convection ($-h/L < 10$), the skewness is probably smaller, and surface friction effects are more important; a Gaussian distribution may then be an approximate description of the w p.d.f.

In the following, we discuss dispersion models for strong and weak convection separately. The division between the two is set somewhat arbitrarily at $-h/L \sim 5$-10, but this is not well defined. If one chooses U/w_* as a stability index, then the strong convection regime should exist for $U/w_* < 6$ (Willis and Deardorff, 1976a).

4.4.1. Strong Convection ($-h/L > 5$-10)

We first consider vertical dispersion from elevated ($z_s/h > 0.1$) and surface ($z_s/h < 0.1$) sources, and then discuss lateral dispersion. Knowledge of the vertical dispersion determines the CWIC field, $C^y(X,Z)$. From the latter we find the GLC assuming a Gaussian crosswind distribution; the GLC along the plume centerline is $C = C^y(X,0)/\sqrt{2\pi}\sigma_y$.

Vertical dispersion from elevated sources

The numerical and laboratory simulations discussed earlier have motivated the development of a new type of dispersion model—the p.d.f. approach—based on the p.d.f. of vertical velocity (Weil and Furth, 1981; Misra, 1982; Venkatram, 1983). Since the latter is non-Gaussian, so also is the predicted vertical concentration distribution.

The conceptual basis of the p.d.f. model is provided by the organized, long-lived nature of the updrafts and downdrafts in the CBL. Effectively, one assumes that passive particles from a fixed source

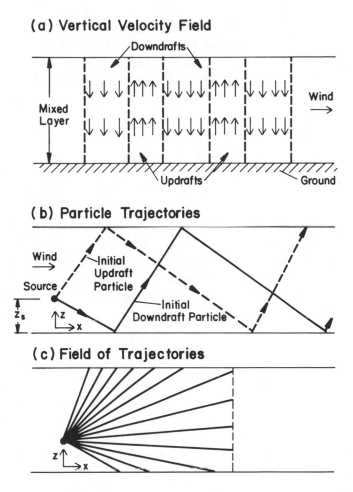

Figure 4.11. Schematic of idealized vertical velocity field and passive particle dispersion in the convective boundary layer.

are emitted into a traveling train of updrafts and downdrafts that are moving with the mean wind speed U (Fig. 4.11a). The vertical velocity (w) in each element, updraft or downdraft, is assumed to be a random variable prescribed by p_w. Note that p_w has dimensions of inverse velocity, and that the probability of w lying in the interval $w - \Delta w/2 < w < w + \Delta w/2$ is $p_w(w)\Delta w$. Considering the definition of probability and the limit as $\Delta w \to 0$, we have

$$\int_{-\infty}^{\infty} p_w(w)dw = 1. \tag{4.38}$$

A key assumption in this model is that the Lagrangian time scale T_L^z of the random vertical velocity is effectively infinite so that a particle remains within an element until it reaches the surface or the capping inversion. There, it reverses its direction and proceeds toward the opposite "boundary," thus behaving somewhat like a bouncing ball between two plates. If we assume that the vertical velocity is uniform with height in a given downdraft (or updraft), then a particle will follow a straightline path (Fig. 4.11b). For a distribution of w governed by p_w, a particle can follow an infinite number of paths (Fig. 4.11c), each having a certain probability of occurrence, $p_w(w)\Delta w$.

The particle trajectories and the associated vertical concentration distribution determined from p_w are for the ensemble-average dispersion, since p_w is an ensemble-average property. Also note that in the above description, we have ignored any small-scale turbulence within an updraft or downdraft that would cause dispersion relative to the large-scale motion (only the latter is depicted in Fig. 4.11). Such dispersion can be incorporated as in the case of buoyant plumes (Sec. 4.5).

A simple calculation of the CWIC, C^y, follows from mass-flux considerations. The flux of particles through an elemental distance Δz at height z in a vertical (y-z) plane of infinite crosswind width is $UC^y(x,z)\Delta z$. This is equal to the emission rate (Q) times the probability of particles lying in the interval $z - \Delta z/2 < z < z + \Delta z/2$. Since the probability is $p_z(z_c)\Delta z$, where p_z is the p.d.f. of particle height z_c, the mass balance is expressed by

$$UC^y\Delta z = Qp_z\Delta z \qquad (4.39)$$

or

$$C^y = \frac{Qp_z}{U} . \qquad (4.40)$$

The task now is to find the p.d.f. of z_c in terms of the p.d.f. of w, which can be done if z_c is a monotonic function of w. The relationship between p_z and p_w is given by

$$p_z = p_w[w(z_c)]\left|\frac{dw}{dz_c}\right| , \qquad (4.41)$$

where the absolute value is taken to ensure that p_z is positive. Writing the argument of p_w as $w(z_c)$ means that we replace w by its equivalent in terms of particle height (z_c) and distance (x). The relationship between w and z_c is found from a differential equation governing the particle trajectory: $dz_c/dx = w(z)/U$, where we permit w to be a function of height (z) in general.

If w is independent of height, the differential equation is simply integrated to yield

$$z_c = z_s + \frac{wx}{U}, \tag{4.42}$$

where it should be remembered that w is a random vertical velocity. The integration to produce Eq. (4.42) follows from the fact that T_L^z is infinitely long, which means that w is determined by its value at the source. $w(z_c)$ is then

$$w = (z_c - z_s)\frac{U}{x} \tag{4.43}$$

and

$$\frac{dw}{dz_c} = \frac{U}{x}. \tag{4.44}$$

Substituting Eqs. (4.41), (4.43), and (4.44) into Eq. (4.40), we have

$$C^y(x,z) = Q\frac{p_w[U(z_c - z_s)/x]}{x}. \tag{4.45}$$

The derivation of C^y from Eqs. (4.40) and (4.41) can be done even if w is a function of z (e.g., see Misra, 1982) or if other factors such as plume buoyancy enter the analysis and modify the trajectory (Sec. 4.5).

To solve for the C^y field, one must choose a function that characterizes the observed p_w. Several authors (Misra, 1982; Baerentsen and Berkowicz, 1984; Weil et al., 1986) have found that a superposition of two Gaussian distributions can provide a good match to observations; i.e.,

$$p_w(w) = \frac{F_1}{\sqrt{2\pi}\sigma_1} \exp\left[\frac{-(w - \overline{w}_1)^2}{2\sigma_1^2}\right] + \frac{F_2}{\sqrt{2\pi}\sigma_2} \exp\left[\frac{-(w - \overline{w}_2)^2}{2\sigma_2^2}\right], \tag{4.46}$$

where $F_1 + F_2 = 1$. The distributions denoted by subscripts 1 and 2 mostly characterize the downdraft and updraft velocities, respectively. If one interprets $\overline{w}_1$ as the mean downdraft velocity and multiplies Eq. (4.46) by $\overline{w}_1$, the result can be compared with the p.d.f.'s found by Lamb. This comparison is shown in Fig. 4.12, where Lamb's p.d.f.'s are unsmoothed versions that have been nondimensionalized using the local mean downdraft velocity, $\overline{w}_d(z)$; the parameters—$\overline{w}_1$, $\overline{w}_2$, σ_1, σ_2, F_1, and F_2, are specified in the figure caption. As can be seen, Eq. (4.46) is a reasonable approximation to Lamb's data.

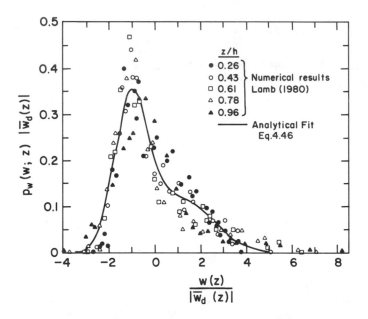

Figure 4.12. Probability density function (p.d.f.) of vertical velocity w at five heights in the convective boundary layer as determined by Lamb (1980, private communication) from Deardorff's (1974) numerical simulations; the p.d.f. and w are made dimensionless by using the mean downdraft velocity w_d at each height. For comparison, Eq. (4.46) is shown for $a_1 = 0.35$, $b_1 = 0.26$, $a_2 = 0.4$, $b_2 = 0.48$, $F_1 = 0.6$, and $F_2 = 0.4$.

The double-Gaussian form for p_w appears to be a suitable representation and offers the flexibility of adjustment to fit other observed p.d.f.'s—in the field (Caughey et al., 1983) or from the laboratory (Deardorff and Willis, 1985). Furthermore, one may wish to adjust the parameters ($\overline{w}_1$, σ_1, etc.) to obtain an optimum fit between the modeled C^y field and observations. Other forms for p_w can be adopted, e.g., those of Venkatram (1983) and Briggs (1985).

To complete the model for C^y, we must include the no-flux boundary conditions at $z = 0$; i.e., particles or material cannot pass through the ground plane. We assume that a similar condition applies at the inversion base, $z = h$. There are three methods for handling this condition: 1) Assume perfect reflection of the velocity at the boundary; i.e., the velocity of a particle coming away from a surface is of the same magnitude but opposite sign as that moving toward the surface (Misra, 1982; Venkatram, 1983). 2) Assume that particles coming away from the surface have velocities "drawn" from the proper distribution—updraft if from the ground or downdraft if from the inversion base (Weil and Furth, 1981). 3) Adopt the "source" solution. We present results here for only the first approach.

The C^y field due to the "direct" plume—that coming directly from the source before reflection at the ground or inversion base—is found by substituting Eq. (4.46) into Eq. (4.45) to obtain

$$
C^y = \frac{QF_1}{\sqrt{2\pi}\sigma_1 x} \exp\left\{\frac{-[(z - z_s)U/x - \overline{w}_1]^2}{2\sigma_1^2}\right\} +
$$
$$
\frac{QF_2}{\sqrt{2\pi}\sigma_2 x} \exp\left\{\frac{-[(z - z_s)U/x - \overline{w}_2]^2}{2\sigma_2^2}\right\} .
$$

(4.47)

The dimensionless CWIC at the surface, $C^y Uh/Q$, can be cast in the following convenient form after multiplying Eq. (4.47) by 2 to account for reflection at $z = 0$:

$$
\frac{C^y Uh}{Q} = \frac{A_1}{X} \exp\left[-\frac{R_1}{2}\left(\frac{Z_s}{a_1 X} - 1\right)^2\right] + \frac{A_2}{X} \exp\left[-\frac{R_2}{2}\left(\frac{Z_s}{a_2 X} + 1\right)^2\right],
$$

(4.48)

where

$$
A_1 = \frac{2F_1}{\sqrt{2\pi}b_1}, \qquad A_2 = \frac{2F_2}{\sqrt{2\pi}b_2}
$$

$$
a_1 = -\overline{w}_1/w_*, \qquad a_2 = \overline{w}_2/w_* \quad (\overline{w}_1 < 0 \text{ and } \overline{w}_2 > 0)
$$

$$
b_1 = \sigma_1/w_*, \qquad b_2 = \sigma_2/w_*
$$

(4.49)

$$
R_1 = (a_1/b_1)^2, \qquad R_2 = (a_2/b_2)^2
$$

$$
Z_s = z_s/h .
$$

To account for multiple reflections of the direct plume at $z = 0$ and $z = h$, we must include "image" sources at $z = \pm(2h - z_s)$, $\pm(2h + z_s)$, $\pm(4h - z_s)$, etc. The resulting nondimensional CWIC is

$$\frac{C^y U h}{Q} = \frac{A_1}{X} \sum_{n=-M}^{M} \exp\left[\frac{-R_1}{2}\left(\frac{2n + Z_s}{a_1 X} - 1\right)^2\right] +$$
$$+ \frac{A_2}{X} \sum_{n=-M}^{M} \exp\left[\frac{-R_2}{2}\left(\frac{2n - Z_s}{a_2 X} - 1\right)^2\right]. \tag{4.50}$$

The number of series terms (M) necessary to obtain a reasonable estimate of $C^y U h/Q$ depends on source height (Z_s) and distance (X). For Z_s up to 0.5 and X out to 2.5, only terms up to $n = \pm 1$ are necessary.

For sources in the lower half of the CBL, the maximum CWIC, C^y_{max}, is determined principally by the first term on the right-hand side of Eq. (4.48). The location, X_{max}, of this maximum can be found by differentiating this term with respect to X and setting the result equal to zero. The X_{max} is then given by

$$X_{max} = \frac{Z_s}{b_1}\left[\left(1 + \frac{R_1}{4}\right)^{1/2} - \frac{R_1}{2}\right] \tag{4.51}$$

and depends both on the mode (or maximum) of the w p.d.f. (i.e., a_1) as well as on the standard deviation, σ_1 or b_1. (For a Gaussian distribution, $R_1 = 0$, and X_{max} reduces to Z_s/b_1 as expected.) For the $a_1(0.35)$ and $b_1(0.26)$ used in the following comparisons, $X_{max} \simeq 2Z_s$.

Figure 4.13 compares the p.d.f. model predictions of C^y with laboratory data (Willis and Deardorff, 1978; Deardorff and Willis, 1984) and numerical simulations (Lamb, 1978, 1982) for source heights of $Z_s \simeq 0.25$ and 0.50. The parameters used in the model $(a_1 = 0.35, a_2 = 0.4, b_1 = 0.26, b_2 = 0.48)$ were chosen as compromise values, i.e., to yield approximate agreement with the laboratory data at both release heights [these parameter values are the same ones used for Eq. (4.46) in Fig. 4.12]. Overall, the fit of the model to the data is reasonably good and especially so on the magnitude and location of C^y_{max}. For reference, we show results for a Gaussian model with $\sigma_z/h = 0.5X$, which is consistent with Lamb's data. The

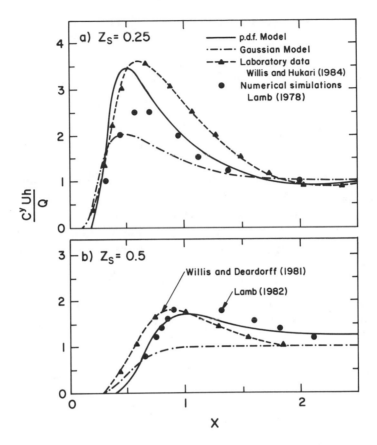

Figure 4.13. Nondimensional crosswind-integrated concentration at the surface versus dimensionless downwind distance for passive releases at two heights in the convective boundary layer.

model peaks at the correct distance, but the magnitude of C^y is too low. This result occurs because of the zero mode, which means that the plume centerline does not descend until a significant number of particle reflections occur at the surface.

Vertical dispersion from surface layer sources

The major problem in dealing with dispersion in the surface layer is the inhomogeneity in the vertical turbulence [Eqs. (4.3), (4.4)]. A few models account for this feature if only in an approximate manner. The first and perhaps simplest is the Gaussian model but with a suitably chosen σ_z. Deardorff and Willis (1975) found that a Gaussian (vertical) distribution was an approximate fit to their laboratory measurements of C^y for $X < 0.5$ and suggested the following formula for σ_z:

$$\frac{\sigma_z}{h} = \left[1.8Z_s^{2/3}X^2 + 0.25X^3 + (b'u_*X/w_*)^2\right]^{1/2}, \tag{4.52}$$

where

$$b' = \frac{0.4 + 0.8(w_*Z_s/u_*X)}{1 + w_*Z_s/u_*X}. \tag{4.53}$$

The first term within the brackets of Eq. (4.52) is the short-time limit of statistical theory and accounts for the initial vertical spread due to the nonzero σ_w at the source height. The second term represents Yaglom's (1972) theory (Eq. 4.32) which must be met as $z_s \to 0$, and the third term is the similarity theory prediction in the neutral stability limit (as $w_*/u_* \to 0$). Briggs (1985) suggested multiplying b' by 1.6 to attain better agreement with field data.

Figure 4.14 compares predictions of $C^y Uh/Q$ at the surface based on the above σ_z with the laboratory measurements of Deardorff and Willis (1975) for $Z_s = 0.067$. The agreement is good over most downwind distances, but for $1 < X < 2$, the model exceeds the data by as much as a factor of 2. This is caused by the plume centerline

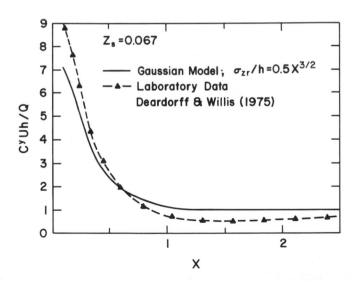

Figure 4.14. Nondimensional crosswind-integrated concentration at the surface versus dimensionless downwind distance for a near-surface source in the convective boundary layer; σ_{zr} is the dispersion based on Yaglom's (1972) result (Eq. 4.32 with $c_1 = 0.75$) as incorporated in Eq. (4.52).

lift-off discussed earlier and the highly non-Gaussian C^y profile that results. For $Z_s \sim 0$, Nieuwstadt's (1980) analysis (Fig. 4.9) shows that the Gaussian model with Yaglom's (1972) σ_z is a good fit to surface observations of C^y out to $X \sim 0.3$.

The eddy-diffusion model or K theory is another way to estimate the short-range vertical dispersion from a surface release (Nieuwstadt and van Ulden, 1978; van Ulden, 1978). In this approach, one solves the diffusion equation assuming that the vertical flux of material is given by $-K\partial C^y/\partial z$, K being the eddy diffusivity that is found from surface layer similarity theory [see Eq. (5.41), Ch. 5]. Van Ulden obtained an analytical expression for C^y that depends on the mean travel speed and height of a plume [see Eqs. (5.42), (5.43)], which are expressed in terms of u_*, L, z_0, and distance. He demonstrated the applicability of this approach out to 800 m from a surface release. More recently, Gryning et al. (1987) successfully applied the model to vertical dispersion from surface layer sources and gave operational methods for implementing it.

In principle, the eddy-diffusion model must fail at large distances where much of the plume is in the mixed layer and is dominated by the large convective eddies, which have a characteristic size $\ell_z \sim h$. For $X > 1$, $\sigma_z \simeq 0.6h$ (Fig. 4.7a), but a necessary condition for applying K theory is $\sigma_z > \ell_z$ (Pasquill and Smith, 1983).

One final approach is an empirical parameterization of C^y based on the laboratory, numerical, and field data discussed earlier. Briggs (1985) adopted this approach and showed that such a parameterization can fit the data quite well over the range of source heights used in the experiments, $0.067 < Z_s < 0.5$. One of the limitations of this approach is the difficulty of extending it to include other effects such as buoyancy, surface depletion, etc.

Lateral dispersion

There is some consensus that a suitable form for σ_y in the PBL is

$$\sigma_y = \sigma_v t f(t/T_L^y) \qquad (4.54)$$

(e.g., Weil, 1985). Here, t can be replaced by x/U when axial diffusion is unimportant; this should be true for $U/w_* > 1.2$ (Willis and Deardorff, 1976b). A convenient function f that satisfies the short and long time limits of statistical theory is

$$f = \frac{1}{(1 + 0.5t/T_L^y)^{1/2}} \qquad (4.55)$$

(Deardorff and Willis, 1975).

In the strong convection limit with $\sigma_v = \alpha_1 w_*$ and T_L^y assumed to be proportional to h/w_*, the following expression results:

$$\frac{\sigma_y}{h} = \frac{\alpha_1 X}{(1 + \alpha_2 X)^{1/2}}, \qquad (4.56)$$

where $\alpha_1 \simeq 0.6$ from Eq. (4.5). There is a fair amount of evidence supporting Eq. (4.56) for $X < 10$. For example, the summary by Briggs (1985) suggested that $\alpha_2 = 2$ provides an approximate lower bound to passive dispersion data [Eq. (4.37)], and $\alpha_2 = 0.6$ is more appropriate for buoyant plumes. Weil and Corio (1985) found that $\alpha_1 = 0.56$, $\alpha_2 = 0.7$ provided a good average fit to σ_y of buoyant plumes. We reiterate that the long-time limit, $\sigma/h \propto X^{1/2}$, is not expected to hold because of the presence of large mesoscale eddies with longer time scales than h/w_*.

4.4.2. Weak to Moderate Convection $(-h/L < 5$–$10)$

Dispersion in this regime can also be divided into surface layer and elevated sources. For surface sources $(z_s/h < 0.1)$, the eddy-diffusion model can be applied bearing in mind its limitation to plumes that are mostly within the surface layer. Operational methods for applying the approach are given by Gryning et al. (1987).

For elevated releases, modeling is on a more tentative basis because of our incomplete knowledge of the turbulence structure, and limited model testing. For the present, the Gaussian plume model may be sufficient. It has been shown to give an adequate description of the GLC field downwind of an elevated source in a wind tunnel simulation for neutral stability (Fackrell and Robins, 1982). The main difficulty in applying the model is accounting for the vertical inhomogeneity of the turbulence on the dispersion parameters. Two simple approaches are available: 1) Assume that the turbulence at the source height is most representative of the ambient field. 2) Calculate the vertical dispersion separately above and below the plume centerline taking into account the inhomogeneity.

Gryning et al. (1987) followed the first approach and estimated σ_y and σ_z from Eq. (4.54) and (4.55). The Lagrangian time scales were assumed to be constant, $T_L^y = 600$ s and $T_L^z = 300$ s, based on earlier empirical results. They determined the turbulence profiles by superposing the velocity variances from the strongly convective and near neutral limits. For σ_w, they gave

$$\frac{\sigma_w^2}{u_*^2} = 1.5\,[-z/(kL)]^{2/3}\exp(-2z/h) + (1.7 - z/h)\,, \qquad (4.57)$$

where the first term on the right-hand side is the convective contribution as parameterized by Baerentsen and Berkowicz (1984), and the second is the result for a near-neutral boundary layer $(-h/L < 1.4)$ obtained by Brost et al. (1982). Brost et al. found that wind tunnel data from a neutral boundary layer agreed well with their field observations. Gryning et al. showed that the Gaussian model with σ_y and σ_z determined as above gave good estimates of C^y for a small sample of field data.

Venkatram and Paine (1985) (see also Ch. 5) followed the second approach, calculating σ_z separately above and below the plume centerline. However, this was done for a σ_w profile characteristic of a slightly stable boundary layer. The approach could be repeated for a profile more representative of a weakly unstable layer, but this remains to be done.

4.5. Applied Dispersion Models: Buoyant Plumes

The conventional Gaussian model of a buoyant plume dispersing in a turbulent wind is divided into a rise regime wherein the plume is dominated by the buoyancy, and a dispersion regime which is governed by ambient turbulence. In the latter, the plume is assumed to be passive and to disperse from an effective stack height $h_e = z_s + \Delta h$, where Δh is the "final" rise. However, casual observation of stack plumes in the CBL shows that this division is an oversimplification of the real process; i.e., plume rise and dispersion (or meandering) by large convective eddies occur simultaneously (see Fig. 3.11).

Recently, much has been learned about buoyant plume dispersion in the CBL from the laboratory simulations of Willis and Deardorff (1983). For example, their experiments clearly demonstrated the superposition of plume rise and the large-scale meander. They also showed that the most important measure of plume buoyancy in the far field was the dimensionless buoyancy flux $F_* = F_b/Uw_*^2 h$, where F_b is the stack buoyancy flux (Ch. 3). For $F_* \sim 0.03$, the plume behaved somewhat like that from a nonbuoyant source after some initial rise. However, for $F_* > 0.1$, the plume rose mostly to the top of the CBL, partially penetrated the capping inversion, and gradually mixed downward in the CBL. That is, the buoyancy-induced velocities were sufficient to overcome the downdraft velocities over considerable distances.

Subsequent measurements by Willis and Hukari (1984) have more clearly demonstrated the highly skewed nature of the vertical concentration distribution. In addition, these measurements as well as those by Deardorff and Willis (1984) showed the substantial reduction in surface concentrations due to plume buoyancy.

In the following, we discuss several models that incorporate some of our current understanding of dispersion in the CBL. These are 1) the Gaussian model with improved formulations of the dispersion parameters (σ_y, σ_z), 2) the p.d.f. model as modified for plume buoyancy, and 3) two other approaches—the impingement model (Venkatram, 1980) and Briggs' (1985) model for highly buoyant effluents. In view of our earlier comments, one may question the discussion of the Gaussian model. We include it because it can approximate the GLC distribution and can deal with buoyant releases and dispersion over a continuous range of stability.

4.5.1. Gaussian Models

A number of σ_y and σ_z expressions based on convective scaling can be applied in the Gaussian model. However, we limit our discussion to models where this has been done, and model predictions have been compared with field observations. Two examples are the PPSP model (Weil and Brower, 1984) and the Danish OML model (Berkowicz et al., 1986).

The PPSP model differs from conventional Gaussian plume models (e.g., CRSTER) in four ways: 1) Briggs' dispersion curves for elevated releases (see Gifford, 1975) replace the PG curves for surface releases; 2) U/w_* is the stability parameter for daytime instead of the Turner (1964) criteria; 3) Briggs' (1975) plume rise formulas for convective conditions (Ch. 3) are included rather than his pre-1975 formulas, which did not address convection; and 4) Briggs' (1984) criteria for estimating plume penetration of elevated inversions are used instead of the "all-or-none" approach of the CRSTER model.

The PPSP model is intended for elevated sources, which are defined as those with $h_e > 0.1h$.

The Briggs curves were given for six "classes" of atmospheric stability ranging from A (most unstable) to F (most stable) and were originally defined in terms of the Turner (1964) criteria. The main difference between them and the PG curves is the σ_z for A stability. The Briggs σ_z is based on elevated releases at the Brookhaven National Laboratory and varies linearly with x, consistent with the short-range behavior for an elevated plume in the CBL (Sec. 4.3). In contrast, the PG σ_z for A stability grows faster than linearly with x, as expected for a surface release. The expected large-time limit, $\sigma_z \propto h$, is simulated in the Gaussian model by the use of image sources above h and below $z = 0$ to satisfy the "no-flux" conditions there.

For small distances ($x < 1$ km), all the Briggs curves conform to the short-range limits of statistical theory, i.e.,

$$\sigma_y = \frac{\sigma_v}{U}x, \qquad \sigma_z = \frac{\sigma_w}{U}x . \qquad (4.58)$$

Using these limits, Weil and Brower redefined the applicability of the A to D (neutral) curves in terms of U/w_*. They adopted interpolation expressions for σ_v and σ_w of the form $\sigma_v = (\sigma_{vn}^2 + \sigma_{vc}^2)^{1/2}$, where σ_{vn} and σ_{vc} represent σ_v in the neutral and strongly convective limits, respectively. The σ_{wn} and σ_{vn} were chosen to agree with Briggs neutral (D) curves: $\sigma_{wn} = 0.06U$ and $\sigma_{vn} = 0.08U$; the convective limits were based on the measurements of Kaimal et al. (1976): $\sigma_{vc} = \sigma_{wc} = 0.56w_*$. Combining the above with the (σ_v, σ_w) interpolation expressions and Eq. (4.58), they found

$$\sigma_y = \left[0.08^2 + (0.56\frac{w_*}{U})^2\right]^{1/2} x$$

$$\sigma_z = \left[0.06^2 + (0.56\frac{w_*}{U})^2\right]^{1/2} x . \qquad (4.59)$$

The stability limits of the dispersion curves were obtained by assuming that a line midway between two neighboring curves divided their regions of applicability; the corresponding U/w_* values were found from Eq. (4.59).

Evaluation of the PPSP model with GLC values of SO_2 measured downwind of Maryland power plants showed that it performed much better than the CRSTER model; the measurements approximated hourly averages along the plume centerline. For example, 69% of the model predictions (C_{pred}) were within factor of 2 of the observed (C_{obs}) concentrations, and the variance (r^2) explained by the model was 0.58; here, r is the correlation coefficient between $\ln C_{obs}$ and $\ln C_{pred}$. For the CRSTER model, only 33% of the predictions met the factor of 2 criterion, and the r^2 was ~ 0. The better performance by the PPSP model was attributed primarily to use of the Briggs A σ_z curve and to the modeled distribution of stability—62%, 30%, 8%, and 0% for the A to D classes, respectively; this distribution is in line with our expectation for daytime. In contrast, the stability distribution given by the CRSTER model was 1%, 19%, 26%, and 53% for the A to D classes; i.e., it was strongly biased toward neutral conditions, which are rather infrequent during daytime.

In another evaluation, Tikvart and Cox (1984) found that the PPSP model overestimated the maximum 1-, 3-, and 24-h average concentrations around the Clifty Creek power plant by factors of $\sim$ 2 to 3. The maximum concentrations were estimated to occur under low $h(<700m)$ and light winds, and when the plume was predicted to be trapped below the elevated inversion. The concentrations were overestimated because the plume was assumed to be passive and to disperse from a source near h if it was unable to penetrate the capping inversion. In contrast, the Willis and Hukari (1984) data for similar conditions (i.e., high F_*) showed that the plume retained its buoyancy over significant distances; this delayed the downward mixing and lowered the maximum concentrations.

The PPSP model was recently modified to account for the high F_* behavior and gave much better results when compared with data from the Clifty Creek, Muskingum River, and Kincaid power plants. The high F_* model is the same as that described below for the p.d.f. model.

Two improvements could be made in the PPSP approach for estimating dispersion parameters. First, the parameters should be given as continuous functions of CBL variables rather than in stability "classes"; this can be achieved using expressions of the form $\sigma_y = \sigma_v t f(t/T_L^y)$ [Eqs. (4.54), (4.55)]. Second, the interpolation formulas for σ_v and σ_w should include u_* in the neutral limit, i.e., $\sigma_{vn}, \sigma_{wn} \propto u_*$, instead of U, and they should account for the local roughness z_0; z_0 can be included by determining u_* from the similarity wind profile (Sec. 4.2.2).

In the OML model (Berkowicz et al., 1986), the dispersion parameters are derived for both surface ($z_s < 0.1h$) and elevated sources and are given as continuous functions of boundary layer variables, which include w_*, h, and u_*. The σ_v and σ_w are expressed as in the PPSP model, i.e., by a superposition of the turbulence components in the neutral (or mechanical) and convective limits. The dispersion parameters are derived assuming that the fluctuating velocities due to convection and to mechanically driven turbulence are uncorrelated. Thus, Berkowicz et al. found $\sigma_z = (\sigma_{zm}^2 + \sigma_{zc}^2)^{1/2}$, where σ_{zm} and σ_{zc} represent the contributions by mechanical and convective turbulence, respectively; a similar form was obtained for σ_y.

σ_{zc} is computed from a rate equation, $d\sigma_{zc}/dt = \sigma_{wc}(z')$, where $t = x/U$ and z' is an effective height at which σ_{wc} is evaluated; σ_{wc} is assumed to be height dependent below $0.1h$ in agreement with free convection scaling, i.e., $\sigma_{wc} \propto z^{1/3}$, and is assumed to be constant, $\sigma_{wc} = 0.57w_*$, above $0.1h$. Thus, for $z_s > 0.1h$, σ_{zc} is $0.57w_*t$. For $z_s < 0.1h$, three expressions are obtained for σ_{zc}, each appropriate in a different distance regime depending on σ_{zc}/z_s and σ_{zc}/h (see Berkowicz et al., 1986).

The mechanical contribution to the dispersion is derived assuming that the turbulence component σ_{wm} is constant throughout the boundary layer, $\sigma_{wm} = 1.1u_*$. The derivation accounts for the strong height variation in the length scale of the mechanically driven turbulence and results in

$$\sigma_{zm} = 1.1u_*t \exp\left(-0.3\frac{u_*t}{z_s}\right) \qquad \text{if } \frac{u_*t}{z_s} < 1$$

$$\sigma_{zm} = 1.1u_*t \exp(-0.3) \qquad \text{if } \frac{u_*t}{z_s} > 1 . \tag{4.60}$$

The σ_y expression is the same for short and tall sources, consistent with the invariance of σ_v with height in the CBL:

$$\sigma_y = \left[\frac{0.25X^2}{1 + 0.9X} + \left(\frac{u_*x}{Uh}\right)^2\right]^{1/2} h . \tag{4.61}$$

In the case of a significant wind direction change Δd over an hour, an effective dispersion σ_{yf} is determined from $\sigma_{yf} = (\Delta d/\sqrt{2\pi})x$. The σ_{yf} is used whenever it exceeds the σ_y given by Eq. (4.61).

Plume rise in the OML model is determined from Briggs' (1975) equations, and the penetration of inversions is found using a slight modification of Briggs' (1984) criteria. The model also includes an empirical correction to reduce the vertical dispersion of highly buoyant plumes.

Evaluation of the OML model with passive tracers emitted from a variety of heights—0.5 m to 200 m—showed that predicted GLCs agreed well with observations. The model also gave reasonable estimates of SF_6 GLCs measured downwind of the Kincaid (Illinois) power plant, thus demonstrating its applicability to buoyant plumes. However, the scatter between C_{obs} and C_{pred} in the buoyant plume comparisons was considerably larger than for the nonbuoyant plumes. This is consistent with the laboratory measurements of concentration fluctuations by Deardorff and Willis (1984) and is related to the greater meander for an elevated, buoyant plume.

4.5.2. P.d.f. Model

The p.d.f. model for buoyant plumes is similar to that for passive releases except that the buoyancy-driven velocity is superimposed on the random convection velocity w (Weil et al., 1986). The model differs for "low" and "high" buoyancy fluxes, which are defined by $F_* < 0.1$ and > 0.1, respectively. This division simplifies the problem analytically and is guided by the laboratory observations discussed earlier as well as a model evaluation with field data.

Low buoyancy fluxes

At low F_*, the conceptual picture is a plume embedded within a field of updrafts and downdrafts that are sufficiently large to displace entire cross sections up and down (Fig. 4.15). In addition, it is assumed that the Lagrangian time scale is infinitely long so that a plume segment remains in the same updraft or downdraft until it reaches the top of the CBL or ground, and w is independent of height. The CWIC, C_b^y, in the buoyant plume is assumed to have a Gaussian distribution about its "instantaneous" centerline:

$$C_b^y = \frac{Q}{\sqrt{2\pi}U\sigma_b} \exp\left(\frac{-(z-z_c)^2}{2\sigma_b^2}\right), \qquad (4.62)$$

where σ_b is the standard deviation of the instantaneous profile. The σ_b arises from buoyancy-induced entrainment and is given by

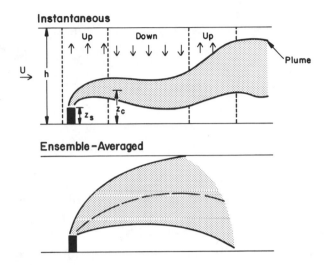

Figure 4.15. Schematic of buoyant plume dispersion in the convective boundary layer for "low" buoyancy fluxes, $F_* < 0.1$, according to the p.d.f. model.

$\sigma_b = \beta' z'_b / \sqrt{2}$, where z'_b is the plume rise and $\beta' = 0.4$ (see Ch. 3). The rise is prescribed by the two-thirds law [Ch. 3, Eq. (3.17)]. By superimposing the vertical displacements due to buoyancy and convection, one finds

$$z_c = z_s + 1.6\frac{F^{1/3}}{U}x^{2/3} + \frac{wx}{U} . \tag{4.63}$$

The ensemble-average CWIC, C^y, is found by integrating over all possible plume heights:

$$C^y = \int C^y_b(x, z, z_c)p_z(z_c)dz_c \tag{4.64}$$

where $p_z = p_w[w(z_c, x)]|dw/dz|$. The only difference between this p_z and that in Sec. 4.4 is the incorporation of plume rise in z_c [Eq. (4.63)].

Combining Eqs. (4.62–4.64) and using the p_z, p_w relationship, Weil et al. found the nondimensional CWIC at the surface, $C^y Uh/Q$, to be

$$\frac{C^y Uh}{Q} = \frac{1.2}{\sqrt{2\pi}\sigma^*_{z1}}\exp\left(\frac{-h^{*2}_1}{2\sigma^{*2}_{z1}}\right) + \frac{0.8}{\sqrt{2\pi}\sigma^*_{z2}}\exp\left(\frac{-h^{*2}_2}{2\sigma^{*2}_{z2}}\right) , \tag{4.65}$$

where

$$\sigma_{zi}^* = [1.3\beta'^2 F_*^{2/3} X^{4/3} + b_i^2 X^2]^{1/2}$$

$$h_i^* = Z_s + 1.6 F_*^{1/3} X^{2/3} + a_i X .$$

(4.66)

The a_i and b_i, where $i = 1$ or 2, are dimensionless coefficients characterizing p_w (Sec. 4.4). In obtaining Eq. (4.65), Weil et al. assumed plume reflection at $z = 0$ to satisfy the no-flux condition there. In Eq. (4.66), the buoyancy terms appear separately from those involving convection so that as $F_* \to 0$, the solution for a nonbuoyant source is recovered.

The zero-flux condition at the top of the CBL is satisfied using a distribution of point sources along $z = h$, where the source strength is chosen to cancel the upward flux of material from the real source. For simplicity, the p_w for these sources is assumed to be Gaussian with a standard deviation $\sigma_w = 0.5 w_*$. A Gaussian distribution in the uppermost parts of the CBL is supported by laboratory data (Deardorff and Willis, 1985) and field measurements (Caughey et al., 1983), although other observations indicate a distribution with positive skewness (Sec. 4.2.1); thus, the form of the distribution at the top of the CBL is not completely clear. For the source distribution at h, the no-flux condition at the surface is satisfied by image sources along $z = -h$; these lead to additional sources at $z = \pm 3h, \pm 5h$, etc.

The total C^y at the ground is given by the sum of contributions from the real source (Eq. 4.65), the sources at $z = h$, and the image sources. The GLC, C, is found from the C^y, assuming a Gaussian crosswind distribution. Thus, the concentration along the centerline at ground level is $C = C^y / \sqrt{2\pi}\sigma_y$, where σ_y is given by Eq. (4.56) with $\alpha_1 = 0.56$ and $\alpha_2 = 0.7$. (The σ_y is the convective limit of "model 1" from Weil and Corio, 1985.)

High buoyancy fluxes

At high F_*, the plume is assumed to rise to the top of the CBL, where it "lofts" or lingers until individual cross sections are mixed to the surface by large-scale downdrafts. The idea, illustrated in Fig. 4.16, is partially inferred from the photographs of laboratory plumes by Willis and Deardorff (1983). "Mix-down" occurs inside only those downdrafts having sufficiently strong vertical velocities to carry plume segments to the surface, and these are randomly dis-

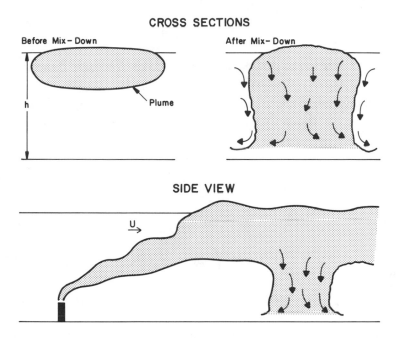

Figure 4.16. Schematic of buoyant plume dispersion in the convective boundary layer for "high" buoyancy fluxes, $F_* > 0.1$, according to the p.d.f. model.

tributed along a plume. Thus, as shown in Fig. 4.16 (side view), one can expect to see plume cross sections mixed to the surface between those that are not.

Prior to mix-down, lofting plumes undergo enhanced lateral spread (Fig. 4.16, cross section) as a result of the pressure difference Δp between the plume and the surrounding air (see Briggs, 1985). The Δp is caused by the density difference $\Delta \rho$, where $\Delta \rho =$ the ambient density minus the plume density; i.e., $\Delta p \propto \Delta \rho g \ell$, where ℓ is the vertical half-width of the plume.

There are three key assumptions in the model. First, the Lagrangian time scale is infinitely long. This means that a plume cross section remains within the same downdraft until it is completely dispersed to the surface. Second, plume segments are mixed to the surface only when the downdrafts have sufficient kinetic energy, $\rho w^2/2$, to overcome the potential energy difference between the plume and the environment, $\Delta \rho g h$. Third, a plume segment within a downdraft is assumed to be uniformly mixed in the vertical at the distance x_m where

$$\Delta \rho g h = \alpha_m \frac{\rho w^2}{2} ; \tag{4.67}$$

i.e., the C^y within the downdraft is equal to Q/Uh. The dimensionless coefficient α_m is empirically found to be 3.1.

The ensemble-averaged CWIC at the surface is Q/Uh times the probability P_m that mix-down has occurred at the distance x or the probability that $x_m < x$; thus,

$$C^y = \frac{Q P_m}{Uh} . \tag{4.68}$$

The P_m is calculated from an integral over the p.d.f. of x_m, p_x:

$$P_m = \int_0^x p_x(x_m) dx_m , \tag{4.69}$$

where p_x is expressed in terms of p_w through the relationship $p_x(x_m) = p_w[w(x_m)]|dw/dx_m|$.

The dependence of w on x_m is obtained from Eq. (4.67) and an expression relating $\Delta \rho$ to the source buoyancy flux F_b. For simplicity, the elevated plume cross section is assumed to be rectangular with half-widths R and ℓ in the lateral and vertical directions, respectively. The ℓ is assumed to be constant prior to mix-down, $\ell = 0.2h$, as inferred from laboratory data (Willis and Deardorff, 1983). The R is given by $\sqrt{2}\sigma_y$ (Briggs, 1975) and σ_y by Briggs' (1985) expression for highly buoyant plumes:

$$\sigma_y = 1.6 \frac{F_b^{1/3}}{U} x^{2/3} . \tag{4.70}$$

By equating the buoyancy flux in the "rectangular plume" to that, πF_b, of the round plume at the source, Weil et al. obtained

$$\frac{\Delta \rho}{\rho} = \frac{\pi F_b}{4 U g \ell R} . \tag{4.71}$$

The vertical velocity required for a uniformly mixed cross section is found by combining Eqs. (4.67), (4.70), (4.71), and $\ell = 0.2h$ with the result

$$\frac{w}{w_*} = A \left(\frac{F_*}{X} \right)^{1/3} , \tag{4.72}$$

where $A = (\pi h / 4.6 \alpha_m \ell)^{1/2}$.

As mentioned earlier, p_w at the top of the CBL is assumed to be Gaussian with $\sigma_w = 0.5w_*$. This assumption, together with Eqs. (4.69) and (4.72), results in the following expression for the dimensionless CWIC at the surface:

$$\frac{C^y U h}{Q} = 1 - \text{erf}\left(\frac{2A}{\sqrt{2}(X/F_*)^{1/3}}\right). \tag{4.73}$$

The finding that the CWIC is a single function of X/F_* is in accord with Briggs' (1985) analysis of field observations [Eq. (4.76) below] and Willis' (1986) summary of laboratory data. However, Eq. (4.73) tends to underestimate these data for $X/F_* > 10$, and this requires further examination. As at low F_*, the GLC is determined from the C^y assuming that the crosswind distribution is Gaussian. The σ_y is given by Eq. (4.70) for $U/w_* > 2$ and by Eq. (4.70) times $0.5U/w_*$ for $U/w_* < 2$; the latter is an empirical correction found by Weil et al. to improve the results.

Evaluation of the p.d.f. model with laboratory measurements of the CWIC near the surface shows good agreement for both non-buoyant and buoyant plumes (Fig. 4.17). For highly buoyant plumes (taken here as $F_* > 0.08$), the model tended to overestimate the CWIC for $X < 1$. This behavior was attributed to an overestimate of σ_y by Eq. (4.70) close to the source and, hence, to an underestimate of $\Delta\rho$; the latter would lead to a more rapid downward mixing of the plume than observed. Evaluation of the model with SF_6 GLCs downwind of the Kincaid power plant (Fig. 4.18) shows fair to good agreement on average. The geometric mean (GM) and geometric standard deviation (GSD) of $C_{\text{pred}}/C_{\text{obs}}$ were 1.1 and 2.1, respectively, and 68% of the predictions were within a factor of 2 of the observations. However, further work is necessary to explain the relatively low variance ($r^2 = 0.34$) and the large scatter. In this regard, emphasis must be placed on concentration uncertainties due to stochastic variability, input data uncertainties, and potential model errors.

For reference, the CRSTER model was compared with the Kincaid data (Fig. 4.19). As can be seen, the results are poorer as demonstrated by greater scatter (GSD = 4.6), fewer predictions within a factor of 2 of C_{obs} (33%), and an r^2 of 0.02. The model also predicted zero concentrations in a number of instances where measureable and

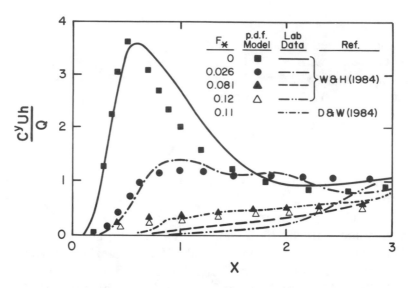

Figure 4.17. Nondimensional crosswind-integrated concentration at the surface versus dimensionless downwind distance for passive and buoyant plumes. W & H (1984) is Willis and Hukari (1984); D & W (1984) is Deardorff and Willis (1984).

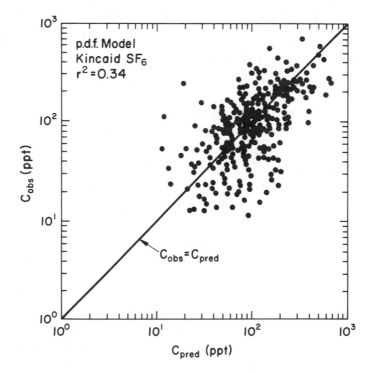

Figure 4.18. Observed versus predicted ground-level SF_6 concentrations for the p.d.f. model at the Kincaid power plant.

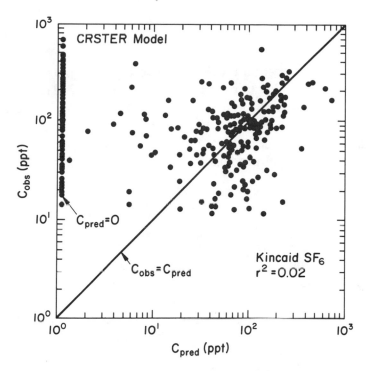

Figure 4.19. Observed versus predicted ground-level SF_6 concentrations for the CRSTER model at the Kincaid power plant.

sometimes high concentrations were observed. The zeroes resulted from predictions of plume penetration of the inversion.

4.5.3. Other Models

The impingement model of Venkatram (1980) is similar to the p.d.f. approach in that plume rise is considered relative to a field of updrafts and downdrafts. However, attention is focused only on plume segments caught in downdrafts since they are most important in determining the GLCs immediately downwind of the source. Venkatram assumes that GLCs occur at some downwind receptor whenever the plume "touches down" or "impinges" on the ground ahead of that receptor. The concentration is found from the source strength, mean wind speed, and the local plume dimensions (at x). The plume is assumed to have a Gaussian crosswind profile and a uniform vertical distribution.

The mean impingement distance $\bar{x}_i$, where the lower plume edge reaches the ground, is found from Briggs' (1975) "touchdown equation":

$$z_s + \frac{F^{1/3}\bar{x}_i^{2/3}}{U} - \frac{\overline{w}_d\bar{x}_i}{U} = 0 . \tag{4.74}$$

In the above, the second term is the rise of lower plume edge due to buoyancy, and the third term is the vertical displacement due to the mean downdraft velocity $\overline{w}_d$. Venkatram assumes that the impingement distance is a random variable following a lognormal distribution, with a mean of $\ln \bar{x}_i$ and an empirically specified standard deviation. The model predicts the GLC along the plume centerline to be

$$C = \frac{Qf_i(x)}{\sqrt{2\pi}U\sigma_y R_z} , \tag{4.75}$$

where R_z is the plume's vertical depth and $f_i(x)$ is the cumulative frequency distribution of the impingement distance. The σ_y and R_z account for buoyancy-induced growth as well as convective turbulence.

Evaluation of this model with SO_2 measurements around two Maryland power plants and the Sudbury (Canada) nickel smelter (Venkatram and Vet, 1981) shows good performance: $\sim 85\%$ of the predicted concentrations are within a factor of 2 of the observed values, and the variance (r^2) explained by the model ranges from ~ 0.6 to 0.8. Further work is necessary to improve the model in two extremes of buoyancy: 1) neutrally buoyant plumes where the model underestimates the maximum concentrations simulated numerically by Lamb (1978) (see Venkatram, 1980), and 2) highly buoyant releases ($F_* > 0.1$–0.2) where a plume can reach the top of the CBL before "touching down"; i.e., the expression for $\bar{x}_i$ is incorrect.

For highly buoyant plumes, Briggs (1985) offered a simple semi-empirical expression to describe the dimensionless CWIC at ground level:

$$\frac{C^y Uh}{Q} = \exp\left[-7(F_*/X)^{3/2}\right] . \tag{4.76}$$

The key thought leading to the above was that strong downward mixing should occur at a distance $x \propto F_b/w_*^3$ or $X \propto F_*$, as suggested by the distance where the rate of lateral spread of a lofting plume is of the order of w_* . This idea led to the notion that the relevant distance variable should be X/F_*. Equation (4.76) is an empirical fit to observations at the Dickerson and Morgantown (Maryland) power plants, and reduces to the correct long-time limit, $C^y Uh/Q = 1$.

The GLC can be obtained from Eq. (4.76) by assuming a Gaussian crosswind distribution with σ_y given by Eq. (4.70).

The Briggs model along with the low F_* component of the p.d.f. approach has been incorporated in a recently developed hybrid dispersion model for tall stacks (Hanna et al., 1986).

4.5.4. Summary

Several models based on improved understanding of the CBL and buoyant plume behavior were described for dispersion applications. My preference is the p.d.f. model because I believe that it best represents the meandering effects of the large convective eddies on plumes, and it reduces to the correct surface CWIC in the limit of zero buoyancy. The model was in reasonable agreement with surface concentration data for both laboratory and full-scale plumes. For high $F_*(> 0.1)$, it needs modification to give an adequate description of the CWIC at large distances, say $X/F_* > 10$.

Another approach is the Gaussian model but with dispersion parameters based on convective scaling (w_*, h) and updated treatments of plume rise and buoyancy effects. Two models—PPSP and OML—were found to give good estimates of GLCs around power plant stacks. However, the PPSP model should be applied only for $F_* < 0.1$ because it tends to overestimate concentrations at higher values for reasons discussed in Sec. 4.5.1. In the high F_* regime, either the p.d.f. or the Briggs (1985) model could be used.

The impingement model is another approach with demonstrated success in predicting GLCs for weakly to moderately buoyant plumes. As discussed, it needs further development to properly model the GLCs in the limits of zero and high ($F_* > 0.1$) buoyancy. For the latter regime, the Briggs model has been shown to give an adequate description of the surface CWIC downwind of two power plants.

References

Baerentsen, J. H., and R. Berkowicz, 1984: Monte Carlo simulation of plume dispersion in the convective boundary layer. *Atmos. Environ.*, **18**, 701–712.

Berkowicz, R., H. R. Olesen, and U. Torp, 1986: The Danish Gaussian air pollution model (OML): Description, test and sensitivity analysis in view of regulatory applications. *Air Pollution Modeling and Its Application V*, C. De Wispelaere, F. A. Schiermeier, and N. V. Gillani, Eds., Plenum, New York, 453–481.

Berkowicz, R., and L. P. Prahm, 1982: Sensible heat flux estimated from routine meteorological data by the resistance method. *J. Appl. Meteor.*, **21**, 1845–1864.

Briggs, G. A., 1975: Plume rise predictions. *Lectures on Air Pollution and Environmental Impact Analyses*, D.A. Haugen, Ed., Amer. Meteor. Soc., Boston, 59–111.

Briggs, G. A., 1984: Plume rise and buoyancy effects. *Atmospheric Science and Power Production*, D. Randerson, Ed., U.S. Dept. of Energy DOE/TIC–27601, 327–366.

Briggs, G. A., 1985: Analytical parameterizations of diffusion: the convective boundary layer. *J. Climate Appl. Meteor.*, **24**, 1167–1186.

Brost, R. A., J. C. Wyngaard, and D. H. Lenschow, 1982: Marine stratocumulus layers. Part II: Turbulence budgets. *J. Atmos. Sci.*, **39**, 818–836.

Carson, D. J., 1973: The development of a dry inversion-capped convectively unstable boundary layer. *Quart. J. Roy. Meteor. Soc.*, **99**, 450–467.

Caughey, S. J., 1982: Observed characteristics of the atmospheric boundary layer. *Atmospheric Turbulence and Air Pollution Modelling*, F. T. M. Nieuwstadt and H. van Dop, Eds., Reidel, 107–158.

Caughey, S. J., M. Kitchen, and J. R. Leighton, 1983: Turbulence structure in convective boundary layers and implications for diffusion. *Bound.-Layer Meteor.*, **25**, 345–352.

Csanady, G. T., 1973: *Turbulent Diffusion in the Environment*. Reidel, 248 pp.

Deardorff, J. W., 1972: Numerical investigation of neutral and unstable planetary boundary layers. *J. Atmos. Sci.*, **29**, 91–115.

Deardorff, J. W., 1974: Three-dimensional numerical study of the height and mean structure of a heated planetary boundary-layer. *Bound.-Layer Meteor.*, 7, 81–106.

Deardorff, J. W., 1982: Structure of convection. *Workshop on the Parameterization of Mixed Layer Diffusion*, Physical Sciences Laboratory, New Mexico State University, Las Cruces, 91–95.

Deardorff, J. W., and G. E. Willis, 1975: A parameterization of diffusion into the mixed layer. *J. Appl. Meteor.*, **14**, 1451–1458.

Deardorff, J. W., and G. E. Willis, 1984: Ground-level concentration fluctuations from a buoyant and a non-buoyant source within a laboratory convectively mixed layer. *Atmos. Environ.*, **18**, 1297–1309.

Deardorff, J. W., and G. E. Willis, 1985: Further results from a laboratory model of the convective planetary boundary layer. *Bound.-Layer Meteor.*, **32**, 205–236.

de Baas, A. F., H. van Dop, and F. T. M. Nieuwstadt, 1986: An application of the Langevin equation for inhomogeneous conditions to dispersion in a convective boundary layer. *Quart. J. Roy. Meteor. Soc.*, **112**, 165–180.

Driedonks, A. G. M., 1982: Models and observations of the growth of the atmospheric boundary layer. *Bound.-Layer Meteor.*, **23**, 283–306.

EPA, 1977: *User's manual for single source (CRSTER) model*. U.S. Environmental Protection Agency, Research Triangle Park, NC, Ref. No. EPA–450/2–77–013, 285 pp.

Fackrell, J. E., and A. G. Robins, 1982: Concentration fluctuations and fluxes in plumes from point sources in a turbulent boundary layer. *J. Fluid Mech.*, **117**, 1–26.

Garratt, J. R., J. C. Wyngaard, and R. J. Francey, 1982: Winds in the atmospheric boundary layer—prediction and observation. *J. Atmos. Sci.*, **39**, 1307–1316.

Gifford, F. A., 1961: Uses of routine meteorological observations for estimating atmospheric dispersion. *Nucl. Safety*, **2**, 47–51.

Gifford, F. A., 1975: Atmospheric dispersion models for environmental pollution applications. *Lectures on Air Pollution and Environmental Impact Analyses*, D.A. Haugen, Ed., Amer. Meteor. Soc., Boston, 35–58.

Gryning, S. E., A. A. M. Holtslag, J. S. Irwin, and B. Sivertsen, 1987: Applied dispersion modelling based on meteorological scaling parameters. *Atmos. Environ.*, **21**, 79–89.

Hanna, S. R., G. A. Briggs, J. Deardorff, B. A. Egan, F. A. Gifford, and F. Pasquill, 1977: AMS workshop on stability classification schemes and sigma curves— Summary of recommendations. *Bull. Amer. Meteor. Soc.*, **58**, 1305–1309.

Hanna, S. R., J. C. Weil, and R. J. Paine, 1986: Plume model development and evaluation, hybrid approach. Electric Power Research Institute, Palo Alto, CA, Report No. D034–500, 550 pp.

Holtslag, A. A. M., and A. P. van Ulden, 1983: A simple scheme for daytime estimates of the surface fluxes from routine weather data. *J. Climate Appl. Meteor.*, **22**, 517–529.

Holzworth, G. C., 1972: Mixing heights, wind speeds, and potential for urban air pollution throughout the contiguous United States. U.S. Environmental Protection Agency, Research Triangle Park, NC, Ref. No. AP–101, 118 pp.

Hunt, J. C. R., 1982: Diffusion in the stable boundary layer. *Atmospheric Turbulence and Air Pollution Modelling*, F. T. M. Nieuwstadt and H. van Dop, Eds., Reidel, 231–274.

Izumi, Y., and S. J. Caughey, 1976: Minnesota 1973 atmospheric boundary layer experiment data report. Air Force Cambridge Research Laboratories, Hanscom AFB, MA, Environmental Research Papers, No. 547, 28 pp.

Kaimal, J. C., J. C. Wyngaard, D. A. Haugen, O. R. Coté, Y. Izumi, S. J. Caughey, and C. J. Readings, 1976: Turbulence structure in the convective boundary layer. *J. Atmos. Sci.*, **33**, 2152–2169.

Lamb, R. G., 1978: A numerical simulation of dispersion from an elevated point source in the convective planetary boundary layer. *Atmos. Environ.*, **12**, 1297–1304.

Lamb, R. G., 1979: The effects of release height on material dispersion in the convective boundary layer. *4th Symposium on Turbulence, Diffusion and Air Pollution*, Amer. Meteor. Soc., Boston, 27–33.

Lamb, R. G., 1982: Diffusion in the convective boundary layer. *Atmospheric Turbulence and Air Pollution Modelling*, F. T. M. Nieuwstadt and H. van Dop, Eds., Reidel, 159–229.

Le Mone, M. A., 1973: The structure and dynamics of horizontal roll vortices in the planetary boundary layer. *J. Atmos. Sci.*, **30**, 1077–1091.

Lilly, D. K., 1968: Models of cloud-topped mixed layers under a strong inversion. *Quart. J. Roy. Meteor. Soc.*, **94**, 292–309.

Lumley, J. L., and H.A. Panofsky, 1964: *The Structure of Atmospheric Turbulence.* Wiley Interscience, New York, 239 pp.

Misra, P. K., 1982: Dispersion of nonbuoyant particles inside a convective boundary layer. *Atmos. Environ.*, **16**, 239–243.

Moninger, W. R., W. L. Eberhard, G. A. Briggs, R. A. Kropfli, and J. C. Kaimal, 1983: Simultaneous radar and lidar observations of plumes from continuous point sources. *21st Conference on Radar Meteorology*, Amer. Meteor. Soc., Boston, 246–250.

Nicholls, S., and C. J. Readings, 1979: Aircraft observations of the structure of the lower boundary layer over the sea. *Quart. J. Roy. Meteor. Soc.*, **105**, 785–802.

Nieuwstadt, F. T. M., 1978: The computation of the friction velocity u_* and the temperature scale T_* from temperature and wind velocity profiles by least-squares methods. *Bound.-Layer Meteor.*, **14**, 235–246.

Nieuwstadt, F. T. M., 1980: Application of mixed-layer similarity to the observed dispersion from a ground-level source. *J. Appl. Meteor.*, **19**, 157–162.

Nieuwstadt, F. T. M., and A. P. van Ulden, 1978: A numerical study of the vertical dispersion of passive contaminants from a continuous source in the atmospheric surface layer. *Atmos. Environ.*, **12**, 2119–2124.

Panofsky, H. A., and J. A. Dutton, 1984: *Atmospheric Turbulence*. Wiley, New York, 397 pp.

Panofsky, H. A., H. Tennekes, D. H. Lenschow, and J.C. Wyngaard, 1977: The characteristics of turbulent velocity components in the surface layer under convective conditions. *Bound.-Layer Meteor.*, 11, 355–361.

Pasquill, F. A., and F. B. Smith, 1983: *Atmospheric Diffusion* (third ed.). Wiley, New York, 437 pp.

Paulson, C. A., 1970: The mathematical representation of wind speed and temperature profiles in the unstable atmospheric surface layer. *J. Appl. Meteor.*, 9, 857–861.

Pennell, W. T., and M. A. Le Mone, 1974: An experimental study of turbulence structure in the fair-weather trade wind boundary layer. *J. Atmos. Sci.*, 31, 1308–1323.

Poreh, M., and J. E. Cermak, 1984: Wind tunnel diffusion in a convective boundary layer. *Bound.-Layer Meteor.*, 30, 431–455.

Sellers, W. D., 1965: *Physical Climatology*. University of Chicago Press, 272 pp.

Singer, I. A., and M. E. Smith, 1966: Atmospheric diffusion at Brookhaven National Laboratory. *Int. J. Air Water Pollut.*, 10, 125–135.

Smith, M. E., 1984: Review of the attributes and performance of 10 rural diffusion models. *Bull. Amer. Meteor. Soc.*, 65, 554–558.

Taylor, G. I., 1921: Diffusion by continuous movements. *Proc. London Math. Soc. Ser. 2*, 20, 196–212.

Tennekes, H., 1973: A model for the dynamics of the inversion above a convective boundary layer. *J. Atmos. Sci.*, 30, 558–567.

Tennekes, H., and A. P. van Ulden, 1974: Short-term forecasts of temperature and mixing height on sunny days. *2nd Symposium on Atmospheric Turbulence, Diffusion and Air Quality*, Amer. Meteor. Soc., Boston, 35–40.

Tikvart, J. A., and W. M. Cox, 1984: EPA's model evaluation program. *4th Joint Conference on Applications of Air Pollution Meteorology*, Amer. Meteor. Soc., Boston, 66–69.

Turner, D. B., 1964: A diffusion model for an urban area. *J. Appl. Meteor.*, 3, 83–91.

Turner, D. B., 1971: *Workbook of Atmospheric Dispersion Estimates*. U.S. Environmental Protection Agency, Office of Air Programs, Ref. No. AP-26, 84 pp.

van Dop, H., F. T. M. Nieuwstadt, and J. C. R. Hunt, 1985: Random walk models for particle displacements in inhomogeneous unsteady turbulent flows. *Phys. Fluids*, 28, 1639–1653.

van Ulden, A. P., 1978: Simple estimates for vertical dispersion from sources near the ground. *Atmos. Environ.*, 12, 2125–2129.

van Ulden, A. P., and A. A. M. Holtslag, 1985: Estimation of atmospheric boundary layer parameters for diffusion applications. *J. Climate Appl. Meteor.*, **24**, 1196–1207.

Venkatram, A., 1980: Dispersion from an elevated source in the convective boundary layer. *Atmos. Environ.*, **14**, 1–10.

Venkatram, A., 1983: On dispersion in the convective boundary layer. *Atmos. Environ.*, **17**, 529–533.

Venkatram, A., and R. Paine, 1985: A model to estimate dispersion of elevated releases into a shear-dominated boundary layer. *Atmos. Environ.*, **19**, 1797–1805.

Venkatram, A., and R. Vet, 1981: Modeling of dispersion from tall stacks. *Atmos. Environ.*, **15**, 1531–1538.

Weil, J. C., 1985: Updating applied diffusion models. *J. Climate Appl. Meteor.*, **24**, 1111–1130.

Weil, J. C., and R. P. Brower, 1983: Estimating convective boundary layer parameters for diffusion applications. Martin Marietta Environmental Center, Columbia, MD, Report No. PPSP–MP–48, 37 pp.

Weil, J. C., and R. P. Brower, 1984: An updated Gaussian plume model for tall stacks. *J. Air Pollut. Control Assoc.*, **34**, 818–827.

Weil, J. C., and L. A. Corio, 1985: Dispersion formulations based on convective scaling. Martin Marietta Environmental Center, Columbia, MD, Report No. PPSP–MP–60, 39 pp.

Weil, J. C., L. A. Corio, and R. P. Brower, 1986: Dispersion of buoyant plumes in the convective boundary layer. *5th Joint Conference of Applications of Air Pollution Meteorology*, Amer. Meteor. Soc., Boston, 335–338.

Weil, J. C., and W. F. Furth, 1981: A simplified numerical model of dispersion from elevated sources in the convective boundary layer. *5th Symposium on Turbulence, Diffusion and Air Pollution*, Amer. Meteor. Soc., Boston, 76–77.

Wieringa, J., 1980: Representativeness of wind observations at airports. *Bull. Amer. Meteor. Soc.*, **61**, 962–971.

Willis, G. E., 1986: Dispersion of buoyant plumes in a convective boundary layer. *5th Joint Conference on Applications of Air Pollution Meteorology*, Amer. Meteor. Soc., Boston, 327–330.

Willis, G. E., and J. W. Deardorff, 1976a: A laboratory model of diffusion into the convective planetary boundary layer. *Quart. J. Roy. Meteor. Soc.*, **102**, 427–445.

Willis, G. E., and J. W. Deardorff, 1976b: On the use of Taylor's translation hypothesis for diffusion in the mixed layer. *Quart. J. Roy. Meteor. Soc.*, **102**, 817–822.

Willis, G. E., and J. W. Deardorff, 1978: A laboratory study of dispersion from an elevated source within a modeled convective planetary boundary layer. *Atmos. Environ.*, **12**, 1305–1312.

Willis, G. E., and J. W. Deardorff, 1981: A laboratory study of dispersion from a source in the middle of the convectively mixed layer. *Atmos. Environ.*, **15**, 109–117.

Willis, G. E., and J. W. Deardorff, 1983: On plume rise within the convective boundary layer. *Atmos. Environ.*, **17**, 2435–2447.

Willis, G. E., and N. Hukari, 1984: Laboratory modeling of buoyant stack emissions in the convective boundary layer. *4th Joint Conference on Applications of Air Pollution Meteorology*, Amer. Meteor. Soc., Boston, 24–26.

Wyngaard, J. C., 1985: Structure of the planetary boundary layer and implications for its modeling. *J. Climate Appl. Meteor.*, **24**, 1131–1142.

Wyngaard, J. C., O. R. Coté, and Y. Izumi, 1971: Local free convection, similarity, and the budgets of shear stress and heat flux. *J. Atmos. Sci.*, **28**, 1171–1182.

Yaglom, A. M., 1972: Turbulent diffusion in the surface layer of the atmosphere. *Izv. Akad. Nauk USSR, Atmos. Ocean. Phys.*, **8**, 333–340.

Yap, D., and T. R. Oke, 1974: Eddy-correlation measurements of sensible heat fluxes over a grass surface. *Bound.-Layer Meteor.*, **7**, 151–163.

CHAPTER 5

Dispersion in the Stable Boundary Layer

Akula Venkatram

5.1. Introduction

In this chapter I use current understanding of micrometeorology to describe dispersion of pollutants in the stable boundary layer (SBL). I emphasize how dispersion works rather than particular formulas that can be used to estimate ground-level concentrations. Whenever it is possible, I will make recommendations on the use of available models for dispersion. However, my main objective is to provide the type of understanding of dispersion that will allow you to tackle a problem without reaching for a handbook on air pollution modeling.

Dispersion is an extremely complex process. Because dispersion is governed by turbulence, whose physics remains largely impenetrable, dispersion models are semi-empirical in the sense that their development almost always involves the fitting of model parameters whose values are not known *a priori*. It is my view that the credibility of a semi-empirical model is closely related to the simplicity of the model structure, and that one should be wary of a complex model with a large number of tunable parameters. I shall attempt to show that it is possible to make a great deal of progress by adopting the philosophy that simpler is better.

In Sec. 5.2 I examine current methods of treating dispersion in the stable boundary layer. In Sec. 5.3 I discuss those aspects of the SBL that are most relevant to dispersion. Section 5.4 discusses the theoretical basis of the dispersion models covered in the following section. In Sec. 5.5, which is the core of this chapter, I describe a set of new, improved models based on recent advances in our understanding of the micrometeorology of the SBL. In Sec. 5.6 I describe methods to estimate the micrometeorological variables used in the proposed models.

5.2. Current Models for Dispersion in the Stable Boundary Layer

The most popular method for estimating dispersion in the stable boundary layer is based on the stability classification system proposed by Pasquill and then modified by Gifford. This PG system, as it is commonly known, relates the dispersion characteristics of the PBL to routinely available meteorological observations. Table 5.1 illustrates the system. Its categories, which range from A to F, are each associated with curves for the dispersion measures σ_y and σ_z. These sigmas are used in a Gaussian formulation to yield ground-level concentrations.

Table 5.1. Stability categories in terms of wind speed, insolation, and state of sky.

Surface wind speed (m s^{-1})	Insolation			Night	
	Strong	Moderate	Slight	Thinly overcast or $\geq$ 4/8 low cloud	$\leq$ 3/8 cloud
< 2	A	A–B	B	–	–
2–3	A–B	B	C	E	F
3–5	B	B–C	C	D	E
5–6	C	C–D	D	D	D
> 6	C	D	D	D	D
	(for A–B take average of values for A and B)				

Strong insolation corresponds to sunny midday in midsummer in England, slight insolation to similar conditions in midwinter. Night refers to the period from 1 h before sunset to 1 h after dawn. The neutral category D should also be used, regardless of wind speed, for overcast conditions during day or night, and for any sky conditions during the hour preceding or following the night as defined above. (From *The Meteorological Magazine*, February, 1961; reproduced by permission of Controller, Her Britannic Majesty's Stationery Office.)

The PG system attempts to account for the effects of shear and buoyancy on turbulence generation through the wind speed at 10 m and the incoming solar radiation. The specification of solar radiation recognizes that the source of buoyant production of turbulence during the day is the surface heat flux, which is driven by incoming solar radiation. At night, the cloudiness is an indirect measure of the incoming thermal radiation which counteracts the radiative cooling of the ground. This radiative cooling, we recall from Ch. 1, directs the heat flux toward the earth's surface, which in turn suppresses the turbulence generated by shear. This suggests that, other things being equal, turbulence levels at night are likely to be higher under cloudy conditions than under cloud-free conditions.

The PG classification system is generally consistent with the turbulent structure of the PBL; however, the correspondence between the PG meteorological variables (wind speed, solar radiation, cloudiness) and the dispersion characteristics is qualitative and far from unique. As we shall see later, we can use our current understanding of the PBL to do a better job of relating dispersion to available meteorological observations.

A major shortcoming of the PG sigma curves is that they are derived from limited observations of dispersion from surface releases. The measurements were restricted to distances less than 1 km, and the sampling time was roughly 3 min; the PG curves are not applicable to substantially different conditions. Current models based on the PG system have ignored this limitation primarily because the theory relating turbulence to the PG meteorological variables was not mature enough to provide guidance for modifying the curves for other dispersion conditions. Our current understanding of micrometeorology allows us to make such modifications; this is a principal reason for seeking a deeper understanding of micrometeorology and dispersion.

I am often told that PG-based models are used only because the micrometeorological inputs required by more advanced models are often not available. I show in this chapter that one can estimate these inputs from the routine observations that are used by the PG system. The point to be stressed here is that a better understanding of the structure of the PBL allows us to make better use of available information.

In the discussion to follow I relate new concepts to familiar ones, and so I shall briefly mention the commonly used sigma curves pro-

posed by Briggs (1973). Table 5.2 shows the analytical fits he derived by combining the PG curves with curves obtained from observations of elevated releases made at Brookhaven National Laboratory
(Singer and Smith, 1966). Briggs attempted to account for the effect
of roughness by increasing the sigmas over urban areas. The curves
are consistent with theoretically expected behavior at small and large
distances from the source. However, σ_z for open-country conditions
appears to level off to a constant value under stable conditions. In
a later section we discuss a new theory proposed by Pearson et al.
(1983) that might explain this anomaly.

Before we get into the topic of dispersion, it is useful to highlight
some relevant features of the structure of the SBL.

Table 5.2. Formulas recommended by Briggs (1973)

a. For $\sigma_y(x)$, m; $10^2 < x < 10^4$ m, open country conditions

Pasquill Type	σ_y (m)	σ_z (m)
A	$.22x(1+.0001x)^{-1/2}$	$.20x$
B	$.16x(1+.0001x)^{-1/2}$	$.12x$
C	$.11x(1+.0001x)^{-1/2}$	$.08x(1+.0002x)^{-1/2}$
D	$.08x(1+.0001x)^{-1/2}$	$.06x(1+.0015x)^{-1/2}$
E	$.06x(1+.0001x)^{-1/2}$	$.03x(1+.0003x)^{-1}$
F	$.04x(1+.0001x)^{-1/2}$	$.016x(1+.0003x)^{-1}$

b. For $\sigma_y(x)$, m and $\sigma_z(x)$, m; $10^2 < x < 10^4$ m, urban conditions

Pasquill Type	σ_y (m)	σ_z (m)
A–B	$.32x(1+.0004x)^{-1/2}$	$.24x(1+.001x)^{1/2}$
C	$.22x(1+.0004x)^{-1/2}$	$.20x$
D	$.16x(1+.0004x)^{-1/2}$	$.14x(1+.0003x)^{-1/2}$
E–F	$.11x(1+.0004x)^{-1/2}$	$.08x(1+.0015x)^{-1/2}$

5.3. The Structure of the SBL

As we saw in Ch. 1, turbulence in the SBL is produced by shear and destroyed by buoyancy and viscous dissipation. This competition between shear and buoyancy results in turbulence intensities that are typically much smaller than those found in the convective boundary layer (CBL). For example, vertical velocity fluctuations are typically of the order of tenths of meters per second. The SBL height h is also small relative to that of the CBL; a typical value is 100 m.

As discussed in Ch. 1, our understanding of the turbulent structure of the SBL is far from satisfactory, in part because it is much less likely to be found in a quasi-steady state than is the CBL; as a result, its structure tends to be constantly evolving in response to changing boundary conditions (Caughey et al., 1979). It is also particularly difficult to measure SBL structure. For example, measurements of turbulent velocities are invariably contaminated by gravity waves, and measurement accuracy is affected by low turbulence levels as well as intermittency (see Caughey and Readings, 1975; Nieuwstadt, 1984a).

Most of what we know about the SBL applies to the situation in which its turbulence is spatially and temporally continuous. The simplest such case is the steady one, which Brost and Wyngaard (1978) argued could be maintained by a constant cooling rate at the surface. In this situation an expression for the height h of the SBL is

$$h = 0.4 \, (u_* L / f)^{1/2} \, . \tag{5.1}$$

This form, first suggested on the basis of dimensional arguments by Zilitinkevich (1972), provides a reasonable estimate for h even when the SBL is not stationary (see Nieuwstadt, 1984b; Venkatram, 1980a). Nieuwstadt (1984b), in discussing rate expressions for h, acknowledged the usefulness of the diagnostic Eq. (5.1), although it does not have theoretical support in other than steady conditions.

A formula for the height of the SBL is valuable in applications, because pollutants released above h are not likely to cause ground-level concentrations. To be able to use Eq. (5.1), we need an estimate of the Monin-Obukhov length L which, in turn, depends on the surface virtual temperature flux $\overline{w\theta}_0$. Because measurements of $\overline{w\theta}_0$ are unlikely to be available at most sites, it is useful to have a method to estimate L from other meteorological measurements. One method is based on the observation that in the stable surface layer $\theta_*(\equiv -\overline{w\theta}_0/u_*)$ shows much smaller variation than either $\overline{w\theta}_0$ or u_*, because these two variables tend to decrease or increase together. Venkatram (1980a) found that θ_* varied slightly around 0.08°C during the Minnesota experiment (Kaimal et al., 1976). Van Ulden and Holtslag (1983) found a similar result in their analysis of data collected at Cabauw, The Netherlands. Their subsequent theoretical analysis showed that although the value of θ_* depended upon the surface energy balance, it varied little with u_*. This suggests that in the absence of more complete information it might be possible to use typical values of θ_* in the expression for the Monin-Obukhov length L. We show in Sec. 5.7 how this simplification also helps in the calculation of u_* and h.

Using an approach identical to that of Venkatram (1980a), Nieuwstadt (1984b) also derived an equation for h in terms of the wind speed at 10 m. He found that this equation provided acceptable estimates of the measurements of h made at Cabauw. In my opinion, therefore, there is sufficient empirical support for the use of a diagnostic equation for h.

Data collected at Minnesota (Caughey et al., 1979) and Cabauw (Nieuwstadt, 1984a) show that the variation of σ_w in the SBL can be represented by

$$\sigma_w \cong 1.3u_*(1 - z/h)^a , \qquad (5.2)$$

where the exponent a varies between 1/2 and 3/4. Equation (5.2) appears to apply only to the high-frequency component of σ_w. As discussed in Ch. 1, the low-frequency component, which includes gravity waves, can lead to an increase of σ_w with height (de Baas and Driedonks, 1985). Because turbulence and gravity waves can have substantially different effects on dispersion, it might be important to separate these two contributions when making measurements of σ_w. However, techniques to do so are still being developed. Therefore, for the present we should use measurements of σ_w if they are available. We discuss the role of σ_w in determining σ_z in a later section.

There is little agreement on the variation of σ_v, an important dispersion variable, in the SBL. As discussed in Ch. 1, surface-layer parameters such as u_* and L do not appear to control σ_v. On the other hand, because σ_v has a substantial low-frequency component, there is reason to believe that mesoscale flows and/or local topographical effects determine its magnitude. In the absence of measurements, a value of σ_v of the order of 1 m s^{-1} is appropriate for the estimation of σ_y.

With this background on the structure of the SBL, we discuss the theoretical basis of the models that have been developed to describe dispersion in the SBL.

5.4. The Theoretical Tools

The theoretical underpinning of most parameterizations of plume spread used in Gaussian models is G. I. Taylor's analysis (Taylor, 1921) of the motion of a fluid element in homogeneous turbulence. In order to follow Taylor's arguments, it is important to define a fluid element (see Hunt, 1982; Saffman, 1960). A fluid element is a control volume of characteristic dimension much larger than the molecular scale but much smaller than the Kolmogorov microscale $= (\nu^3/\epsilon)^{1/4}$, where ν is the kinematic viscosity and ϵ is the dissipation rate of turbulent kinetic energy. The motion of the centroid of this fluid element corresponds to the average over the molecules enclosed by the control volume. The size of the fluid element implies that it can be regarded as part of the fluid continuum, and its centroid responds to all scales of turbulent motion.

It is important to note that molecules can move in and out of the fluid element. Therefore, the concentration of the fluid element is not conserved unless the molecular diffusivity of the pollutant is zero. In describing pollutant dispersion, it is clear that we have to deal with the motion of molecules rather than that of fluid particles. However, most theoretical analyses (with notable exceptions such as Saffman, 1960) are restricted to the motion of fluid elements. Under neutral and unstable conditions, fluid elements (or particles) move away from the source on average, and their statistics of spread are not sensitive to molecular effects. Under stable conditions, buoyancy can restrict the vertical motion of fluid particles, and it becomes necessary to distinguish between molecules and fluid elements (see Hunt, 1982). We shall not make that distinction here, however.

We derive Taylor's result for plume spread using methods that I believe illustrate the physics better than the purely mathematical derivations given in standard textbooks. For simplicity, assume that the total displacement Y of a particle consists of four discrete steps:

$$Y = y_1 + y_2 + y_3 + y_4 , \qquad (5.3)$$

where each of the displacements y occurs in a small time interval Δt so that

$$y_1 = v_1\Delta t, \ y_2 = v_2\Delta t \ , \ \text{and so on.} \tag{5.4}$$

Squaring (5.3) we obtain

$$Y^2 = y_1^2 + y_2^2 + y_3^2 + y_4^2 + 2(y_1y_2 + y_3y_4 + y_2y_3) \\ + 2(y_1y_3 + y_2y_4) + 2y_1y_4 \ , \tag{5.5}$$

or

$$Y^2 = \sum_{i=1}^{4} y_i^2 + 2\sum_{i=1}^{3} y_iy_{i+1} + 2\sum_{i=1}^{2} y_iy_{i+2} + 2\sum_{i=1}^{1} y_iy_{i+3} \ . \tag{5.6}$$

It is now easy to see the pattern in the terms of Eq. (5.6). We can now generalize the equation for N displacements

$$Y^2 = \sum_{i=1}^{N} y_i^2 + 2\sum_{i=1}^{N-1} y_iy_{i+1} + 2\sum_{i=1}^{N-2} y_iy_{i+2} + \cdots + 2\sum_{i=1}^{1} y_iy_{i+N-1} \ . \tag{5.7}$$

If we substitute Eq. (5.4) in Eq. (5.7) and average over an ensemble of particles, we get

$$\langle Y^2 \rangle = \Delta t^2 \sum_{i=1}^{N} \langle v_i^2 \rangle + 2\Delta t^2 \sum_{i=1}^{N-1} \langle v_iv_{i+1} \rangle + 2\Delta t^2 \sum_{i=1}^{N-2} \langle v_iv_{i+2} \rangle + \cdots \\ + 2\Delta t^2 \sum_{i=1}^{1} \langle v_iv_{i+N-1} \rangle \ . \tag{5.8}$$

In homogeneous turbulence the statistics of the velocity v are independent of position and the autocorrelation terms depend only upon the lag times, so that we can write

$$\langle v_i^2 \rangle = \sigma_v^2 \tag{5.9}$$

and

$$\langle v_iv_{i+n} \rangle \equiv \sigma_v^2 \, R_L^y(n\Delta t) \ , \tag{5.10}$$

where $R_L^y(t)$ is the Lagrangian autocorrelation function.
   With the definitions

$$n\Delta t \equiv t \quad \text{and} \quad N\Delta t \equiv T \tag{5.11}$$

we can rewrite Eq. (5.8) as

$$\langle Y^2 \rangle = \Delta t T \sigma_v^2 + 2\sigma_v^2 \sum_{n=1}^{N} R_L^y(t)(T-t)\Delta t . \qquad (5.12)$$

In the limit $\Delta t \to 0$, the first term on the right-hand side disappears, the sum becomes an integral, and we obtain

$$\langle Y^2 \rangle = 2\sigma_v^2 \int_0^T (T-t)R_L^y(t)dt . \qquad (5.13)$$

The textbook derivation (Csanady, 1973), which is much shorter, uses mathematical manipulations that are not always clear to a student.

Note that $R_L^y(t)$ measures the particle's ability to remember its velocity between 0 and t. At small times, $R_L^y(t)$ is close to unity; at large times, we expect the particle velocities to be uncorrelated with the initial velocity so that $R_L^y \to 0$ as t becomes large. The terms "small" and "large" times become meaningful only in the context of the Lagrangian time scale defined by

$$T_L^y = \int_0^\infty R_L^y(t)dt . \qquad (5.14)$$

Then

$$R_L^y(t) \to 1 , \quad t \ll T_L^y \qquad (5.15a)$$

$$R_L^y(t) \to 0 , \quad t \gg T_L^y . \qquad (5.15b)$$

Now we are in a position to look at the asymptotic behavior of $\langle Y^2 \rangle$ in Eq. (5.13). For small t we can take $R_L^y = 1$ so that

$$\langle Y^2 \rangle \cong 2\sigma_v^2 \int_0^T (T-t)dt \qquad (5.16a)$$

$$= \sigma_v^2 T^2 . \qquad (5.16b)$$

To see what happens at large T, let us rewrite Eq. (5.13) as follows:

$$\langle Y^2 \rangle = 2\sigma_v^2 T \int\limits_0^T (1 - \frac{t}{T}) R_L^y(t) dt \; . \qquad (5.17)$$

When T is large, the term (t/T) is close to zero at finite values of R_L^y so that

$$\langle Y^2 \rangle \cong 2\sigma_v^2 T \int\limits_0^\infty R_L^y(t) dt \qquad (5.18a)$$

$$= 2\sigma_v^2 \; T \; T_L^y \; . \qquad (5.18b)$$

It is customary to write $\langle Y^2 \rangle$ as σ_y^2 and the corresponding variance in the z direction as σ_z^2. See Appendix A for further discussion on the limits of Taylor's equation.

The results derived thus far are strictly applicable only to homogeneous flows, but they are more generally useful in providing guidance for theories of dispersion in inhomogeneous flows. Like so many other turbulence theories, they are semi-empirical in the sense that they rely on observations to complete them.

A useful result for describing vertical dispersion in the surface layer can be derived by restating Eq. (5.18) in terms of the z direction:

$$\frac{d\sigma_z^2}{dt} = 2K_z \; , \qquad (5.19)$$

where the eddy diffusivity K_z is defined by the equation

$$K_z \equiv \sigma_w^2 T_L^z \; . \qquad (5.20)$$

This equation, applicable at large times $(t >> T_L^z)$, would not be particularly useful if it were only a definition. However, we can show that K_z behaves like the eddy diffusivity that appears in the diffusion equation. To make this connection, we use the observation that the concentration distribution downwind of a source in homogeneous turbulence has a Gaussian shape. Then, for a continuous release Q, the concentration C is given by (Csanady, 1973)

$$C(x,y,z) = \frac{Q}{2\pi U \sigma_y \sigma_z} \; \exp\left(-\frac{y^2}{2\sigma_y^2}\right) \exp\left(-\frac{z^2}{2\sigma_z^2}\right) \; , \qquad (5.21)$$

where the point of release is $(0, 0, 0)$. It can be easily shown that this $C(x,y,z)$ is a solution of the diffusion equation (Csanady, 1973)

$$U\frac{\partial C}{\partial x} = K_y\frac{\partial^2 C}{\partial y^2} + K_z\frac{\partial^2 C}{\partial z^2} , \qquad (5.22)$$

where the diffusion coefficients are defined by

$$K_y = \frac{U\,d\sigma_y^2}{2\,dx} ; K_z = \frac{U\,d\sigma_z^2}{2\,dx} . \qquad (5.23)$$

This is the same form as Eq. (5.19) if we notice that $t = x/U$. The diffusion Eq. (5.22) is a statement of mass conservation in which the fluxes in y and z have been replaced by the flux-gradient relationships

$$F_y = -K_y\frac{\partial C}{\partial y} ; F_z = -K_z\frac{\partial C}{\partial z} . \qquad (5.24)$$

What can we conclude from all these arguments? We have shown that

1) under special conditions ($t \gg T_L^{\tilde{z}}$), dispersion can be described in terms of an eddy diffusivity that depends upon flow properties, and
2) this eddy diffusivity satisfies the flux-gradient relationship given by Eq. (5.24).

This suggests that, under these conditions, the concentration field can be obtained from the diffusion equation in which the eddy diffusivity is independent of the travel time from the source. In general, however, it does depend on the turbulence properties of the flow.

Another description of the region of validity of eddy diffusivity is given by $\sigma_z \gg \sigma_w T_L^{\tilde{z}}$. With $T_L^{\tilde{z}} \sim \ell_w/\sigma_w$, where ℓ_w is an appropriate length scale of turbulence, this tells us that the concept of eddy diffusivity is useful when the scale of the plume, σ_z, is much greater than the scale of turbulence, ℓ_w. An equivalent statement is that Fickian diffusion (associated with eddy diffusivity) exists when the scale of concentration variation is much larger than the turbulence scale.

Eq. (5.23) is a general relationship that assumes nothing about the form of σ_y and σ_z. Note that in general

$$\frac{d\sigma_z^2}{dt} = 2\sigma_w^2\int_0^t R_L^{\tilde{z}}(\zeta)d\zeta = 2K_z(t) . \qquad (5.25)$$

Therefore, the effective K_z depends upon the travel time t from the source. This means that K_z at the same point in space will have different values for two plumes from sources that are displaced along the direction of the wind; i.e., it depends on the geometry of the source distribution. Thus, the eddy diffusivity for the pollutant field emanating from a line source could differ from that for temperature and water vapor diffusing in the same flow, since the latter have an effectively infinite area source at the surface. In general, then, the eddy diffusivity for point or line source diffusion differs from that traditionally used by micrometeorologists to describe heat, momentum, and moisture transfer in the boundary layer, although in many applications the difference is not great.

This points out the weakness of the concept of eddy diffusivity to describe dispersion. In spite of this, the eddy diffusivity is useful because it allows us to extend our understanding of the transport of heat and momentum to that of matter. As a first guess we can assume that K_z is equal to that for heat or momentum, and then test this assumption by comparing the predicted concentration distribution with the observed.

5.5. Dispersion Models

5.5.1. Surface Releases

There has been much progress in the prediction of dispersion in the surface layer. This might seem strange, given that our theoretical understanding of dispersion applies only to homogeneous flows; however, although mean profiles vary near the surface, many turbulence statistics are relatively uniform there. Furthermore, a number of carefully planned experiments (Project Prairie Grass, Greenglow, Hanford-30) have allowed workers to develop working models for dispersion. See Horst et al. (1979) for an analysis of the data collected during these experiments.

Recent models of dispersion near the surface have attempted to exploit our understanding of the micrometeorology of the surface layer (Ch. 1). Because the vertical spread of a surface release is controlled by eddies of roughly the same size as the plume, the eddy diffusivity Eq. (5.19) is a good starting point:

$$\frac{d\sigma_z^2}{dt} = 2K_z \; . \tag{5.26}$$

In order to make progress we assume that K_z is similar to K_h or K_m so that (Businger, 1973)

$$K_z = ku_*z/(1 + \beta z/L) \,, \tag{5.27}$$

where $\beta = 4.7$. This equation depends upon z, which suggests that we should evaluate it at some height representative of σ_z. It is reasonable to assume that $z \sim \sigma_z$, which is consistent with the notion that plume spread in the surface layer is governed by eddies whose length scales are proportional to z. Then

$$\sigma_z \frac{d\sigma_z}{dt} \sim K_h(\sigma_z/L) \,. \tag{5.28}$$

Rather than integrate Eq. (5.28), let us consider the asymptotic forms for small and large σ_z. For small σ_z, $K_h \sim u_*\sigma_z$ so that

$$\sigma_z \sim u_*t \,. \tag{5.29}$$

What is the travel time t? Let us write the formal expression

$$\sigma_z \sim u_*x/U_e \,, \tag{5.30}$$

where U_e is some effective velocity. If we assume that U_e is also associated with the dilution of the plume, then

$$\overline{C}^y (x,0) \sim \frac{Q}{\sigma_z U_e} \sim \frac{Q}{u_*x} \,, \tag{5.31}$$

which shows that the crosswind-integrated concentration falls off as x^{-1}. Although the arguments leading up to (5.31) are not rigorous, the result is essentially correct.

Because the largest eddies in the surface layer scale with L (Ch. 1), one expects $K_z \sim u_*L$ when $\sigma_z > L$, and we can write

$$\sigma_z \frac{d\sigma_z}{dt} \sim u_*L \tag{5.32a}$$

or

$$U_t \frac{d\sigma_z^2}{dx} \sim u_*L \,, \tag{5.32b}$$

where U_t is a transport velocity. It is reasonable to assume that U_t corresponds to the wind evaluated at some $z = \alpha\sigma_z$, where α

is a constant. When $z \gg L, U(z) \sim u_* z/L$ so that $U_t \sim u_* \sigma_z/L$. Substituting this expression for U_t in Eq. (5.32b) and integrating, we find

$$\sigma_z \sim L^{2/3} x^{1/3} . \tag{5.33}$$

Now $\overline{C}^y(x, 0)$ is

$$\overline{C}^y = \frac{Q}{U_t \sigma_z} \sim \frac{1}{u_* L^{1/3} x^{2/3}} . \tag{5.34}$$

Notice that although σ_z grows as $x^{1/3}$, $\overline{C}^y$ falls off at the faster rate of $x^{2/3}$ because of the increase of U_t with σ_z. This also shows that the conventional use of a wind speed measured at an arbitrary height is bound to lead to errors in the estimation of concentrations from ground-level sources.

Equations 5.31 and 5.34 can be rewritten by defining $\overline{C} = \overline{C}^y u_* L/Q$ and $\overline{x} = x/L$,

$$\overline{C} \sim \overline{x}^{-1} \quad , \overline{x} \ll 1 \tag{5.35a}$$

$$\overline{C} \sim \overline{x}^{-2/3} , \overline{x} \gg 1 . \tag{5.35b}$$

The implied constants of proportionality in Eq. (5.35) can be obtained by fitting them to observations. This was done by Venkatram (1982), who obtained

$$\overline{C} = \overline{x}^{-1} \quad ; \overline{x} \leq 1.4 \tag{5.36a}$$

$$\overline{C} = 0.89 \overline{x}^{-2/3} ; \overline{x} > 1.4 . \tag{5.36b}$$

Figure 5.1 shows the fit obtained with Eq. (5.36); 98% of the observations (Prairie Grass) lie within a factor of 2 of the model predictions.

An empirical equation that does an acceptable job of explaining the variation of $\overline{C}^y$ over the range of observed $\overline{x}$ is

$$\overline{C} = 1.25 \overline{x}^{-0.83} \tag{5.37}$$

or

$$\overline{C}^y = 0.38 Q/u_*^{1.34} x^{0.83} . \tag{5.38}$$

Note again that the use of Eq. (5.38) requires only an estimate of u_*, which can be obtained from a measurement of wind speed at a single level and information on the roughness length (see Venkatram, 1980a).

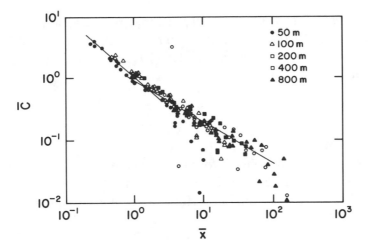

Figure 5.1. Plot of $\overline{C} = \overline{C}^y u_* L/Q$ against $\bar{x} \equiv x/L$ for Prairie Grass data presented by Horst et al. (1979). The intersecting straight lines correspond to $\overline{C} \sim \bar{x}^{-1}$ and $\overline{C} \sim \bar{x}^{-2/3}$.

The centerline ground-level concentration C is

$$C = \overline{C}^y/\sqrt{2\pi}\sigma_y . \qquad (5.39)$$

Because the behavior of σ_v under stable conditions is sensitive to local topography and mesoscale turbulence, and to averaging time, it is risky to estimate σ_y with micrometeorological variables such as u_* or L. More analysis of existing observations is required before we can recommend "typical" values of σ_v under different conditions. If near-ground-level measurements of σ_v are available, σ_y can be estimated from

$$\sigma_y = \sigma_v t/(1 + x/x_0)^{1/2} , \qquad (5.40)$$

where $x_0 = 10^4$ m according to Briggs (1973). Evidently, however, there is little theoretical or empirical basis for this choice of x_0 (G. Briggs, personal communication). The travel time t is x/U, where U presumably corresponds to the same height as σ_v. We can write $\sigma_v/U \cong \sigma_\theta$ for small σ_θ; this approximation is useful because σ_θ can be measured directly with bivanes. Note that $\tan \sigma_\theta \neq \sigma_v/U$ when σ_θ is large. In the absence of site-specific data, the formula $\sigma_y = 0.1x$ should be adequate for rough estimates.

Our analysis of concentrations from ground-level releases provides no information on the vertical distribution. Our use of local

K_z to model the growth of σ_z suggests that this vertical distribution can be obtained from a diffusion equation. This is what van Ulden (1978) and Nieuwstadt and van Ulden (1978) assumed in solving the equation

$$U\frac{\partial \overline{C}^y}{\partial x} = \frac{\partial}{\partial z}\left(K_z \frac{\partial \overline{C}^y}{\partial z}\right), \qquad (5.41)$$

where $\overline{C}^y$ is the crosswind integrated concentration, U is described by the similarity profile (Businger, 1973), and K_z is taken to be K_h. Van Ulden (1978) obtained a useful analytical approximation to the solution of Eq. (5.41),

$$\overline{C}^y(x,z) = \overline{C}^y(x,0) \exp\left[-(0.66z/\overline{z})^{1.5}\right]. \qquad (5.42)$$

Here $\overline{C}^y(x,0)$ is given by

$$\overline{C}^y(x,0) = 0.73/\overline{U}\,\overline{z}. \qquad (5.43)$$

The mean particle height $\overline{z}$ and its associated velocity $\overline{U}$ can be easily obtained from implicit equations that involve u_*, L, and z_0. One possible short-cut is to use Eq. (5.38) rather than Eq. (5.43). Then $\overline{z}$, the only unknown, can be obtained from the implicit equation

$$\overline{z}[\ln(0.6z/z_0) + 4.7\overline{z}/L] = 0.92u_*^{0.34}x^{0.83}. \qquad (5.44)$$

This concludes the section on surface releases. We will now treat models for dispersion of releases in the upper part of the stable boundary layer.

5.5.2. Elevated Releases

The main difficulty in estimating dispersion in the SBL is vertical inhomogeneity of the mean and turbulent velocities. Before suggesting methods for dealing with this problem, let us first consider plumes whose vertical extent is small compared with the scales of the velocity and temperature profiles. Although these plumes do not cause high ground-level concentrations in flat terrain, they are important because they can impact on the sides of hills.

Hunt (1982) pointed out that the motion of fluid elements in stable flows has features that have to be given special consideration. A fluid element that does not exchange mass or momentum with its surroundings can travel through a vertical distance $\sim \sigma_w/N$ only if its initial velocity is σ_w, where N is the Brunt-Väisälä frequency. The autocorrelation function for such a fluid element will have negative loops corresponding to this oscillatory motion. Csanady (1973) and Hunt (1982) showed that this can lead to a Lagrangian time scale T_L that is zero; therefore, σ_z will stop growing after it reaches a maximum of σ_w/N. Pearson et al. (1983) developed a theory that indicates that the subsequent molecular diffusion leads to very slow growth beyond this maximum value. Because the observational evidence for this behavior is not conclusive, we assume that it is not necessary to account explicitly for molecular effects.

We base our model for vertical dispersion on the assumption that the relevant length scale ℓ_s is proportional to σ_w/N (see Ch. 1) so that

$$\ell_s = \gamma^2 \sigma_w/N , \qquad (5.45)$$

where γ is the proportionality constant that appears in the theory of Pearson et al. (1983). Unlike them, we assume that γ is independent of molecular diffusion. To determine its value, we apply Eq. (5.45) to the surface layer in the limit of very stable stratification. Note that ℓ_s is related to the eddy diffusivity K_z and the Lagrangian time scale T_L^z through

$$K_z \equiv \sigma_w \ell_s = \gamma^2 \sigma_w^2/N ; \quad \ell_s = \sigma_w T_L^z . \qquad (5.46)$$

To derive a surface-layer eddy diffusivity K_z in terms of σ_w and N, we start with the expression for the mean temperature gradient during very stable conditions (Businger, 1973):

$$\frac{d\Theta}{dz} = \frac{\beta \theta_*}{kL} . \qquad (5.47)$$

If we substitute Eq. (5.47) into the identity $\overline{w\theta_0} \equiv -\theta_* u_*$ we find

$$\overline{w\theta_0} = -\frac{k}{\beta} L u_* \frac{d\Theta}{dz} \equiv -K_h \frac{d\Theta}{dz} . \qquad (5.48)$$

Notice from Eq. (5.48) that the eddy diffusivity for heat K_h is

$$K_h \equiv \frac{k}{\beta} L u_* .$$ (5.49)

If we substitute for $\overline{w\theta}_0$ from Eq. (5.48) into the definition for the Monin-Obukhov length and solve for L, noting that $N^2 = g\frac{d\Theta}{dz}/T_0$, we find

$$L = \frac{\sqrt{\beta}}{k} \frac{u_*}{N} .$$ (5.50)

If we now put Eq. (5.50) for L into Eq. (5.49) and use the relationship $\sigma_w = bu_*$ in the surface layer, where b is a constant, the expression for K_h becomes

$$K_h = \frac{1}{\sqrt{\beta b^2}} \frac{\sigma_w^2}{N} .$$ (5.51)

We now have written K_h for the surface layer in terms of σ_w and N. Comparing Eq. (5.46) with Eq. (5.51) yields

$$\gamma^2 = \frac{1}{\sqrt{\beta b^2}} .$$ (5.52)

With $\beta = 4.7$ and $b = 1.3$ (Businger, 1973), γ is 0.52, and the length scale ℓ_s for stable conditions becomes

$$\ell_s = \gamma^2 \sigma_w / N ; \gamma = 0.52 .$$ (5.53)

When N is small, ℓ_s can become large enough to require consideration of the effect of the earth's surface on the length scale. To account for this, let us consider the expression for the neutral length scale ℓ_n in the absence of stratification:

$$\ell_n = \alpha z_r ,$$ (5.54)

where z_r is the height of release. To estimate the constant α, we notice that K_h for neutral stratification is

$$K_h = \sigma_w kz / [b\phi_h(0)] ,$$ (5.55)

which implies that

$$\ell_n = kz/[b\phi_h(0)] \tag{5.56}$$

or

$$\alpha = k/b\phi_h(0) . \tag{5.57}$$

If we take $k = 0.35, b = 1.3$, and $\phi_h(0) = 0.74, \alpha$ is 0.36.

We now write a formula for ℓ that interpolates between the limits ℓ_n and ℓ_s:

$$1/\ell = 1/\ell_n + 1/\ell_s . \tag{5.58}$$

As discussed in Ch. 1 this expression, or one much like it, has been used by other investigators (see Delage, 1974; Brost and Wyngaard, 1978; Hunt et al., 1983, for example). Our formula for σ_z can now be written as

$$\sigma_z = \sigma_w t/(1 + t/2T_L)^{1/2} , \tag{5.59a}$$

where

$$T_L = \ell/\sigma_w . \tag{5.59b}$$

Notice that Eq. (5.59a) is a convenient way of interpolating between the linear and square-root growth rates of σ_z.

We now turn to the performance of this model for σ_z. Figure 5.2 shows the predicted curve $\sigma_z/\sigma_w t$ versus t/T_L together with observations of σ_z made during an experiment conducted at Cinder Cone Butte, Idaho (Lavery et al., 1982) in 1980 under the auspices of the USEPA Complex Terrain Model Development Study. Details of the methods used to derive σ_z are given in Venkatram et al. (1984). The variables constituting T_L were evaluated at the release height z_r. The model provides a good description of the observations, but at large t/T_L there is some indication that the growth of σ_z is smaller than $t^{1/2}$; this suggests that the theory of Pearson et al. (1983) will have to be examined more carefully. For the time being, I would recommend the use of Eq. (5.59) with Eq. (5.58).

The lateral spread σ_y can be represented as

$$\sigma_y = \sigma_v t/(1 + t/2T_L^y)^{1/2} . \tag{5.60}$$

We know little about estimating T_L^y (or, equivalently, ℓ_v/σ_v, where ℓ_v is a length scale). Both ℓ_v and σ_v are controlled by mesoscale variables which, in most cases, are unrelated to the micrometeorology of the site (see Ch. 1). For the present, we have to rely on empirical estimates of T_L^y. Draxler (1976) suggested a value of 500 s. One has

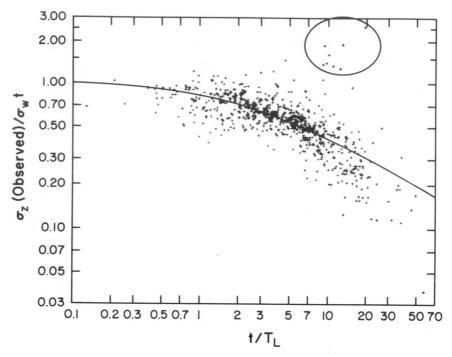

Figure 5.2. Variation of $\sigma_z/\sigma_w t$ with t/T_L. Circled points are outliers.

to be careful about using a constant value because T_L^y depends upon ℓ_v as well as σ_v. If we draw guidance from studies on dispersion in the convective boundary layer (Venkatram, 1980b), it is not unreasonable to assume that T_L^y is very large so that

$$\sigma_y = \sigma_v t \ . \tag{5.61}$$

One might want to use Eq. (5.60) with $T_L^y = 500$ s for regulatory applications that require conservatism in the estimation of concentrations.

The crosswind shear in the SBL can also affect σ_y. In Appendix B we show that the shear contribution to σ_y is given by

$$\sigma_{ys}^2 = \alpha^2 S^2 \sigma_z^2 t^2 \ , \tag{5.62}$$

where α is a constant, and S is the constant shear (dV/dz) of the cross-wind velocity. Note that when S is large during stable conditions, σ_z is likely to be small. This suggests that the effects of shear on σ_y are not likely to be significant until t becomes large enough.

A formula for σ_{ys} often used in practice is based on Smith's (1965) statistical analysis:

$$\sigma_{ys}^2 = \frac{1}{6} S^2 \sigma_w^2 T_L^z t^3 , \qquad (5.63)$$

which can be rewritten for large t as

$$\sigma_{ys}^2 = \frac{1}{12} S^2 \sigma_z^2 t^2 . \qquad (5.64)$$

If θ is the wind direction, it is easy to see that

$$\frac{d\theta}{dz} \cong \frac{1}{U} \frac{dV}{dz} = \frac{1}{U} S . \qquad (5.65)$$

Substituting Eq. (5.65) into Eq. (5.64), we find that

$$\sigma_{ys}^2 = \frac{1}{12} U^2 \left(\frac{d\theta}{dz}\right)^2 \sigma_z^2 t^2 . \qquad (5.66)$$

Writing $t = x/U$ and $\Delta\theta \equiv \frac{d\theta}{dz} \sigma_z$, this becomes

$$\sigma_{ys}^2 = \frac{1}{12} (\Delta\theta)^2 x^2 ,$$

the form often used in practice. The total σ_y is obtained by combining the "turbulence" σ_{yt} with σ_{ys} to obtain an effective σ_y as follows:

$$\sigma_y^2 = \sigma_{yt}^2 + \sigma_{ys}^2 . \qquad (5.67)$$

For problems involving buoyant plumes, it is necessary to add another component corresponding to buoyancy-generated turbulence. This component is generally assumed to be proportional to the plume rise, which is given by (Briggs, 1975)

$$\Delta h = 2.6(F/UN^2) , \qquad (5.68)$$

where F is the buoyancy parameter of the stack in question. Then the buoyancy-related spread σ_{yb} is

$$\sigma_{yb} = \frac{\beta}{\sqrt{2}} \Delta h , \qquad (5.69)$$

where β is the entrainment coefficient, which is usually taken to be $\cong 0.6$. This spread also contributes to σ_z so that

$$\sigma_z^2 = \sigma_{zb}^2 + \sigma_{zt}^2 , \qquad (5.70)$$

where $\sigma_{zb} = \sigma_{yb}$.

The discussion thus far has dealt with cases in which σ_z is small enough that meteorological variables can be evaluated at effective source height. How do we estimate ground-level concentrations caused by elevated releases when the plume has to disperse through a highly inhomogeneous layer? The next subsection provides some suggestions.

5.5.3. Dispersion in a Shear-Dominated Boundary Layer

Elevated releases can cause high ground-level concentrations even during stable conditions if the winds are high enough (Venkatram and Paine, 1985). Under these conditions, the shear-generated turbulence can be vigorous enough to bring the plume down close to the stack. To model such a plume it is necessary to account for the vertical variation of the dispersion variables such as σ_w and U; one approximate way of doing this is to use averages of these variables over the layer of interest. However, this procedure, apart from being without theoretical support, gives rise to questions on the depth of the region of the SBL that needs to be averaged. Because σ_z grows with x, this averaging layer has to be also a function of x. This method, in my opinion, is bound to be cumbersome; I think a more acceptable model can be built on the basis of some ideas suggested by Moore (1975).

Figure 5.3 shows the plume behavior Moore assumed in deriving his equation for the ground-level concentration $C(x,y)$. The upper and lower parts of the plume are associated with σ_{zu} and $\sigma_{z\ell}$, respectively. Their growth rates are taken to reflect the different levels of turbulence encountered by these plume segments. The maximum concentration in the plume is assumed to remain at a constant level h_e, the effective stack height. The expression for $C(x,y)$ then follows immediately:

$$C(x,y) = \frac{Q}{\pi U_T \sigma_y \bar{\sigma}_z} \exp\left(-\frac{h_e^2}{2\sigma_{z\ell}^2}\right) \exp\left(-\frac{y^2}{2\sigma_y^2}\right) , \qquad (5.71)$$

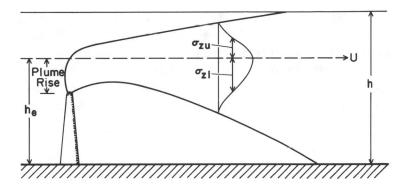

Figure 5.3. Simplified picture of vertical dispersion in the neutral boundary layer.

where

$$\bar{\sigma}_z = (\sigma_{zu} + \sigma_{zl})/2 \tag{5.72}$$

and U_T is the average velocity between the surface and h_e. For a power-law profile,

$$U_T = \frac{U(h_e)}{1 + p}, \tag{5.73}$$

where p is the exponent of the profile. Notice from Eq. (5.71) that the centerline concentration is diluted by an average σ_z whereas the ground-level concentration is governed by the vertical spread of the lower plume σ_{zl}.

In Moore's formulation, the sigmas were related to eddy diffusivities averaged over the upper and lower parts of the SBL. Because the specification of eddy diffusivity is an uncertain exercise, Venkatram and Paine (1985) used the following equations to model sigmas:

$$\frac{d\sigma_{zu}}{dt} = \sigma_w(z = h_e + \sigma_{zu}) \tag{5.74a}$$

$$\frac{d\sigma_{zl}}{dt} = \sigma_w(z = h_e - \sigma_{zl}). \tag{5.74b}$$

The growth rates depend upon σ_w evaluated at the "edges" of the plumes. In the SBL, in which σ_w decreases with height, the rate of increase of σ_{zu} slows down with travel time. On the other hand, the growth rate of σ_{zl} increases as the plume descends into regions of higher turbulence.

Venkatram and Paine (1985) substituted Eq. (5.2) for σ_w into Eq. (5.74) and obtained analytical expressions for the sigmas. The model of Eq. (5.71) was then used to estimate concentrations around two power plants situated in the midwestern U.S. As can be seen from Fig. 5.4, the model performed reasonably well, and it does represent a significant improvement over the EPA-recommended CRSTER model.

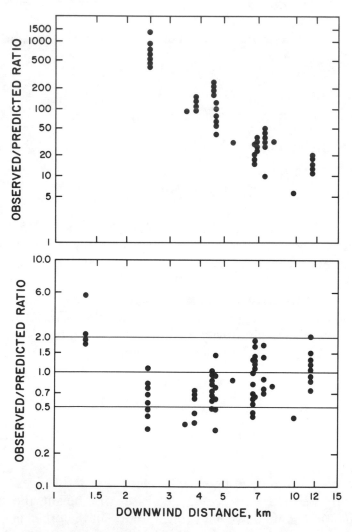

Figure 5.4. Observed/predicted concentration ratio vs. downwind distance (km). Upper panel, CRSTER; lower panel, high wind model.

A shortcoming of Eq. (5.74) is that it does not incorporate the dispersion time scales that are likely to become important at long travel times. One way of improving this formulation is to express the growth laws by

$$\frac{d\sigma_z}{dx}\left(1 + \frac{\sigma_z}{\ell}\right) = \frac{\sigma_w}{U} .$$

(5.75)

This form suggested by Moore (personal communication) satisfies the theoretically expected linear and square-root asymptotic growth rates. The length scale ℓ in Eq. (5.75) should be that given in Eq. (5.58).

We see that our experience with modeling dispersion of elevated releases in the SBL is very limited. Much more work needs to be done before we can recommend an appropriate model.

5.6. Computation of Micrometeorological Variables

In a recent paper, van Ulden and Holtslag (1985) provided an excellent account of practical methods of estimating the micrometeorological variables required by modern dispersion models. Rather than describing these methods here, I shall show how one can estimate the required inputs from the type of information used in the PGT classification. We assume that the only information available to us is

1) wind speed U_r at z_r,
2) temperature Θ_r at z_r,
3) an estimate of the surface roughness z_0,
4) an estimate of the cloud cover.

We begin our analysis by using the following empirical equation for θ_* (van Ulden and Holtslag, 1985):

$$\theta_* = 0.09 \, (1 - 0.5C^2) ,$$

(5.76)

where C is the fractional cloud cover. As pointed out earlier, this empirical equation for θ_* will vary from site to site. This then allows us to write the Monin-Obukhov length L as

$$L = Au_*^2 ,$$

(5.77a)

where

$$A = \frac{T_0}{g} \frac{1}{\theta_* k} .$$

(5.77b)

Substituting Eq. (5.77a) in the surface layer similarity profile (Businger, 1973) we obtain the following formula for u_* (Venkatram, 1980a):

$$u_* = C_{DN}U_r\left\{\frac{1}{2} + \frac{1}{2}\left[1 - \left(\frac{2U_0}{C_{DN}^{1/2}U_r}\right)^2\right]^{1/2}\right\} ,$$ (5.78)

where

$$C_{DN} = k/\ln(z_r/z_0)$$ (5.79a)

and

$$U_0^2 = \beta z_r/kA .$$ (5.79b)

This estimate of u_* allows us to estimate the SBL height h from

$$h = 0.4\left(\frac{u_*L}{f}\right)^{1/2}$$ (5.80a)

$$= 0.4\left(\frac{Au_*^3}{f}\right)^{1/2} ,$$ (5.80b)

where f is the Coriolis parameter. We can now estimate the profile of σ_w in the SBL from (Nieuwstadt, 1984a)

$$\sigma_w = 1.3u_*(1 - z/h)^{3/4} .$$ (5.81)

We are also in a position to calculate the velocity profile (van Ulden and Holtslag, 1985) from

$$U(z) = \frac{u_*}{k}\left[\ln\left(\frac{z}{z_0}\right) - \psi_M(z/L) + \psi_M\left(\frac{z_0}{L}\right)\right] ,$$ (5.82a)

where

$$\psi_M(z/L) = -17[1 - \exp(-0.29z/L)] .$$ (5.82b)

Holtslag (1984) found that Eq. (5.82) provided good estimates of the wind speed up to heights of 200 m even though h was below 200 m.

The turning of the wind with height can be estimated from an empirical equation suggested by van Ulden and Holtslag (1985):

$$D(z)/D(h) = d_1[1 - \exp(-d_2z/h)] ,$$ (5.83)

where $d_1 = 1.58$ and $d_2 = 1.0$. Here $D(z)$ and $D(h)$ are turning angles in degrees. $D(h)$ can be taken to be $\cong 35°$ as a typical value.

We next estimate the temperature profile within the SBL. To do this, we use Wetzel's (1982) results, which indicate that the temperature profile is linear through the bulk of the SBL. This suggests the applicability of the log-linear similarity profile for heights up to $z = h$. This means that we can calculate $\Theta(z)$ from

$$\phi_h = \frac{d\Theta}{dz} \frac{kz}{\theta_*} = 0.74 + 4.7 \frac{z}{L} . \qquad (5.84)$$

Integrating, we find

$$\frac{\Theta(z) - \Theta_r}{\theta_*} = \frac{0.74}{k} \ln \left(\frac{z}{z_r}\right) + \frac{4.7}{k}(z - z_r) . \qquad (5.85)$$

Note that θ_* is obtained from Eq. (5.76). Above $z = h$, we can follow Wetzel's suggestion and use a quadratic function to interpolate between $\Theta(h)$ and Θ_a, the temperature of the mixed layer at the end of the previous day.

We have shown how we can use rather limited information to derive the micrometeorological variables relevant to dispersion. This should serve to blunt the criticism often directed at more advanced models: that they require inputs that are not usually available. This is clearly not true. In fact, our knowledge of SBL physics permits us to make the best use of the information we have, including that available for the PG classification system.

References

Briggs, G. A., 1973: Diffusion estimation for small emissions. 1973 Annual Report, Air Resources Atmos. Turb. and Diffusion Lab., Environmental Res. Lab., Report ATDL-106, USDOC-NOAA.

Briggs, G. A., 1975: Plume rise predictions. *Lectures on Air Pollution and Environmental Impact Analysis*, Amer. Meteor. Soc., Boston, 59–111.

Brost, R. A., and J. C. Wyngaard, 1978: A model study of the stably stratified planetary boundary layer. *J. Atmos. Sci.*, **35**, 1427–1440.

Businger, J. A., 1973: Turbulent transfer in the atmospheric surface layer. *Workshop on Micrometeorology*, D. A. Haugen, Ed., Amer. Meteor. Soc., Boston, 67–100.

Caughey, S. J., and C. J. Readings, 1975: An observation of waves and turbulence in the earth's boundary layer. *Bound.-Layer Meteor.*, **9**, 279–296.

Caughey, S. J., J. C. Wyngaard, and J. C. Kaimal, 1979: Turbulence in the evolving stable boundary layer. *J. Atmos. Sci.*, **36**, 1041–1052.

Csanady, G. T., 1973: *Turbulent Diffusion in the Environment*. Reidel, Dordrecht, 248 pp.

de Baas, A. F., and A. G. M. Driedonks, 1985: Internal gravity waves in a stably stratified boundary layer. *Bound.-Layer Meteor.*, **31**, 303–323.

Delage, Y., 1974: A numerical study of the nocturnal atmospheric boundary layer. *Quart. J. Roy. Meteor. Soc.*, **100**, 351–364.

Draxler, R. R., 1976: Determination of atmospheric diffusion parameters. *Atmos. Environ.*, **10**, 99–105.

Holtslag, A. A. M., 1984: Estimates of diabatic wind speed profiles from near surface weather observations. *Bound.-Layer Meteor.*, **29**, 225–350.

Horst, T. W., J. C. Doran, and P. W. Nickola, 1979: Evaluation of empirical atmospheric diffusion data. Battelle Pacific Northwest Laboratories, Richland, WA, NRC Fin. No. B2086, 137 pp.

Hunt, J. C. R., 1982: Diffusion in the stable boundary layer. *Atmospheric Turbulence and Air Pollution Modelling*, F. T. M. Nieuwstadt and H. van Dop, Eds., Reidel, Dordrecht, 231–274.

Hunt, J. C. R., J. C. Kaimal, J. E. Gaynor, and A. Korrell, 1983: Observations of turbulence structure in stable layers at the Boulder Atmospheric Laboratory. Studies of Nocturnal Stable Layers at BAO, Report No. 4, January, 1983, 1–52. Available from NOAA/ERL, Boulder, CO.

Kaimal, J. C., J. C. Wyngaard, D. A. Haugen, O. R. Coté, Y. Izumi, S. J. Caughey, and C. J. Readings, 1976: Turbulence structure in the convective boundary layer. *J. Atmos. Sci.*, **33**, 2152–2169.

Lavery, T. F., A. Bass, D. G. Strimaitis, A. Venkatram, B. R. Green, P. J. Drivas, and B. A. Egan, 1982: EPA Complex Terrain Model Development: First Milestone Report—1981. EPA-600/3-82-036, Research Triangle Park, NC, 304 pp.

Moore, D. J., 1975: A simple boundary layer model for predicting time mean ground-level concentrations of material emitted from tall chimneys. *Proc. Inst. Mech. Eng.*, **189**, 33–43.

Nieuwstadt, F. T. M., 1984a: The turbulent structure of the stable nocturnal boundary layer. *J. Atmos. Sci.*, **41**, 2202–2216.

Nieuwstadt, F. T. M., 1984b: Some aspects of the turbulent stable boundary layer. *Bound.-Layer Meteor.*, **30**, 31–54.

Nieuwstadt, F. T. M., and A. P. van Ulden, 1978: A numerical study on the vertical dispersion of passive contaminants from a continuous source in the atmospheric surface layer. *Atmos. Environ.*, **12**, 2119–2114.

Pasquill, F., 1975: The dispersion of materials in the atmospheric boundary layer— the basis for generalization. *Lectures in Air Pollution and Environmental Impact Analysis*, Amer. Meteor. Soc., Boston, 1–34.

Pearson, H. J., J. S. Puttock, and J. C. R. Hunt, 1983: A statistical model of fluid-element motions and vertical diffusion in a homogeneous stratified turbulent flow. *J. Fluid Mech.*, **129**, 219–249.

Saffman, P. G., 1960: On the effect of the molecular diffusivity in turbulent diffusion. *J. Fluid Mech.*, **8**, 273–283.

Singer, I. A., and M. E. Smith, 1966: Atmospheric dispersion at Brookhaven National Laboratory. *Int. J. Air Water Pollution*, **10**, 125–135.

Smith, F. B., 1965: The role of wind shear in horizontal diffusion of ambient particles. *Quart. J. Roy. Meteor. Soc.*, **91**, 318–329.

Taylor, G. I., 1921: Diffusion by continuous movements. *Proc. London Math. Soc.*, Ser. 2, **20**, 196–211.

van Ulden, A. P., 1978: Simple estimates for vertical diffusion from sources near the ground. *Atmos. Environ.*, **12**, 2125–2129.

van Ulden, A. P., and A. A. M. Holtslag, 1983: The stability of the atmospheric surface layer during nighttime. *6th Symposium on Turbulence and Diffusion*, Amer. Meteor. Soc., Boston, 257–260.

van Ulden, A. P., and A. A. M. Holtslag, 1985: Estimation of atmospheric boundary layer parameters for diffusion applications. *J. Climate Appl. Meteor.*, **24**, 1196–1207.

Venkatram, A., 1980a: Estimating the Monin-Obukhov length in the stable boundary layer for dispersion calculations. *Bound.-Layer Meteor.*, **19**, 481–485.

Venkatram, A., 1980b: Dispersion from an elevated source in the convective boundary layer. *Atmos. Environ.*, **14**, 1–10.

Venkatram, A., 1982: A semi-empirical method to compute concentrations associated with surface releases in the stable boundary layer. *Atmos. Environ.*, **16**, 245–248.

Venkatram, A., and R. Paine, 1985: A model to estimate dispersion of elevated releases into a shear-dominated boundary layer. *Atmos. Environ.*, **19**, 1797–1805.

Venkatram, A., D. Strimaitis, and D. DiCristofaro, 1984: A semi-empirical model to estimate vertical dispersion of elevated releases in the stable boundary layer. *Atmos. Environ.*, **18**, 823–928.

Wetzel, P. J., 1982: Toward parameterization of the stable boundary layer. *J. Appl. Meteor.*, **21**, 7–13.

Zilitinkevich, S. S., 1972: On the determination of the height of the Ekman boundary layer. *Bound.-Layer Meteor.*, **3**, 141–145.

Appendix A
Interpretation of Taylor's Equation

Although Taylor's equation (Eq. 5.13) applies strictly to dispersion in homogeneous turbulence, it is frequently used to describe field observations of plume spread. In view of this, I believe that a physical interpretation of the equation will provide guidance on its use. Recall that Taylor's equation describes the variance of particle positions as a function of travel time from a fixed point of release. The meaning of the equation is best explained through an idealized experiment in which particles are released one at a time from a source (time intervals between releases are not relevant) into a "steady" turbulent flow field that is described by the three velocity components,

$$V = (U + u, v, w) .\qquad\text{(A1)}$$

In Eq. (A1), U is a constant and u, v, and w are fluctuating turbulent velocities that satisfy

$$\bar{u} = \bar{v} = \bar{w} = 0 ,\qquad\text{(A2)}$$

where the averages are taken over time. A particle released in this flow is assumed to acquire the velocity V so that its motion in Lagrangian coordinates (X, Y, Z) is described by

$$\frac{dX}{dt} = U + u \qquad\text{(A3a)}$$

$$\frac{dY}{dt} = v \qquad\text{(A3b)}$$

$$\frac{dZ}{dt} = w .\qquad\text{(A3c)}$$

Here the coordinate system is fixed to the release point. Note that u, v, and w are evaluated at the position of the particle (X, Y, Z). Now the position of the particle at a time T from the moment of release is given by

$$X_i(T) = UT + \int_0^T u_i(r, t)dt \qquad\text{(A4a)}$$

$$Y_i(T) = \int_0^T v_i(r, t)dt \qquad\text{(A4b)}$$

and

$$Z_i(T) = \int_0^T w_i(r,t)dt \; , \tag{A4c}$$

where

$$r = (X_i, Y_i, Z_i) \; . \tag{A4d}$$

In Eqs. (A4), i tags the particle we are observing. We can write similar equations for all the particles (an infinity, in principle) that can be released from the source. We can now derive an *average* particle position at time T by averaging over all the particles released from the source. This is the *ensemble* average we use below.

Any one of the integrals in Eq. (A4) represents the average of the fluctuating velocities along the trajectory taken by the particle during the time interval $(0, T)$. Although the average consists of the sum of a number of positive and negative velocities, it need not be zero for any given particle trajectory. However, the integral over an infinite number of trajectories is equivalent to an average over all time, which we know is zero by assumption. Thus, we see that *single-point* velocity statistics defined over trajectories taken by particles released from a fixed point are equivalent to statistics over time. We use this result in the following discussion.

The average particle position is given by

$$\langle X(T) \rangle = UT \tag{A5a}$$

$$\langle Y(T) \rangle = \langle Z(T) \rangle = 0 \; , \tag{A5b}$$

where the angle brackets denote averaging over particles.

We saw earlier (Eq. 5.18a) that the average variance of the horizontal position of the particle is given by

$$\langle Y^2 \rangle = 2\sigma_v^2 T \int_0^T (1 - \frac{t}{T}) R_L^y(t)dt \tag{A6}$$

where $R_L^y(t)$ is the Lagrangian autocorrelation function defined by

$$R_L^y(t) = \langle v(\tau + t)v(\tau) \rangle / \sigma_v^2 \; . \tag{A7}$$

An important time scale T_L^y is associated with R_L^y,

$$T_L^y = \int_0^\infty R_L^y(t)dt \; . \tag{A8}$$

The Lagrangian time scale T_L^y represents the time over which the velocity of a particle is correlated with itself. It is roughly the time over which a particle maintains its initial velocity before it undergoes a turbulent "collision" and changes its velocity.

The asymptotic limits of Eq. (A6) can be stated in terms of the T_L^y as follows:

$$\langle Y^2 \rangle = \sigma_y^2 = \sigma_v^2 T^2; \quad T << T_L^y \qquad \text{(A9a)}$$

$$= 2\sigma_v^2 T T_L^y; \quad T >> T_L^y \qquad \text{(A9b)}$$

These limits can be derived easily from Eq. (A6). However, it is instructive to derive them directly by using semi-quantitative reasoning that relies on our understanding of the physics.

Let us begin with the short-time limit. For $T << T_L^y$, a given particle maintains its velocity at release so that we can write

$$Y_i = v_i T . \qquad \text{(A10)}$$

Squaring both sides, and noticing that $\langle v_i^2 \rangle = \sigma_v^2$ yields the earlier result,

$$\sigma_y^2 = \sigma_v^2 T^2 . \qquad \text{(A11)}$$

In order to derive the large-time limit, we divide the travel time T into intervals of duration T_L^y. This allows us to treat each of these time intervals independently. The velocity is taken to be constant in each of them; furthermore, these velocities are uncorrelated with each other because they are separated by time intervals that are at least as long as the Lagrangian time scale T_L^y. Denoting the velocity of the i-th particle in the j-th time interval by v_{ij}, the total distance traveled by the particle is

$$Y_i = \sum_{j=1}^{N} v_{ij} T_L^y , \qquad \text{(A12)}$$

where N is the number of intervals in the travel time T. By definition $N = T/T_L^y$. If we square both sides of (A12) we obtain

$$Y_i^2 = (T_L^y)^2 \left(\sum_{j=1}^{N} v_{ij}^2 + 2 \sum_{j \neq k} \sum v_{ij} v_{ik} \right) . \qquad \text{(A13)}$$

The second term on the right-hand side of the equation represents the products of the velocities in the different time intervals within T. As mentioned earlier, these terms disappear when we average over all particles to find

$$\langle Y^2 \rangle = N\sigma_v^2 (T_L^y)^2 \tag{A14}$$

$$= \sigma_v^2 T T_L^y \text{ with } N = T/T_L^y . \tag{A15}$$

We have lost the factor 2 in our approximate analysis (see Eq. A9b). However, this exercise does provide a clear understanding of the important role played by the Lagrangian time scale T_L^y.

We have stressed the two limits of plume growth primarily because they are often used to describe observed plume behavior. One commonly used form for σ_y that satisfies these two limits is

$$\sigma_y = \sigma_v T/(1 + T/2T_L^y)^{1/2} . \tag{A16}$$

The time scale T_L^y is treated as an empirical parameter that is derived by fitting (A16) to observations (see Draxler, 1976). Alternatively, we can postulate an expression for T_L^y in terms of a known length scale ℓ as follows:

$$T_L^y = \alpha \ell / \sigma_v . \tag{A17}$$

The new empirical parameter is α. Venkatram et al. (1984) used this approach in their analysis of σ_z observations from stable conditions.

Appendix B
Effects of Shear on σ_y

The approach used to derive the two limits of plume growth can also be used to examine the effects of vertical shear on horizontal spread. As we did earlier, let us describe the motion of a particle in a flow with a constant cross-wind shear $S = dV/dz$,

$$\frac{dY}{dt} = v + \frac{dV}{dz}Z \tag{B1a}$$

$$\frac{dZ}{dt} = w . \tag{B1b}$$

For convenience, we have dropped the subscript i that denotes the particle of interest. If the travel time is much less than the Lagrangian time scales for v and w, we can integrate (B1) to obtain

$$Y = vT + \frac{1}{2}SwT^2 \tag{B2a}$$

$$Z = wT . \tag{B2b}$$

Squaring both sides of the equations and averaging, we find

$$\sigma_y^2 = \sigma_v^2 T^2 + S\langle vw \rangle T^3 + \frac{1}{4}S^2\sigma_w^2 T^4 \tag{B3a}$$

and

$$\sigma_z^2 = \sigma_w^2 T^2 . \tag{B3b}$$

In homogeneous turbulence, $\langle vw \rangle = 0$, and the shear contribution to σ_y becomes

$$\sigma_{ys}^2 = \frac{1}{4}S^2\sigma_z^2 T^2 , \tag{B4}$$

where we have used the expression for σ_z (Eq. B3b).

Notice that the shear contribution is proportional to σ_z, which suggests that large shears during stable conditions are unlikely to be very important because σ_z is small in stable stratification. Under these conditions, T_L^z is also likely to be small so that Eq. (B4) will not be applicable to travel times of interest.

The result for large travel times compared with T_L^z is more useful. We can derive the appropriate equation using the earlier technique of dividing the travel time into intervals of length T_L^z. Let us denote the vertical distance traveled in each of these intervals by Δz_i, where i denotes the time interval. We can express Δz_i as

$$\Delta z_i = w_i T_L^z , \tag{B5}$$

so that

$$\langle (\Delta z_i)^2 \rangle = \langle w_i^2 \rangle (T_L^z)^2 = \sigma_w^2 (T_L^z)^2 . \tag{B6}$$

The total distance Z_i traveled by a particle in i time intervals can be written as

$$Z_i = \sum_{j=1}^{i} \Delta z_j , \tag{B7a}$$

With these preliminaries we are now in a position to derive an expression for σ_{ys}. To do so we will look at only the shear term in Eq. (B1a),

$$\frac{dY}{dt} = SZ . \tag{B8}$$

We can integrate this equation as follows:

$$Y_{i+1} = Y_i + SZ_i T_L^z, \quad i = 1 \text{ to } N - 1 . \tag{B9}$$

Recall that i denotes each of the time intervals of duration T_L^z in T. Using Eq. (B7a), Eq. (B9) can be rewritten as

$$Y_{i+1} = Y_i + ST_L^z \sum_{j=1}^{i} \Delta z_j; \quad i = 1 \text{ to } N - 1 . \tag{B10}$$

Equations (B10) relate the horizontal displacements in successive time intervals. They can be more explicitly written as

$$Y_N = Y_{N-1} + ST_L^z \sum_{i=1}^{N-1} \Delta z_i \tag{B10a}$$

$$Y_{N-1} = Y_{N-2} + ST_L^z \sum_{i=1}^{N-2} \Delta z_i \tag{B10b}$$

and so on, until

$$Y_1 = 0 . \tag{B10c}$$

If we add these N equations we obtain the following expression for Y_N:

$$Y_N = ST_L^z \left(\sum_{i=1}^{N-1} \Delta z_i + \sum_{i=1}^{N-2} \Delta z_i + \cdots + \Delta z_1 \right) \quad \text{(B11a)}$$

$$= ST_L^z [(N-1)\Delta z_1 + (N-2)\Delta z_2 + \cdots \Delta z_{N-1}] \quad \text{(B11b)}$$

If we square both sides of the equation and take an ensemble average, we obtain terms of the form $\langle \Delta z_i \Delta z_j \rangle$. Because, by assumption, velocities in different time intervals are uncorrelated, Eq. (B6) implies

$$\langle \Delta z_i \cdot \Delta z_j \rangle = \delta_{ij} \sigma_w^2 (T_L^z)^2 . \quad \text{(B12)}$$

This result allows us to write

$$\sigma_{ys}^2 = S^2 \sigma_w^2 (T_L^z)^4 [(N-1)^2 + (N-2)^2 + \cdots + 1] . \quad \text{(B13)}$$

The sum within the parentheses can be readily expressed as

$$\sum_{i=1}^{N-1} k^2 = \frac{(N-1)(N)(2N-1)}{6} \quad \text{(B14a)}$$

$$\simeq N^3/3 \text{ for large } N . \quad \text{(B14b)}$$

Then Eq. (B13) becomes

$$\sigma_{ys}^2 = S^2 \sigma_w^2 (T_L^z)^4 \frac{N^3}{3} . \quad \text{(B15)}$$

Because $N = T/T_L^z$, Eq. (B15) can be rewritten as

$$\sigma_{ys}^2 = \alpha S^2 \sigma_w^2 T_L^z T^3 , \quad \text{(B16)}$$

where α is a constant whose value has to be determined by exact analysis (see Smith, 1965, for example) rather than the approximate one we have used.

If we notice that $\sigma_z^2 \sim \sigma_w^2 T T_L^z$ for large T/T_L^z, we can rewrite (B16) as

$$\sigma_{ys}^2 = \alpha^2 S^2 \sigma_z^2 T^2 . \quad \text{(B17)}$$

This equation has the same form as that for small T/T_L^z. The shear contribution to σ_y that is used in air pollution modeling practice is based on Eq. (B17) (Pasquill, 1975).

CHAPTER 6

Topics in Applied Dispersion Modeling

Akula Venkatram

6.1. Introduction

The preceding chapters have laid the foundation for the material presented in this chapter. Here, I shall show how the ideas described earlier can be used to develop practical air pollution models. I shall focus on three topics that are considered important in applied dispersion modeling. The first is dispersion in coastal areas. The recent interest in this subject has been stimulated by offshore drilling for oil. Most of the models developed to estimate the impact of these offshore activities use dispersion parameterizations that are strictly applicable to sources over land. In this chapter I shall emphasize the need to relate the dispersion to the physics of the over-water boundary layer, which differs from that over land. A realistic dispersion model must also treat the sharp change in boundary layer structure as air flows from water to land. I discuss methods to incorporate this transition into models and also discuss estimating the height of the coastal boundary layer, one of the critical inputs to a shoreline dispersion model.

The second topic in this chapter is dispersion in complex terrain, where my emphasis will be on the behavior of plumes during stable conditions; during these conditions elevated plumes can impact on hills, causing high concentrations. The regulatory interest in this problem has given rise to several research projects, the largest being the Complex Terrain Model Development program sponsored by the United States Environmental Protection Agency. The models I describe stem from those I helped to develop during that project.

My third topic, model evaluation, is crucially important in dispersion modeling. Although the American Meteorological Society has sponsored workshops on this topic (Fox, 1981; 1984) to gain consensus on the subject, we still seem to be struggling with some basic issues. Therefore my treatment of the topic is necessarily incomplete, and also reflects a personal view. I believe that the current tendency to rely primarily on statistical measures to quantify model performance is likely to impede the application of recent advances in our understanding of dispersion and micrometeorology. In this chapter I emphasize the need to tie model evaluation to the iterative process of using discrepancies between predictions and observations to improve the model. Statistical manipulation of the residuals between predictions and observations should be only one of many steps involved in model evaluation.

6.2. Dispersion in the Coastal Boundary Layer

6.2.1. The Boundary Layer Over the Water Surface

Moisture evaporating from the ocean plays an important role in the overlying boundary layer. Most of the radiative energy input to the water surface goes into evaporation rather than increasing the temperature of the water. What little energy is absorbed gives rise to much smaller variations of the temperature than over land because of the relatively high specific heat of water. In the absence of advection, the oceanic boundary layer would change little over diurnal time scales.

Most of the dynamic activity in this boundary layer is driven by the temperature and moisture differentials between the water and the air masses advected over it. Because the properties of these air masses have little to do with local radiation inputs, the Pasquill-Gifford (PG) classification should not be used to characterize dispersion over water. Over land, by contrast, diurnal radiative forcing plays the major role in most instances; that is why the PG system based on radiation and wind speed applies there.

This implies that in the absence of direct measurements of turbulence it is necessary to use information on weather patterns and climatology to characterize dispersion over water. A systematic method of doing so has not yet been developed because dispersion over water has not been of wide concern until recently. For the time being, we will have to require at least single-level measurements of wind speed, temperature, and humidity to infer turbulence aloft.

Studies performed over the ocean (e.g., Lenschow et al., 1980) indicate that our understanding of the structure of the PBL over land can be transferred to that over water if we account for the effects of moisture on density. Because water vapor is lighter than air, the injection of water vapor into the boundary layer leads to a decrease in the density of the layer. This process can be simulated by increasing the temperature of the air to produce this smaller density. Thus, we can ignore the presence of water vapor if we use a "virtual" temperature T_v,

$$T_v = T(1 + 0.61q) , \tag{6.1}$$

where q is the specific humidity of the air, and 0.61 is simply the ratio of the molecular weights of air and water vapor minus 1. The moisture "correction" to the temperature T is usually very small because q rarely exceeds 10^{-2}. However, because Δq can be of the same order as q, both contributions to

$$\Delta T_v \simeq \Delta T + 0.61 T \Delta q \tag{6.2}$$

can be of the same order. The Monin-Obukhov length becomes

$$L = -\frac{T_{vo}}{g} \frac{u_*^3}{k \overline{w \theta}_0} , \tag{6.3}$$

where (as mentioned in Ch. 1) $\overline{w \theta}_0$ is understood to be the surface flux of virtual temperature. With these definitions, we can apply the similarity profiles obtained over land to flows over water.

Under the Monin-Obukhov scaling hypothesis the similarity profiles of the SBL should be the same over land and over water, and what evidence we have supports this. However, the empirical relationships $L \sim u_*^2$ and $h \sim u_*^{3/2}$ (see Ch. 5) have not been tested against observations over water surfaces. The usefulness of such formulas clearly suggests the need for further study in this area.

In using surface measurements to infer heat and momentum fluxes, it is useful to remember that the roughness of the ocean depends upon wave height which, in turn, is a function of the surface shear stress and the gravitational force. Kitaigorodskii (1973) suggested the following equation for z_0:

$$z_0 = 0.035 u_*^2 / g . \tag{6.4}$$

For typical values of u_*, the roughness length z_0 is of the order of millimeters over water. This, as we shall see, simplifies the calculation of u_* which, under neutral conditions, is given by

$$u_*^2 = C_{DN}(U_r - U_w)^2 , \qquad (6.5)$$

where U_r is the wind at a reference height z_r, U_w is the speed of the water surface (usually much smaller than U_r), and the neutral drag coefficient C_{DN} is

$$C_{DN} = [k/\ln(z_r/z_0)]^2 . \qquad (6.6)$$

Because z_r is typically of the order of meters, the ratio (z_r/z_0) is so large that it varies little at the measurement heights of interest. Therefore, C_{DN} to a good approximation can be taken to be invariant with height, and we can use the following empirical equation suggested by Smith (1980):

$$C_{DN} = (0.61 + 0.63U_{10})10^{-3} , \qquad (6.7)$$

where U_{10} is the wind speed (in m s^{-1}) at 10 m. However, as stated earlier the precise height of measurement is not important. To calculate the friction velocity u_* during diabatic conditions, we can use the equations proposed by Louis (1979) as approximations to those derived from the similarity profiles.

Although there have been several studies of the structure of the boundary layer over the ocean, few have examined dispersion in the light of the observed turbulent structure. Most studies of dispersion over water have characterized observed dispersion using the PG classification scheme. This, as we have seen earlier, has very little merit. In my opinion, there is a strong need for dispersion experiments that characterize plume spread with micrometeorological variables such as u_*, L, and h.

6.2.2. Shoreline Fumigation

The model

The literature contains several models for shoreline fumigation, the most recent being those of Deardorff and Willis (1982), Misra (1980) and van Dop et al. (1979). In my opinion, the most practical models are the last two. The discussion here follows Misra (1980). The model illustrates a general method of dealing with dispersion in situations in which the boundary layer undergoes a transition in properties.

Figure 6.1 shows a schematic of shoreline fumigation. The plume is initially embedded in the stable onshore flow; fumigation is associated with the entrainment of the plume from the stable layer into the growing thermal internal boundary layer (TIBL). In order to estimate ground-level concentrations, we will treat the surface of the TIBL as a set of sources that emit material into the TIBL.

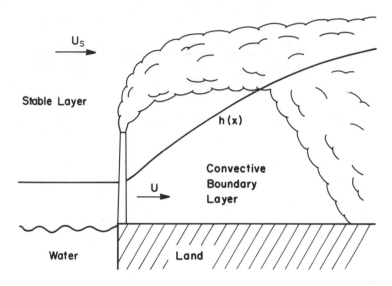

Figure 6.1. Schematic of shoreline fumigation.

Figure 6.2 shows the idealization used to derive the model. We first need an equation for the source strength $dQ(x', y', z')$ associated with the infinitesimal arc AB on the surface of $h(x)$. If we assume homogeneity in the y direction, we see that the flux of material through AB is the sum of the fluxes through AC and CB. Now

$$\text{Downward flux through } CB = K_s \frac{\partial C_s}{\partial z} \Delta x' \Delta y' , \quad (6.8)$$

where the subscript s refers to the stable layer. From our earlier discussion (see Ch. 5), we can write the eddy diffusivity K_s as

$$K_s = \frac{1}{2} d\sigma_{zs}^2 / dt , \quad (6.9)$$

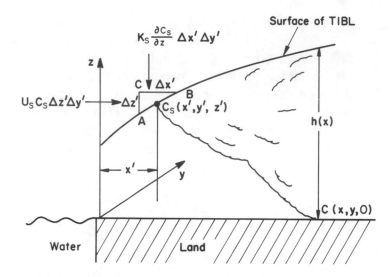

Figure 6.2. Geometry used to derive the expression for the ground-level concentration during fumigation.

where σ_{zs} is the plume spread in the stable layer, and t is the time of travel from the source.

Now the flux through AC is simply the advection of pollutant, or $U_s C_s \Delta z' \Delta y'$. Therefore, the source strength associated with the entrainment of material through AB is

$$dQ(x',y',z') = K_s \frac{\partial C_s}{\partial z} \Delta z' \Delta y' + U_s C_s \Delta z' \Delta y' . \qquad (6.10)$$

Notice that

$$z' = h(x',y') \text{ and } \Delta z' = \frac{dh}{dx} \Delta x' \qquad (6.11)$$

so that Eq. (6.10) can be rewritten as

$$dQ = \left(K_s \frac{\partial C_s}{\partial z} + U_s C_s \frac{dh}{dx} \right) \Delta x' \Delta y' . \qquad (6.12)$$

To evaluate Eq. (6.12), we need an expression for C_s. It is reasonable to assume that the concentration distribution in the stable layer is Gaussian:

$$C_s(x',y',z') = \frac{Q}{2\pi U_s \sigma_{ys} \sigma_{zs}} \exp\left[-\frac{(h_e - z')^2}{2\sigma_{zs}^2} - \frac{y'^2}{2\sigma_{ys}^2} \right] . \qquad (6.13)$$

where h_e is the effective stack height. In writing Eq. (6.13) we have assumed that material entrained into $h(x)$ cannot affect the concentration in the stable layer, so there is no reflection term in Eq. (6.13). We can now evaluate dQ by putting $z' = h(x')$ in Eq. (6.13).

To calculate the concentration associated with $dQ[x',y',h(x')]$ at $(x,y,0)$, we assume that the pollutant is well mixed in the vertical by the time it travels through the distance $x - x'$. Clearly, this is likely to be incorrect if $x - x'$ is small compared with the mixing distance $Uh(x)/w_*$. Later we suggest a method to account for this error.

Taking the crosswind distribution to be Gaussian, the contribution $dC(x,y,0|x',y',h)$ associated with dQ is

$$dC(x,y,0|x',y',h) = \frac{dQ[x',y',h(x')]}{U\sqrt{2\pi}\ \sigma_y(x,x')h(x)} \exp\left[-\frac{(y-y')^2}{2\sigma_y^2}\right],$$
(6.14)

where the unsubscripted variables refer to the TIBL. The spread $\sigma_y(x,x')$ refers to the incremental growth between x and x'. Then $C(x,y,0)$ becomes

$$C(x,y,0) = \int dC[x,y,0|x',y',h(x')] .$$
(6.15)

With Eqs. (6.12), (6.13), and (6.14), Eq. (6.15) can be written as

$$C(x,y,0) = \frac{Q}{2\pi h(x)} \int_0^x \frac{1}{\sigma_{ye}U} \exp\left(-\frac{p^2}{2}\right) \frac{dp}{dx'} \exp\left(-\frac{y^2}{2\sigma_{ye}^2}\right) dx' ,$$
(6.16)

where

$$p(x') = \frac{h(x') - h_e(x')}{\sigma_{zs}(x')}$$
(6.17a)

and

$$\sigma_{ye}^2 = \sigma_{ys}^2(x') + \sigma_y^2(x,x') .$$
(6.17b)

Notice that this formulation allows for the variation of the effective height $h_e(x')$ with distance from the stack. Both Misra (1980) and van Dop et al. (1979) obtained Eq. (6.16) using slightly different derivations. However, they differed in their interpretation of $\sigma_y^2(x,x')$. Misra (1980) assumed that the dispersion in the stable layer and the TIBL are completely independent so that

$$\sigma_y^2(x,x') = \sigma_y^2(x - x') \ . \tag{6.18}$$

This equation implies that a particle entrained at x' forgets its history and behaves as if it were released from a new point source at $x',y',h(x')$. On the other hand, van Dop et al. (1979) assumed that plume spread in the TIBL corresponds to particle release at $x = 0$. This implies that

$$\sigma_y^2(x,x') = \sigma_y^2(x) - \sigma_y^2(x') \ . \tag{6.19}$$

Because $\sigma_y^2(x,x')$ of Eq. (6.19) is always smaller than that of (6.18), Misra's model will predict smaller concentrations than the model of van Dop et al. (1979). I prefer Eq. (6.19) to Eq. (6.18) because it reduces to the correct form when the crosswind spread in the stable layer is the same as that in the TIBL. In this limit, we find from Eqs. (6.17b) and (6.19) that

$$\sigma_{ye}^2(x) = \sigma_{ys}^2(x) \ . \tag{6.20}$$

When large-scale wind meandering controls σ_y, Eq. (6.20) is likely to be valid. That is, the rate of growth of σ_y in the TIBL will not be affected by convective turbulence. Then Eq. (6.16) reduces to the simple form

$$C(x,y,0) = \frac{Q}{2\sqrt{2\pi}\sigma_y h} \left[1 + \mathrm{erf}\left(\frac{h(x) - h_e(x)}{\sqrt{2}\sigma_{zs}}\right) \right] \exp\left(-\frac{y^2}{2\sigma_y^2}\right) \ . \tag{6.21}$$

This equation is similar to one derived by Lyons and Cole (1973) except that they used an enhanced σ_y to account for the increased spread in the TIBL. Note that when $h \gg h_e$ so that the plume is completely entrained into the TIBL, erf (the error function) becomes unity, and Eq. (6.21) reduces to the familiar expression for the well mixed boundary layer.

The most serious problem with this model for shoreline fumigation is the assumption of instantaneous vertical mixing. Laboratory results presented by Deardorff and Willis (1982) indicate that the time required for material released at the top of the boundary layer to reach the surface is about $2h/w_*$. This means that the model presented here could overestimate concentrations. An approximate way of accounting for incomplete vertical mixing is to modify the upper limit of the integral to an x^* defined by

$$x^* = x - 4hU/w_* = x - 4UT_m \; ; \; T_m \equiv h/w_* \; , \tag{6.22}$$

where I have assumed that it takes a time of about $4h/w_*$ for the material to mix through the depth of the boundary layer. This formulation allows only part of the material entrained between 0 and x to contribute to the ground-level concentration at x.

Some applications

Misra and Onlock (1982) tested Eq. (6.16) together with Eq. (6.18) against observations of ground-level concentrations made during experiments conducted in 1978/1979 around the Nanticoke Generating Station situated on the shore of Lake Erie. They used lidar observations to give estimates of σ_{ys} and σ_{zs}, and they modeled σ_z in the TIBL with equations proposed by Lamb (1978) on the basis of his numerical experiments. The boundary-layer height $h(x)$ was estimated with a few measurements from an acoustic sounder. The surface heat flux, required for the calculation of w_*, was taken to be a fraction of the incoming solar radiation.

Figure 6.3 shows that the model performed well against hourly averaged ground-level concentrations of SO_2 associated with the generating station. The model does not overpredict concentrations as one would expect in light of the assumption of instantaneous vertical mixing. One reason for this could be that most of the observations were made at distances at which most of the plume was entrained into the TIBL, and therefore any assumptions on vertical mixing would not be important. In any event, this indicates that Eq. (6.16) is a good candidate for a practical, easy-to-implement model for shoreline fumigation.

One of the inputs required by the model is w_*, which depends on the surface virtual temperature flux $\overline{w\theta}_0$. Because $\overline{w\theta}_0$ varies with distance from the shoreline, it is useful to define an effective flux $\overline{w\theta}_e$ as follows. Let us assume that a constant $\overline{w\theta}_e$ gives rise to the growth of the TIBL between 0 and x. Then a simple thermal energy balance results in

$$h^2(x) - h^2(0) = \frac{2\overline{w\theta}_e \alpha x}{\gamma U} \; , \tag{6.23}$$

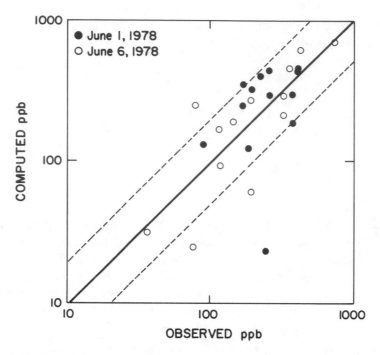

Figure 6.3. Comparison of observed SO_2 concentrations with predictions from the shoreline fumigation model. From Misra and Onlock (1982).

where α is a constant related to the entrainment at the top of the TIBL, and γ is the gradient of the potential temperature in the stable layer. Details of the derivation are given in the next section. From Eq. (6.23), $\overline{w\theta_e}$ becomes

$$\overline{w\theta_e} = \frac{[h^2(x) - h^2(0)]\gamma U}{2x\alpha} \, .$$ (6.24)

For simplicity, let us assume that $h(x) \gg h(0)$ so that

$$\overline{w\theta_e} = \frac{h^2(x)\gamma U}{2x\alpha} \, .$$ (6.25)

Then the effective w_* is

$$w_* \equiv \left(\frac{g}{T_0}\overline{w\theta_e}h\right)^{1/3} = h\left(\frac{g}{T_0}\frac{\gamma U}{2x\alpha}\right)^{1/3} \, .$$ (6.26)

Then the mixing time scale h/w_* becomes

$$T_m = h/w_* \equiv \left(\frac{T_0}{g} \frac{2x\alpha}{\gamma U} \right)^{1/3} . \tag{6.27}$$

Notice that an estimate of T_m required in Eq. (6.22) does not depend upon $\overline{w\theta}_0$. Furthermore, w_* is only a weak function of γ and U, which suggests that the variable of most importance in the use of the shoreline fumigation model is the TIBL height $h(x)$.

6.2.3. The TIBL Height

Introduction

In this section, the thermal internal boundary layer (TIBL) refers to the boundary layer that forms when stable air flows from water onto warmer land. The temperature differential between the land and the air results in upward heat flux, which in turn gives rise to a convectively dominated boundary layer that grows with distance from the shoreline.

In the next subsection, we postulate a simplified picture of the TIBL and then derive the equations that govern its growth. We show that these equations constitute the basis for most existing models of the TIBL.

The model

Figure 6.4 shows a schematic of the physical situation. We assume that the internal boundary layer begins to develop when the heat flux at the surface is directed upward. Notice that the convective activity can be initiated over the water surface. The initial boundary layer height at the land-water boundary allows for this possibility. The following derivation assumes that the TIBL growth occurs under quasi-steady conditions. This implies that the time of travel x/U, where U is the mixed-layer-average wind speed, is much less than the time scale of the temperature change of the land surface.

In our analysis, the origin of the coordinate system is fixed at the land-water boundary. Conditions are taken to be homogeneous in y, so that the problem is two-dimensional. The temperature profile within the TIBL is idealized as follows. Except in the shallow superadiabatic surface layer, the potential temperature is taken to be uniform in the vertical. The TIBL is capped by a sharp inversion, which is idealized by a first-order discontinuity in temperature. This temperature jump is surmounted by the remnants of the stable onshore flow.

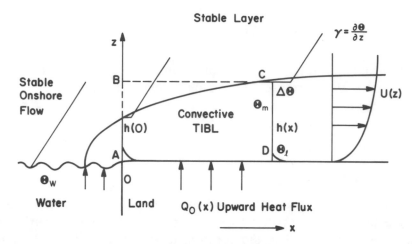

Figure 6.4. Schematic of the development of the thermal internal boundary layer.

The equation governing the growth of the TIBL can be derived by applying a thermal energy balance to the control volume ABCD shown in Fig. 6.4. If we neglect vertical motion, we see that the conservation of enthalpy requires

(Energy outflow through CD) − (Energy inflow through AB) = Heat flux through AD:

$$\rho_a C_p \left[\int_0^{h(x)} U(x,z)\Theta(x,z)dz - \int_0^{h(x)} U(0,z)\Theta(0,z)dz \right] = \int_0^x Q_0(x)dx .$$

$$(6.28)$$

Here ρ_a is the air density, C_p is the specific heat of air at constant pressure, $U(x,z)$ is the horizontal velocity, and $\Theta(x,z)$ is the potential temperature. Notice that this formulation can accommodate arbitrary temperature and velocity profiles. However, in order to develop a model for $h(x)$ that can be used in practical applications we have to make some simplifications. We begin by assuming that the velocity is uniform in the vertical so that we simplify Eq. (6.28) to

$$\int_0^{h(x)} \Theta(x,z)dz - \int_0^{h(x)} \Theta(0,z)dz = \frac{1}{\rho_a C_p U} \int_0^x Q_0(x)dx , \qquad (6.29)$$

where U is an "effective" velocity whose relationship to a measured velocity has to be determined empirically.

The computation of the second integral on the left-hand side of Eq. (6.29) is illustrated in Fig. 6.5. Note that the potential temperature profile is taken to be piecewise linear. Thus, there is no need

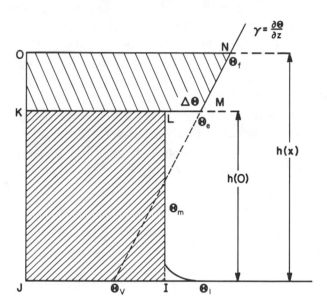

$$\int_0^{h(x)} \theta(0,z)\, dz = \text{area of rectangle IJKL} + \text{area of trapezoid KLMNO}$$

$$= \text{IJ} \cdot \text{IL} + \frac{(\text{KM} + \text{NO}) \cdot (\text{OK})}{2}$$

$$= \theta_m h(0) + \frac{(\theta_m + \Delta\theta + \theta_f)(h(x) - h(0))}{2}$$

$$= (\theta_e - \Delta\theta)h(0) + \frac{(\theta_e - \Delta\theta + \Delta\theta + \theta_f)(h(x) - h(0))}{2}$$

$$= (\theta_v + \gamma h(0) - \Delta\theta)h(0) + \frac{(2\theta_v + \gamma h(0) + \gamma h(x))(h(x) - h(0))}{2}$$

$$= \theta_v h(0) + \gamma h^2(0) - \Delta\theta h(0) + \theta_v(h(x) - h(0)) + \frac{\gamma}{2}(h^2(x) - h^2(0))$$

$$= \gamma h^2(0) - \Delta\theta h(0) + \theta_v h(x) + \frac{\gamma}{2}(h^2(x) - h^2(0)) .$$

Now assume that $\Delta\theta = F\gamma h(0)$ (entrainment assumption). Then

$$\int_0^{h(x)} \theta\, dz = \gamma h^2(0)(1 - F) + \theta_v h(x) + \frac{\gamma}{2}(h^2(x) - h^2(0)) .$$

Figure 6.5. Computation of second integral on the left-hand side of Eq. (6.29).

to assume that the gradient γ is constant with height. A knowledge of the potential temperature profile of the onshore stable flow allows us to compute γ and the associated Θ_v. Note that Θ_v is *not* equal to the water temperature Θ_w. Venkatram (1977a) emphasizes that Θ_v is the temperature that is relevant to the TIBL height formulation. Because the gradient of the surface-based inversion over water is usually much larger than that of the temperature profile aloft, $(\Theta_\ell - \Theta_w) > (\Theta_\ell - \Theta_v)$. Therefore, in practical applications, the water temperature Θ_w would be an underestimate of Θ_v.

It should be stressed that the entrainment assumption $\Delta\Theta = F\gamma h$ has not been validated with observations. F is taken to be a constant only because our understanding of the dynamics of entrainment at the inversion is poor. This uncertainty in the value F might, however, not be important because the expression for the TIBL height is relatively insensitive to F (see Venkatram, 1977b). From Fig. 6.5, it is seen that

$$\int_0^{h(x)} \Theta(0,x)dz = \gamma h^2(0)(1-F) + \Theta_v h(x) + \frac{\gamma}{2}\left[h^2(x) - h^2(0)\right] . \quad (6.30)$$

We can immediately write the second integral at x by setting $h(0) = h(x)$ in Eq. (6.30),

$$\int_0^{h(x)} \Theta(x,z)dz = \gamma h^2(1 - F) + \Theta_v h(x) . \quad (6.31)$$

Substituting Eqs. (6.30) and (6.31) into Eq. (6.29), we find

$$\frac{\gamma}{2}(1 - 2F)[h^2(x) - h^2(0)] = \frac{1}{\rho_a C_p U}\int_0^x Q_0(x)dx \quad (6.32)$$

or

$$h^2(x) = h^2(0) + \frac{2}{\gamma(1 - 2F)U}\int_0^x \overline{w\theta}_0(x)dx \quad (6.33a)$$

where the surface temperature and heat fluxes are related by

$$\overline{w\theta}_0 \equiv Q_0/\rho_a C_p . \quad (6.33b)$$

Notice the appearance of the square of the initial height $h(0)$ in Eq. (6.33a).

We can recast Eq. (6.33a) in an alternative form by differentiating it with respect to x,

$$U\gamma(1 - 2F)h\frac{dh}{dx} = \overline{w\theta}_0(x) .\qquad (6.34)$$

Equation (6.34) can be used to relate the value of F to the familiar ratio of the inversion flux $\overline{w\theta}_i(x)$ to $\overline{w\theta}_0(x)$. The inversion flux is (Tennekes, 1973)

$$\overline{w\theta}_i = -\Delta\Theta U\frac{dh}{dx}\qquad (6.35a)$$

$$= -F\gamma h U\frac{dh}{dx} .\qquad (6.35b)$$

Then

$$\frac{\overline{w\theta}_i}{\overline{w\theta}_0} = -\frac{F}{(1 - 2F)} \equiv -R .\qquad (6.36)$$

If we use the estimate $R = 0.2$ from Tennekes (1973), F becomes 1/7. As mentioned earlier, current estimates (and understanding) of the entrainment parameter are rather uncertain, and should be used cautiously.

Because measurements of $\overline{w\theta}_0(x)$ are not likely to be routinely available, the usefulness of Eq. (6.33) hinges upon our ability to estimate it from routine observations. The next section discusses some of the more popular parameterizations for $\overline{w\theta}_0$.

An expression for the TIBL height

In deriving a useable equation for the TIBL, one of two assumptions is usually made, namely 1) the surface flux $\overline{w\theta}_0$ does not vary with x or, 2) the land temperature Θ_ℓ does not vary with x. Because there is little theoretical or observational justification for either, we can test their correctness only by examining the expressions for $h(x)$ that result from them.

If one assumes that $\overline{w\theta}_0$ is constant, Eq. (6.33a) can be reduced to

$$h^2(x) = h^2(0) + \frac{2x}{\gamma(1 - 2F)U}\overline{w\theta}_0 .\qquad (6.37)$$

The equations suggested by Weisman (1976) and Plate (1971) correspond to different values of F in Eq. (6.37). As it stands, the most
valuable information conveyed by Eq. (6.37) is that

$$h \sim A \, x^{1/2} \qquad\qquad (6.38)$$

if $\overline{w\theta}_0$ is indeed reasonably invariant with distance. Some observations
(Raynor et al., 1975; Kerman et al., 1982) do suggest this behavior
at least at small distances from the shoreline. However, note that a
constant $\overline{w\theta}_0$ does not allow the TIBL to reach an equilibrium height;
$h(x)$ will keep on growing, which is inconsistent with the observations
of Kerman et al. (1982). It is difficult to accept the hypothesis that
$\overline{w\theta}_0$ is constant when the properties of the TIBL vary with distance
from the shoreline. In any event, no study has been conducted to
test this hypothesis.

Even if Eq. (6.37) were reasonable, it could not be used in prac-
tical applications because direct measurements of $\overline{w\theta}_0$ are not likely
to be available. Multiple-level measurements of temperature and
velocity to use surface-layer similarity relationships are also not routine. Note that the use of similarity relationships assumes horizontal
homogeneity, which often cannot be justified in the evolving TIBL.

In order to use Eq. (6.37) we have to develop a parameterization
for $\overline{w\theta}_0$ in terms of variables that can be readily measured. Because
the surface flux $\overline{w\theta}_0$ is driven by the temperature differential between
the heated land and the onshore flow, it is reasonable to postulate

$$\overline{w\theta}_0 = A \, u_* \, (\Theta_\ell - \Theta_w) , \qquad\qquad (6.39)$$

where the water temperature Θ_w is used to characterize the temperature of the air. The empirical factor A is likely to be site-specific, and
will depend upon the method used to calibrate its value. Substitution
of Eq. (6.39) into Eq. (6.37) results in an equation that is similar in
form to that suggested by Raynor et al. (1975). If we assume that
u_* is some fraction of U, we can rewrite Eq. (6.49) in the form

$$h^2(x) = h^2(0) + A \, \frac{(\Theta_\ell - \Theta_w)}{\gamma} \, x , \qquad\qquad (6.40)$$

where we have absorbed the unknown empirical factors into A. Notice that even in this simple form, Eq. (6.37) is not transferable from

site to site because A will depend upon site characteristics as well as the way Θ_ℓ and Θ_w are defined.

As mentioned before, the major deficiency with formulations exemplified by (6.40) is that the boundary-layer depth h continues to grow at all x. This is inconsistent with observations (see Kerman et al., 1982), which indicate that h grows initially and then reaches an equilibrium height farther inland. In the next section, we examine models that allow for this leveling off in the TIBL height.

Other models of the TIBL height

The model described in the last section assumes that $\overline{w\theta}_0$ is constant over land. Because there is little observational evidence to support this assumption, it is useful to examine the consequences of other possible assumptions about the behavior of $\overline{w\theta}_0(x)$. Venkatram (1977a,b) assumed that Θ_ℓ is constant. This causes the heat flux to reach its maximum value at the land-water boundary where the temperature contrast between the air and land is greatest; $\overline{w\theta}_0(x)$ then decreases as the upward heat flux warms the air as it passes over the land. It can be easily shown that for small x, this variation of $\overline{w\theta}_0(x)$ gives rise to the square-root behavior

$$h^2(x) = h^2(0) + \frac{Au_*(\Theta_\ell - \Theta_w)x}{U\gamma}, \qquad (6.41)$$

where the symbols have their previous meaning. The decrease of $\overline{w\theta}_0(x)$ with increasing x causes $h(x)$ to reach an equilibrium value h_e

$$h_e \sim (\Theta_\ell - \Theta_w)/\gamma . \qquad (6.42)$$

Although this behavior is qualitatively consistent with observations, the model has two shortcomings, namely 1) the unrealistic assumption of constant Θ_ℓ, and 2) its reliance on the empirical factor A, which is a function of the definitions of Θ_ℓ and Θ_w.

We can construct a more realistic model for the TIBL height by paying more attention to the observed behavior of $h(x)$, and not specifying the precise variation of $\overline{w\theta}_0(x)$. Observations indicate that for small $x, h(x) \sim x^{1/2}$, and $h(x) \sim h_e$ at large x. A plausible interpolation between these limits is

$$h^2(x) = h^2(0) + \left[h_e^2 - h^2(0)\right](1 - e^{-x/D}) . \qquad (6.43)$$

In Eq. (6.43), the equilibrium height h_e is the boundary-layer height far from the shoreline where the upward heat flux is driven by the incoming solar radiation. We can write an expression for the relaxation distance D by noticing that it measures the distance over which the air flowing from the water adjusts to the heat flux over the land. The greater this heat flux, the greater is D. Because h_e is controlled by the heat flux over land, it is reasonable to assume that

$$D = h_e/\beta , \qquad (6.44)$$

where β is an empirical constant. When Eq. (6.44) is substituted into Eq. (6.43) and $h(x)$ is expanded for small x,

$$h^2(x) \sim \beta h_e x \ [\text{neglecting } h(0)] , \qquad (6.45)$$

we now recover the observed behavior $h \sim x^{1/2}$. At large distances $x \gg D, h = h_e(t)$.

The simplicity of Eq. (6.43) is very appealing. The only unknown is $h_e(t)$, which can be estimated with a variety of methods. If measurements (rawinsonde) are not available, we can calculate $h_e(t)$ from the heat flux at the surface and the morning temperature profile. The heat flux can, in turn, be related to the incoming solar radiation. Note that this relationship is reasonable only far from the shoreline where the solar radiation is the driving force. A useful formulation for $h_e(t)$ is (Carson and Smith, 1974)

$$h_e^2(t) = h_e^2(0) + \frac{2}{\gamma(1 - 2F)} \int_0^t \overline{w\theta}_0(t)dt . \qquad (6.46)$$

where $\overline{w\theta}_0$ is the surface temperature flux far inland, γ is the potential temperature gradient above the mixed layer, and F is the entrainment factor referred to earlier. The flux $\overline{w\theta}_0$ can be related to the incoming solar radiation $S(t)$ through

$$\rho_a C_p \overline{w\theta}_0 = BS(t), \qquad (6.47)$$

where B is an empirical constant that ranges from 0.2 to 0.4 depending upon the surface. Alternatively, $\overline{w\theta}_0$ can be estimated from surface similarity relationships that are valid for the homogeneous boundary layer that occurs inland. Because we have experience with making reliable estimates of h_e, the suggested equation for $h(x)$ is more useful than other formulations that are based on ill-defined quantities such as Θ_l, Θ_w, u_*, and U. However, like the other models, it has to be tested thoroughly against observations before it can be used.

6.3. Dispersion in Complex Terrain

6.3.1. Introduction

Our present state of understanding of dispersion and flow around complex terrain does not allow us to recommend one model that is generally applicable to the wide range of problems associated with dispersion in complex terrain. The models that we examine in this chapter apply to a narrow range of physical situations. We shall concentrate on stable atmospheric conditions, when plumes are expected to cause high concentrations by impacting on hillsides. We shall also look at situations in which stable stratification depresses the plume centerline toward the hill and thus leads to a magnification of the ground-level concentration over that in flat terrain. Although our examination of dispersion in complex terrain is far from exhaustive, it does deal with situations in which concentrations are expected to be high and are thus important from a regulatory viewpoint. For a more complete treatment of the subject, the reader is referred to Schiermeier (1984) and Hunt et al. (1979).

6.3.2. The Effects of Complex Terrain on Mean Flow and Turbulence in the SBL

The concept of the dividing streamline height plays a crucial role in the description of flow and dispersion in complex terrain. It is defined implicitly by the equation

$$\frac{1}{2}U^2(H_c) = \int_{H_c}^{H} N^2(z)(H - z)dz \,, \qquad (6.48)$$

where N is the Brunt-Väisälä frequency defined by

$$N^2 = \frac{g}{T_0}\frac{d\Theta}{dz} \,. \qquad (6.49)$$

The left-hand side of Eq. (6.48) is the kinetic energy of the fluid at H_c; the right-hand side is the potential energy gained by the fluid in rising through the height $H - H_c$ to the top of the hill at $z = H$.

This energy argument suggests a two-layer flow about a three-dimensional obstacle. The fluid below H_c does not have enough energy to surmount the hill, and thus tends to flow around the obstacle. The fluid above H_c flows over the top of the hill. In a series of experiments, Snyder et al. (1983) demonstrated the usefulness of this concept in describing plume behavior in stratified flows. They showed that plumes released below H_c remained horizontal before impinging upon the surface of the obstacle. A plume released above H_c rises over the hill.

For constant U and N^2, Eq. (6.48) reduces to

$$H_c = H(1 - \text{Fr}) , \qquad (6.50)$$

where the Froude number Fr is defined by

$$\text{Fr} = U/NH . \qquad (6.51)$$

Notice that region 2 (see Fig. 6.6) has a depth $H - H_c = \text{Fr}H = U/N$. It is now easy to see that the concept of the dividing streamline is related to the idea that a fluid with velocity U can rise only through a distance U/N.

The fluid below H_c can remain roughly horizontal as it flows around a three-dimensional body. However, for a two-dimensional body, this flow can become blocked, and the effects of blocking can be transmitted upstream. This implies a readjustment of the flow to surmount the initial potential energy barrier. Our understanding of these effects is incomplete, and it is not clear that the computed H_c can be used to delineate two regions of flow as around three-

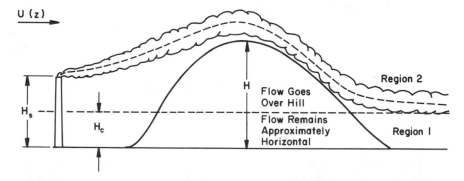

Figure 6.6. Schematic diagram of plume behavior in stable flow around a terrain obstacle.

dimensional bodies. For our purposes, we assume that the concept of H_c also applies to two-dimensional hills.

In order to describe plume behavior in the flow above H_c, it is necessary to have an understanding of the effects of complex terrain on the mean flow and turbulence. When the flow goes over the top of the hill, it speeds up. This is accompanied by a convergence of streamlines in the vertical direction. At the same time, the streamlines in the crosswind direction spread out. Near the surface of the hill, all three components of turbulence are increased above their free-stream values. This increase is roughly proportional to the slope of the hill.

What are the consequences of these hill-induced changes on dispersion of plumes? The convergence of streamlines depresses the plume centerline and at the same time leads to a decrease in σ_z. The crosswind spread σ_y is increased by streamline divergence. The increased wind speed decreases the concentration while the increased turbulence counteracts the decrease in σ_z caused by streamline convergence. Thus, complex terrain induces changes that can either increase or decrease the concentration on the hill's surface. The combined effect of these changes on the ground-level concentration can be examined only through a model that incorporates the complex terrain phenomena we have mentioned. The model described in the next section attempts to parameterize these hill effects. This approach is a major departure from conventional models in which terrain correction factors are used primarily to lower the effective stack height over the hill. This, of course, produces the desired effect of increasing the ground-level concentration. Because there is little theoretical justification for this procedure, the validity of commonly used complex terrain dispersion models is highly suspect.

6.3.3. Complex-Terrain Models

Introduction

Figure 6.7 shows the idealized flow that forms the basis of the dispersion model to be described here. Flow below H_c is assumed to remain horizontal, and the portion of the plume that is embedded in this flow wraps around the hill. The concentrations associated with this plume are taken to be that corresponding to direct impaction on the hill. This is a critical assumption that is based on the work of Hunt et al. (1979), whose analysis can be explained as follows. As the flow approaches the hill, the wind transporting the plume

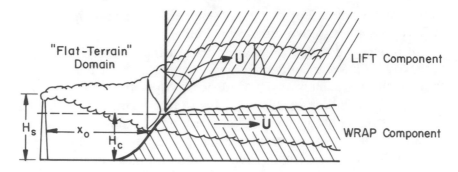

Figure 6.7. Idealized stratified flow about hills.

decreases as the streamlines diverge near the stagnation point. For most practical purposes these changes can be assumed to occur over distances that are small compared with the distances from the source to the receptor. This means that the plume behaves as if the hill were not present until it almost impacts on the hill. Because the rapid changes in the impaction zone can be considered to be pure deformation, the concentration is not affected. The net result is that we can ignore the presence of the hill in estimating the concentration at the stagnation point.

The model for the dispersion of the plume segment embedded in the flow above H_c assumes that only the plume segment above H_c surmounts the hill. The presence of H_c effectively lowers the plume. However, this lowering corresponds to only that part of the plume material that is above H_c. The consequences of this will be examined later.

The WRAP model

The model applicable to plume dispersion below H_c will be referred to as the WRAP model. This terminology has its origin in the picture of a plume wrapping around the hill when the release height is below H_c.

Figure 6.8 shows the idealized geometry considered in the formulation of WRAP. We shall consider only the estimation of the maximum concentration, which we assume to occur at the stagnation point A. The assumption here is that the concentrations on the hill's surface are primarily associated with the one streamline, the stagnation streamline, which carries plume material onto the surface. All other streamlines are deflected away from the hill. If we assume

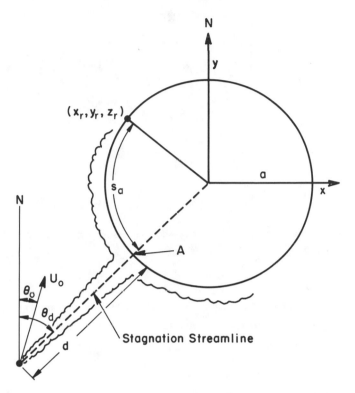

Figure 6.8. Idealized plume dispersion in the region of horizontal flow.

that dispersion across streamlines is small, we need consider only the plume associated with the direction of the stagnation streamline θ_d.

In order to derive an expression for the maximum concentration C_m at A, we assume that the time of travel from the source to A is small compared with the Lagrangian time scale of the horizontal velocity fluctuations. This is consistent with the large wind meandering observed during stable conditions. It implies that "particles" originating from the source travel along straight lines for the source-receptor distances of interest. In order to formulate the model for C_m, let us first take a more general approach and then specialize the resulting expression to the case at hand.

Figure 6.9 shows the geometry we shall consider in our derivation. Our assumption of straight-line trajectories implies that the particles emitted at Q into the angle $\Delta\theta$ will all pass through the arc CD. Now let us write a mass balance for CD:

$$Q \times \text{fraction of particles emitted into } \Delta\theta = \overline{C}^z(r,\theta)U_\theta r\Delta\theta, \quad (6.52)$$

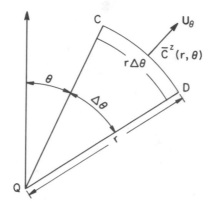

Figure 6.9. Geometry used to derive the expression for the concentration at the stagnation point A.

where $\overline{C}^z$ is the vertically integrated concentration at (r,θ). The fraction f emitted into $\Delta\theta$ can be readily expressed as

$$f = P(\theta)\Delta\theta ,\tag{6.53}$$

where $P(\theta)\Delta\theta$ is the probability that the horizontal wind blows into $\Delta\theta$ around the direction θ. Substituting (6.53) into (6.52), we find

$$\overline{C}^z(r,\theta) = \frac{QP(\theta)}{U_\theta r} .\tag{6.54}$$

We have assumed that the wind speed U_θ does not vary along r. This is consistent with our hypothesis that we need not consider the presence of the hill in estimating the impaction concentration.

The expression for $C(r,\theta,z)$ can be readily written

$$C(r,\theta,z) = \frac{QP(\theta)}{U_\theta r}S(z) ,\tag{6.55}$$

where $S(z)$ is an appropriate shape function. If we assume that $S(z)$ is Gaussian, we find

$$C(r,\theta,z) = \frac{QP(\theta)}{2U_\theta r\sqrt{\pi}}\left\{\exp\left[-\frac{(H_s - z)^2}{2\sigma_z^2}\right] + \exp\left[-\frac{(H_s + z)^2}{2\sigma_z^2}\right]\right\} ,\tag{6.56}$$

where H_s is the source height, and we have assumed reflection at the ground.

C_m can now be simply written as

$$C_m = C(d, \theta_d, z) . \tag{6.57}$$

Notice that this formulation for the impaction concentration does not require a mean wind direction or speed. Modelers who have dealt with highly variable winds can appreciate the advantage of this formulation.

We now provide a correction to account for cases in which the source-receptor distance is too large for us to assume straight-line trajectories of particles. To do so, we use the result (see Ch. 5)

$$\sigma_y = \sigma_v t / (1 + t/2T_{Lv})^{1/2} . \tag{6.58}$$

The linear growth rate corresponds to the small-time straight-line trajectories of particles. At longer times, particles forget their initial velocities and their spread is now described by the square-root law. This suggests the following modification to (6.55):

$$C_m = \frac{Q\, P(\theta_d)(1 + t/2T_{Lv})^{1/2} S(z)}{U_\theta d} . \tag{6.59}$$

Notice that if $P(\theta)$ is Gaussian, we can rewrite Eq. (6.59) as follows:

$$C_m = \frac{Q}{\sqrt{2\pi}\sigma_\theta d U_\theta} \exp\left[-\frac{(\theta_d - \theta_0)^2}{2\sigma_\theta^2}\right]\left(1 + \frac{t}{2T_{Lv}}\right)^{1/2} S(z) . \tag{6.60}$$

Here θ_0 is the mean wind direction. If we notice that $\sigma_\theta d \cong \sigma_v t$, Eq. (6.60) reduces to the familiar form

$$C_m = \frac{Q}{\sqrt{2\pi}\sigma_y U_\theta} \exp\left[-\frac{(\theta_d - \theta_0)^2}{2\sigma_\theta^2}\right] S(z) . \tag{6.61}$$

Therefore, the primary justification for the correction in Eq. (6.59) is that it reduces to the correct form when $P(\theta)$ is Gaussian.

Equation (6.59) forms the basis for estimating the concentrations on the hill's surface when the plume is released below H_c. To calculate the concentration at any receptor, x_r, y_r, z_r (see Fig. 6.8), we simply use the distance, $d + s_a$, measured along the hill to estimate σ_z.

The LIFT model

The LIFT model, a name suggested by the lifting of the plume over the hill, applies to flow above H_c. To develop this model, we idealize the flow shown in Fig. 6.7 as that shown in Fig. 6.10. The distance between the source to the point at which H_c intersects the hill surface is denoted by x_0. The plume above H_c is assumed to encounter the hill at x_0. This is shown as a sharp transition of the hill height to H_c. The flow above H_c is taken to follow the contour of the hill. This allows us to represent irregular terrain by a flat surface if we measure distance along the contours.

In order to calculate the concentration at a distance x^*, we assume that flow and dispersion properties undergo a discontinuous change at x_0. The properties assigned to the modified flow will, in principle, correspond to the average changes between 0 and x^*. Note that all modified flow variables are signified by $*$.

We have seen earlier that the hill causes a depression of the streamlines in the z-direction and an expansion of the streamlines in the y-direction. These changes are accompanied by a speed-up of the wind and an increase in the turbulence levels. Although these changes occur simultaneously, we assume that the distortion between 0 and x^* can be represented as a step change at x_0. The distorted plume is then allowed to disperse into an environment typified by U^*, σ_{y^*} and σ_{z^*}. These changes are illustrated in Figs. 6.10a and 6.10b. Note that plume distortion can displace the plume centerline to one side of the original centerline at x_0. However, this is not relevant to the analysis, and we take the displacement to be entirely in the vertical.

To proceed with our analysis, we focus attention on the plume element A and its mapping A^* (Fig. 6.10b) on the distorted plume. To trace the evolution of the distorted plume over the hill, we represent it by a set of point sources whose strength dQ^* is given by

$$dQ^* = C^*(0, y^*, z^*)U^*dy^*dz^* , \qquad (6.62)$$

where $(0, y^*, z^*)$ are the coordinates of a plume element A^*. We now assume that the distortion of the plume can be represented by

The Transformation:

$$z = z^*/T_z + H_c ; \quad y = y^*/T_y ;$$
$$x = x^* + x_0$$

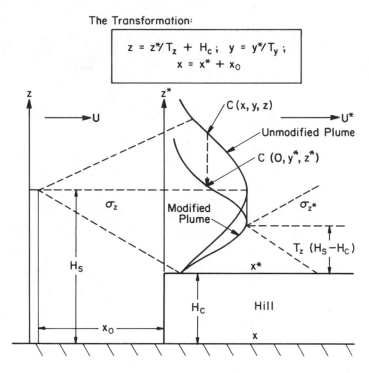

Figure 6.10a. Idealization of flow and dispersion above H_c. Modified variables are signified by $*$.

constant scaling factors T_z and T_y. This then allows us to calculate C^* (see Fig. 6.10a) from

$$C^*(0,y^*,z^*) = C(x_0,y^*/T_y,z^*/T_z + H_c) , \qquad (6.63)$$

where C is the concentration of the unmodified plume; we take C to be the Gaussian form

$$C(x_0,y,z) = \frac{Q}{2\pi\sigma_y\sigma_z U} \left\{ \exp\left[-\frac{(H_s - z)^2}{2\sigma_z^2} \right] + \exp\left[-\frac{(H_s + z)^2}{2\sigma_z^2} \right] \right\} . \qquad (6.64)$$

The hill-surface concentration at $(x^*,y^*,0)$ associated with dQ^* is again taken to be given by the Gaussian equation with reflection at $z^* = 0$,

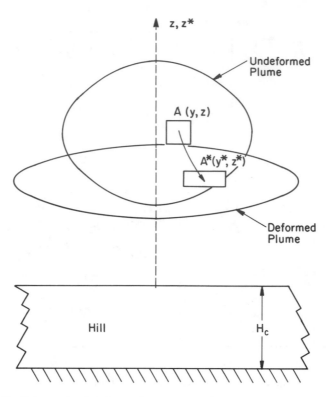

Figure 6.10b. Schematic of deformation of the plume at x_o. A maps onto A*. View is looking into mean wind.

$$d^2 C^*(x^*, y^*, 0) = \frac{dQ^*}{\pi U^* \sigma_{y*} \sigma_{z*}} \exp\left(-\frac{z^{*2}}{2\sigma_{z*}^2}\right) \exp\left(-\frac{y^{*2}}{2\sigma_{y*}^2}\right) . \quad (6.65)$$

Then the $C^*(x^*, y^*, 0)$ is given by the integral

$$C^*(x^*, y^*, 0) = \int_{-\infty}^{\infty} \int_0^{\infty} d^2 C(x^*, y^*, 0) . \quad (6.66)$$

Before we can perform the integration, we need to interpret the modified plume spread σ_{z*} and σ_{y*}. If $H_c = 0$ and the flow is not modified, Eq. (6.66) represents a method of calculating the ground-level concentration from a concentration profile at x_0. We also know that if we do the mathematics correctly, the result from (6.66) in this case should be identical to the concentration calculated directly from the point source at $x = 0$. The only way of guaranteeing this identity is to express the σ_z that appears in Eq. (6.66) as

$$\sigma_{z\bullet}^2(x^*) = [\sigma_z^2(x_0 + x^*) - \sigma_z^2(x_0)]T_{\sigma z}^2 \;, \tag{6.67}$$

where $T_{\sigma z}$ represents the effect of the hill on plume spread. It is a gross parameterization for the combined effects of changes in turbulence levels, wind speed, and other meteorological variables that control σ_z. A similar equation applies to $\sigma_{y\bullet}$.

We are now in a position to obtain a definite result from Eq. (6.66). Substituting (6.63) to (6.66), and performing the integration yields

$$
\begin{aligned}
C^*(x^*, y^*, 0) = \frac{Q}{\sqrt{2\pi}\sigma_{ze}U} \Bigg\{ &\exp\left[-\frac{1}{2}\left(\frac{H_s - H_c}{\sigma_{ze}}\right)^2\right] \\
&\times \mathrm{erfc}\left(\frac{H_c - H_s}{\sqrt{2}\sigma_{ze}}\frac{\sigma_{z\bullet}}{T_z\sigma_{zo}}\right) \\
&+ \exp\left[-\frac{1}{2}\left(\frac{H_s + H_c}{\sigma_{ze}}\right)^2\right] \\
&\times \mathrm{erfc}\left(\frac{H_c + H_s}{\sqrt{2}\sigma_{ze}}\frac{\sigma_{z\bullet}}{T_z\sigma_{zo}}\right) \Bigg\} \\
&\times \exp\left(-\frac{y^{*2}}{2\sigma_{ye}^2 T_y^2}\right)
\end{aligned}
\tag{6.68}
$$

where erfc is the error function complement.

Here $\sigma_{zo} = \sigma_z(x_0)$, and the effective σ_{ze} is

$$\sigma_{ze}^2 = \sigma_{zo}^2 + \sigma_{z\bullet}^2/T_z^2 \tag{6.69}$$

and

$$\sigma_{ye}^2 = \sigma_{yo}^2 + \sigma_{y\bullet}^2/T_y^2 \tag{6.70}$$

$$\sigma_{y\bullet}^2 = [\sigma_y^2(x_0 + x^*) - \sigma_y^2(x_0)]T_{\sigma y}^2 \;, \tag{6.71}$$

where T_y represents the deformation factor in the y direction and $T_{\sigma y}$ is the correction factor for plume spread.

We shall find it convenient to refer to T_z, T_y, etc., collectively as T factors. It would seem that in order to use the dispersion model, all we have to do is to calculate them; however, this is not easy because the theory to do so is far from developed. It is only for the simplest of hill shapes that we can even make estimates for the T factors. So how do we use the dispersion model that we have developed? My answer is that the model should be looked upon as a framework to analyze observations. The T factors should be derived semi-empirically from observations. This is not too different from the adjustment of terrain correction factors to get the correct concentrations. However, in our case, the more elaborate framework of the dispersion models allows for a more justifiable adjustment of the T factors.

6.3.4. Application of the Dispersion Model

Simplification of the model

Before illustrating the applications of the complex terrain model, it is useful to describe its limitations. As it is now formulated, it does not account for the vertical inhomogeneity of the stable boundary layer. We know that both σ_w and U vary substantially across the depth of the SBL. These variations by themselves will necessitate major modifications to predictions from models that assume homogeneity. Until we have a model that performs well in stable conditions over *flat* terrain, it is difficult to develop a model that has to include the additional complications of complex terrain. Hunt (1982) pointed out that complex terrain will increase the gradients of temperature and velocity of the incoming flow. Note that the velocity must vanish at $z = H_c$ while the remaining flow goes over the top of the hill. This shear is not included in the LIFT model. This is a shortcoming because shear by itself (Hunt, 1982) can depress the plume centerline and thus increase the concentration on the hill.

The flow distortion, the wind speed-up, and increased turbulence over the hill have been modeled in LIFT with T factors that are assumed to be uniform across the plume. Clearly this cannot be physically accurate when we know that these effects have their largest magnitudes close to the surface of the hill and decrease with height. These limitations suggest that the application of the model is not a simple matter of determining values for the T factors. The basic *structure* of the model might require modification in order to explain observations.

The model that will be tested against observations is a modification of the structure presented earlier. Like other dispersion models, it is semi-empirical in the sense that its formulation draws heavily upon our experience with describing observations. Its basic theoretical framework is based upon the equations for the LIFT and WRAP models.

Our model is best explained by considering Eq. (6.70) that describes the LIFT model. When H_c is equal to zero, the expression for the ground-level concentration becomes

$$C^*(x^*,y^*,0) = \frac{Q}{\pi U \sigma_{ye}\sigma_{ze}} \exp\left(-\frac{H_s^2}{2\sigma_{ze}^2}\right) \exp\left(-\frac{y^2}{2\sigma_{ye}^2}\frac{1}{T_y^2}\right) . \quad (6.72)$$

I rewrite this equation using the following definitions:

$$\sigma_{zh}^2 = T_z^2\sigma_z^2(x_0) + T_{\sigma z}^2[\sigma_z^2(x) - \sigma_z^2(x_0)] \quad (6.73a)$$

$$\sigma_{yh}^2 = T_y^2\sigma_y^2(x_0) + T_{\sigma y}^2[\sigma_y^2(x) - \sigma_y^2(x_0)] , \quad (6.73b)$$

where the subscript h refers to hill. We expect $T_z < 1$ (compression of streamlines) and $T_y > 1$. Both $T_{\sigma z}$ and $T_{\sigma y}$ are likely to be greater than 1, signifying the increase in turbulence over the hill. However, the appearance of the increased wind speed in the expressions for the sigmas might force these T factors to be close to 1.

The definitions (Eqs. 6.73) for the hill sigmas are consistent with the general observation that σ_z decreases while σ_y increases when the plume passes over the hill. With (6.73), Eq. (6.72) becomes

$$C_x(x,y,0) = \frac{QT_yT_zT_u}{\pi U_h\sigma_{zh}\,\sigma_{yh}} \exp\left(-\frac{H_s^2T_{hz}^2}{2\sigma_{zf}^2}\right) \exp\left(-\frac{y^2}{2\sigma_{yh}^2}\right) , \quad (6.74)$$

where we have replaced U by U_h/T_u; T_u is a wind speed-up factor, and the subscript f signifies unmodified, flat-terrain values. The terrain correction factor T_{hz} is now defined by

$$T_{hz} = T_z\sigma_{zf}/\sigma_{zh} . \quad (6.75)$$

I change the notation in order to reduce the equations to a form that is similar to that used in available complex terrain models. Mass continuity requires that

$$T_zT_yT_u = 1 , \quad (6.76)$$

so that (6.74) becomes

$$C_h(x,y,0) = \frac{Q}{\pi U_h\sigma_{zh}\,\sigma_{yh}} \exp\left[-\left(\frac{H_sT_{hz}}{\sqrt{2}\sigma_{zf}}\right)^2\right] \exp\left(-\frac{y^2}{2\sigma_{yh}^2}\right) . \quad (6.77)$$

We could use Eq. (6.77) if we knew how to calculate the hill variables. Although such a calculation might be possible for simple hill shapes (see Strimaitis et al., 1985), there is no doubt that the complex terrain encountered in real problems will require elaborate theories with its host of inevitable approximations. This suggests that we should simplify Eq. (6.77) and then rely on empiricism to develop a practical complex terrain model.

A plausible assumption is

$$U_h \sigma_{yh} \sigma_{zh} = U_f \sigma_{yf} \sigma_{zf} \; . \tag{6.78}$$

The rationale is that the changes in the variables on the left-hand side of the equation compensate for each other, and the error involved in replacing them with the corresponding flat-terrain variables is likely to be small. Then Eq. (6.77) becomes

$$C_h(x,y,0) = \frac{Q}{\pi U_f \sigma_{yf} \sigma_{zf}} \exp\left[-\frac{1}{2}\left(\frac{H_s T_{hz}}{\sigma_{zf}}\right)^2\right] \exp\left(-\frac{y^2}{2\sigma_{yh}^2}\right) \; . \tag{6.79}$$

Equation (6.79) is very similar to that used in current practice, except that the terrain correction factor T_{hz} is now

$$T_{hz} = T_z \sigma_{zf} / \sigma_{zh} \; . \tag{6.80}$$

When x is close to x_0, and the hill effects are unimportant,

$$T_{hz} \to 1 \quad \text{as expected} \; , \tag{6.81}$$

and when x is much larger than x_0 so that hill effects dominate, we find

$$T_{hz} \to T_z / T_{\sigma z} = T_r \; . \tag{6.82}$$

With the above definition for T_r, (6.80) can be rewritten as

$$T_{hz} = T_r / [1 - R^2(1 - T_r^2)]^{1/2} \; , \tag{6.83a}$$

where

$$R \equiv \sigma_{zf}(x_0) / \sigma_{zf}(x) \; . \tag{6.83b}$$

This form allows us to collapse T_z and $T_{\sigma z}$ into one factor, T_r, which can be determined by fitting model predictions to observations. However, its value, as well as its variation with relevant inputs, should be guided by available theory.

Notice that the calculation of the off-centerline concentration still requires σ_{yh}. However, the maximum centerline concentration depends only upon one empirical factor, T_r. If the sigmas and U are based on measurements made close to the hill, we can assume that they include some of the effects of the hill. We can then use them as hill variables in Eq. (6.77). In fact, if short-term measurements of wind speed and direction are available at a tower close to the hill, we can estimate $P(\theta)$ and hence use Eq. (6.55):

$$C_h(r,\theta,0) = \frac{QP(\theta)}{\sqrt{2\pi}U_\theta r\sigma_{zf}} \exp\left[-\frac{1}{2}\left(\frac{H_s T_{hz}}{\sigma_{zf}}\right)^2\right], \qquad (6.84)$$

where r is the distance from the source to the receptor and θ is the direction of the line joining the source to the receptor.

We are now in a position to write the expression for C_h when H_c is finite:

$$C_h(x,0,0) = \frac{Q}{2\pi U_f \sigma_{yf}\sigma_{zf}}\left\{\exp\left[-\frac{1}{2}\left(\frac{(H_s-H_c)T_{hz}}{\sigma_{zf}}\right)^2\right]f_u\right.$$
$$\left. + \exp\left[-\frac{1}{2}\left(\frac{(H_s+H_c)T_{hz}}{\sigma_{zf}}\right)^2\right]f_\ell\right\}, \qquad (6.85)$$

where

$$f_u = \text{erfc}\left[\frac{(H_c-H_s)T_{hz}}{\sqrt{2}\sigma_{zf}}\frac{(1-R^2)^{1/2}}{RT_r}\right] \qquad (6.86a)$$

and

$$f_\ell = \text{erfc}\left[\frac{(H_c+H_s)T_{hz}}{\sqrt{2}\sigma_{zf}}\frac{(1-R^2)^{1/2}}{RT_r}\right]. \qquad (6.86b)$$

At $x = x_0$, $C_h(x_0,0,0)$ reduces to the impingement concentration

$$C_h(x_0, 0, 0) = \frac{Q}{2\pi U_f \sigma_{yf} \sigma_{zf}} \left\{ \exp\left[-\frac{(H_s - H_c)^2}{2\sigma_{zf}^2} \right] \right.$$
$$\left. + \exp\left[-\frac{(H_s + H_c)^2}{2\sigma_{zf}^2} \right] \right\} . \tag{6.87}$$

This means that a plume with its centerline close to H_c can give rise to concentrations that are approximately equal to the centerline concentration. An examination of the preceding equations shows that complex terrain will always cause an amplification of the corresponding concentration over flat terrain. In designing this feature of our model, we were guided by results obtained in the EPA wind tunnel by Thompson and Snyder (1984).

Equation (6.85) then represents the model for concentrations that occur at receptors above H_c. For receptors below H_c, we will use the impingement calculation of Eq. (6.55). Notice that if the release is below H_c, the maximum concentration will also occur below H_c as long as the centerline of the plume is directed along the stagnation streamline. When the release is above H_c, the maximum concentration on the hill will occur above H_c.

Model testing

One of the objectives of model testing is to find out whether it is possible to derive a semi-empirical value of T_r that will yield acceptable model predictions.

The first data set we use is derived from observations made during the first Small Hill Impaction Study (SHIS #1), which was conducted at Cinder Cone Butte (CCB), a roughly axisymmetric, isolated 100-m-tall hill in the Snake River Basin near Boise, Idaho. The field program consisted of ten flow visualization experiments and 18 multi-hour tracer gas experiments conducted during stable conditions. Two tracer gases (SF_6 and CF_3Br) were released from various locations around the hill using a mobile release crane, and concentrations on the hill surface were measured by a network of samplers. A comprehensive meteorological data base using data from six towers, a tethersonde, and free balloons was compiled. The data include wind turbulence and other meteorological parameters derived specifically for the tracer gas release heights as well as hourly average emission rates and ground-level concentration data. The modeling studies described here were performed with the SF_6 data base.

We chose 79 experimental hours for model testing. Of these, 27 hours corresponded to cases in which the maximum observed concentrations occurred below H_c, and the remaining 52 hours contained the maximum observed concentrations above H_c. The model was tested in the following manner:

(1) For cases with maximums below H_c, we used the impingement model to calculate the concentrations at all receptors below H_c. We used the stagnation angle in the probability density $P(\theta)$ in Eq. (6.55). The maximum observed was then paired with the maximum of the predicted concentrations. We did not pair in space because of the acknowledged difficulty in determining the plume centerline, especially in complex terrain situations. The average of the top two and five observed concentrations was also compared with the corresponding predictions.

(2) We used a similar procedure for maximums above H_c except that we used Eq. (6.85) to estimate concentrations. The Gaussian crosswind distribution was replaced by $P(\theta)/r$ as in the impingement calculation, except that θ corresponded to the receptor angle.

The second data set was obtained from the Small Hill Impaction Study 2 (SHIS #2), which was conducted during the fall of 1982 at the Hogback Ridge (HBR). HBR is a long, roughly 100-m-tall ridge 15 miles west of Farmington, New Mexico. The location is characterized by frequent, stable easterly winds at night. The experiments were similar to those performed at CCB and included the following:

- releases east of HBR of two tracer gases (SF_6 and CF_3Br) and oil fog using a mobile 150-ft crane and a tower as source platforms;

- meteorological measurements using instrumented towers, acoustic sounders, anemometers, and tethersondes;

- ground-level tracer gas concentrations;

- lidar measurements; and

- photographs and videotapes.

We selected 35 hours for model testing, 28 hours of which contained maximum observed concentrations below H_c; the remaining 7 consisted of cases with maximum observed concentrations above H_c. These concentrations were paired with predictions using the procedure described earlier for data from SHIS #1.

The two data sets for SHIS #1 and SHIS #2 are representative of conditions in two broad classes of complex terrain, namely three-dimensional hills and two-dimensional ridges. Therefore, a model tested against these data is likely to be of general value in estimating dispersion in complex terrain.

All the meteorological inputs required by the model were evaluated at source height. The vertical spread σ_z was calculated with the equation presented in the chapter on dispersion in the stable boundary layer (Venkatram et al., 1984). T_r was taken to be 0.4. This choice is based on theoretical work (Strimaitis et al., 1985) that indicates that $T_z \simeq 0.6$ at the crest of a two-dimensional ridge. Taking $T_{\sigma z} > 1$, $T_r = 0.4$ is not unreasonable.

It is seen from Fig. 6.11 that although the scatter is not small, there is reasonable correspondence between model predictions and observations at CCB. The model performs better at HBR in SHIS #2, as seen in Fig. 6.12. In general, our complex terrain model provides better estimates of observations at HBR than at CCB. This is to be expected because slight errors in the model inputs have a bigger effect on whether the plume reaches a receptor on a three-dimensional hill than on a two-dimensional ridge. The performance of the complex terrain dispersion model is encouraging. However, it is necessary to evaluate it against a much larger set of data before it can be considered to be acceptable. This evaluation will inevitably result in modifications in the parameters and the structure of the model.

6.3.5. Future Work

It should be clear from the preceding discussion that our understanding of dispersion in complex terrain is far from adequate. The models described in this chapter represent a start toward placing complex terrain modeling on sounder footing, but we still need to go some way before these models become suitable for operational use. However, I believe that the model described in Sec. 6.3.4 is a good foundation for a model that can be used in regulatory applications.

Future work in complex-terrain modeling must address an effect that has not been discussed thus far. This is related to the surface boundary layer in which gradients of concentrations are likely to be different from that predicted by the Gaussian equation. This layer will be more important in complex terrain than in flat terrain because

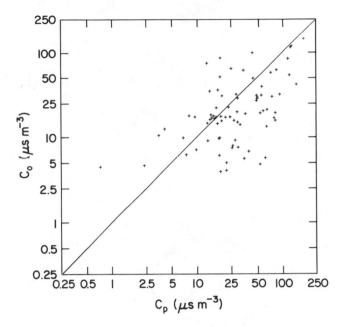

Figure 6.11a. Predicted vs. observed maximum concentrations at Cinder Cone Butte, Idaho. Concentrations are normalized by emission rates.

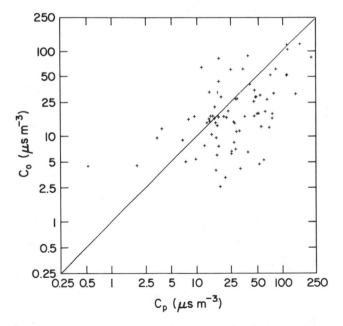

Figure 6.11b. Predicted vs. average of observed top two concentrations at Cinder Cone Butte.

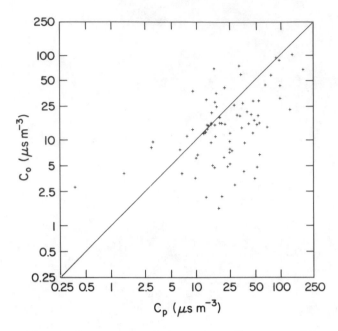

Figure 6.11c. Predicted vs. average of observed top five concentrations at Cinder Cone Butte.

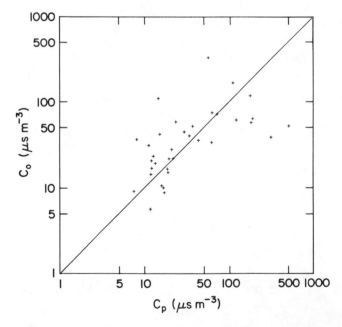

Figure 6.12a. Predicted vs. observed maximum concentrations at Hogback Ridge, New Mexico.

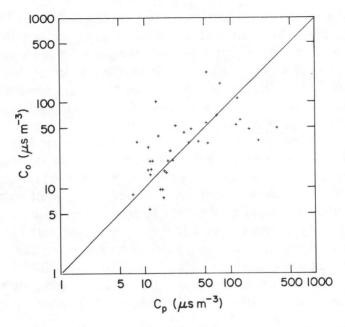

Figure 6.12b. Predicted vs. average of observed top two concentrations at Hogback Ridge.

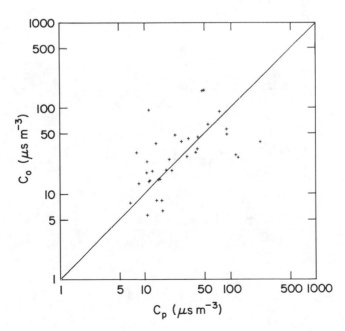

Figure 6.12c. Predicted vs. average of observed top five concentrations at Hogback Ridge.

the major effect of a hill is to bring the plume close to the hill's surface. This allows the surface boundary layer to have increased access to the high concentrations near the plume centerline. It can, in effect, reach into the high concentrations and bring them down to the surface. I believe that this is the major mechanism for the amplification of the flat terrain concentration in complex terrain. It also explains the popularity of the empirical correction of lowering the plume height over complex terrain. The terrain correction factor applied only to the plume height is an approximate representation of the boundary layer effect. This is why I think that the model of Sec. 6.3.4 has some promise. However, if we want to account explicitly for the surface boundary layer, we have to take another approach. My ideas on a possible approach are presented here.

The effect of the boundary layer is to mix down the material through a height ℓ from the hill's surface. To model this we need to derive an expression for the concentration at any receptor height z measured from the hill's surface. A little algebra yields $C_h(x,y,z)$,

$$C_h(x,y,z) = \frac{Q}{4\pi U_h \sigma_{yh} \sigma_{zh}} \exp\left(-y^2/2\sigma_{yh}^2\right) (T_1 + T_2 + T_3 + T_4) ,$$

(6.88)

where

$$T_1 = \exp\left[-\frac{(H_m + z)^2}{2\sigma_{zh}^2}\right] \left\{ \mathrm{erfc}\left[\frac{H_m F}{\sqrt{2}\sigma_{zh}} \left(1 - \frac{z}{H_m} \frac{1}{F^2}\right)\right] \right\} \quad (6.89a)$$

$$T_2 = \exp\left[-\frac{(H_m - z)^2}{2\sigma_{zh}^2}\right] \left\{ \mathrm{erfc}\left[\frac{H_m F}{\sqrt{2}\sigma_{zh}} \left(1 + \frac{z}{H_m} \frac{1}{F^2}\right)\right] \right\} \quad (6.89b)$$

$$T_3 = \exp\left[-\frac{(H_p + z)^2}{2\sigma_{zh}^2}\right] \left\{ \mathrm{erfc}\left[\frac{H_p F}{\sqrt{2}\sigma_{zh}} \left(1 - \frac{z}{H_p} \frac{1}{F^2}\right)\right] \right\} \quad (6.89c)$$

$$T_4 = \exp\left[-\frac{(H_p - z)^2}{2\sigma_{zh}^2}\right] \left\{ \mathrm{erfc}\left[\frac{H_p F}{\sqrt{2}\sigma_{zh}} \left(1 + \frac{z}{H_p} \frac{1}{F^2}\right)\right] \right\} , \quad (6.89d)$$

where

$$F \equiv \left[\sigma_{zf}^2(x) - \sigma_{zf}^2(x_0)\right]^{1/2} / T_z \sigma_{zf}(x_0) \qquad (6.90)$$

and

$$H_m = T_z(H_c - H_s) \qquad (6.91a)$$

$$H_p = T_z(H_c + H_s) . \qquad (6.91b)$$

We have indicated earlier that the surface boundary layer mixes the concentration profile through a height $\ell(x)$ that can be likened to an internal boundary layer that erodes the profile aloft. Then it is reasonable to assume that the modified concentration on the hill's surface is

$$\tilde{C}_h(x,y,0) = [C_h(x,y,0) + C_h(x,y,\ell)]/2 , \qquad (6.92)$$

where C_h is the concentration given by Eq. (6.88).

In order to use Eq. (6.92), we need an expression for the mixing height $\ell(x)$. A plausible equation suggested by Miyake (described in Panofsky and Dutton, 1984) is

$$\frac{d\ell}{dx} = \frac{u_*}{U} , \qquad (6.93)$$

where the right-hand side is evaluated at $z = \ell$. On substituting the similarity expression for U under stable conditions, Eq. (6.93) results in the following implicit equation for $\ell(x)$:

$$\ell\left[\ln\left(\frac{\ell}{z_0}\right) - 1\right] + \frac{\beta\ell^2}{2L} = k(x - x_0) . \qquad (6.94)$$

Here $\beta = 4.7$, k is the von Kármán constant, z_0 is the roughness length, and L is the Monin-Obukhov length. We have shown earlier (Venkatram, 1980) and in Ch. 5 how L can be estimated from empirical relationships.

I have taken some trouble to describe the possible effects of the surface boundary layer because complex terrain models can be improved considerably. Until these improvements are made, I would recommend the use of the plume height terrain correction factor described in Sec. 6.3.4. As noted before, it is similar to that used in popular complex terrain models. However, it has much better theoretical justification in the sense that it is derived from a general model of dispersion in complex terrain.

6.4. Model Evaluation

6.4.1. Introduction

Most modelers would agree upon the following attributes of a good model:
(1) It should incorporate a realistic description of the physical processes that govern the system being modeled.
(2) It should provide adequate estimates of available observations made on the relevant system.

The first attribute provides the confidence necessary to apply the model beyond the range of observations used to test the model.

Air pollution models are, in general, only broad descriptions of the complex processes, such as turbulence, that govern dispersion. This means that it is very difficult to provide objective measures of the two attributes of model performance. Because our understanding of the processes of dispersion is rather incomplete, it is possible to have a multitude of views on what constitutes an adequate description of these processes. These gaps in our understanding also give rise to relatively large deviations between model predictions and observations.

In the past few years, the modeling community in the United States has attempted to crystallize its thoughts on model evaluation by holding two workshops on the subject (Fox 1981, 1984). One of the products of these workshops was a listing of statistical procedures to describe the residuals between model predictions and observations. Although several other recommendations were made, these statistical techniques were looked upon as the major result by the modelers. It has now become fashionable to compute all sorts of statistical measures that are supposed to quantify performance. In my opinion, the reason for this is that it is easy to compute statistics and compare numbers (which are often meaningless). On the other hand, it is much more difficult to examine the science that governs models. At the present, model evaluation has become synonymous with the computation of performance statistics. I intend to emphasize the need to get away from this narrow view of model testing. My ideas on model evaluation are summarized as follows:

(1) It is a series of tests to determine whether the model is an adequate representation of the real system being modeled. These tests are specific to the model at hand and are best designed by the model builders in conjunction with other scientists interested in model development.

(2) Evaluation is a process in which the model is improved on the basis of the comparison between model predictions and observations. This interaction between comparison and model improvement will give rise to additional testing procedures. Therefore, model evaluation is an evolutionary process that is an integral part of the model development process.

(3) Model evaluation is *not* an application of specified statistical tests, the results of which pass or fail the model. While statistical measures are useful in describing the performance of the model, they are not a major component of model evaluation. On the other hand, the success of a model evaluation program is critically dependent upon the scientific judgment of the model builders.

(4) The product of the evaluation process is a summary of the results from a battery of tests performed on the model. This product will help to determine the acceptability of the model to the interested scientific community.

Because model evaluation consists of a host of tests that have to be tailored to the model at hand, it is not possible to be very specific about the subject. The best I can do is to discuss the set of principles that ought to guide the process of model evaluation. I shall also discuss the topic of "inherent uncertainty," which plays an important role in examining the performance of models.

6.4.2. Steps in Model Evaluation

The basic steps involved in model evaluation are

(1) examination of the structure of the model,
(2) sensitivity analysis, and
(3) testing of model predictions against observations.

As pointed out earlier, it is important to examine the assumptions involved in formulating the model being evaluated. This step is extremely important because the relatively large expected deviation between model predictions and observations makes discrimination between different models an ambiguous exercise. Without such an examination it is often possible to make one model look better than the other simply by choosing the appropriate statistical measures. Modelers who have participated in some of the recent evaluation "Olympics" can appreciate this. It is recognized that it is difficult to judge what constitutes good science. However, I am reasonably sure that a group of experts can readily reach consensus on whether a given model contains adequate science.

Sensitivity analysis is important because it identifies the critical inputs of the model. This information can be used to allocate resources to a measurement program. In testing the model, we would concentrate on the inputs to which the model is most sensitive. We would then be able to find out whether nature displays the same sensitivity to these inputs. It is necessary to test the model over the widest possible range of model predictions because of the unavoidably large expected deviation between model predictions and observations.

The comparison of model predictions with observations is the core of model evaluation. It is often not realized that this comparison is not a simple matter of computing statistics of the residuals. It is important to understand the relationship between model predictions and observations before getting into descriptive statistics. This relationship is explored in the next few sections.

6.4.3. The Relationship Between Model Prediction and Observations

Inherent uncertainty

The past few years have seen a steady improvement in our understanding of the physics of the planetary boundary layer (see Nieuwstadt and van Dop, 1982). Air quality models that have incorporated these advances do show marked improvement over existing models. In spite of these improvements in model performance, it is discouraging to find that the deviations between model predictions and observations are still relatively large. This is illustrated in the results obtained in a model development project sponsored by EPRI (Weil et al., 1986). Figure 6.13 compares the performance of two models. The first model, commonly referred to as CRSTER, is based on the Pasquill-Gifford sigma curves developed at least 20 years ago. It is recommended by the EPA for regulatory applications. The second

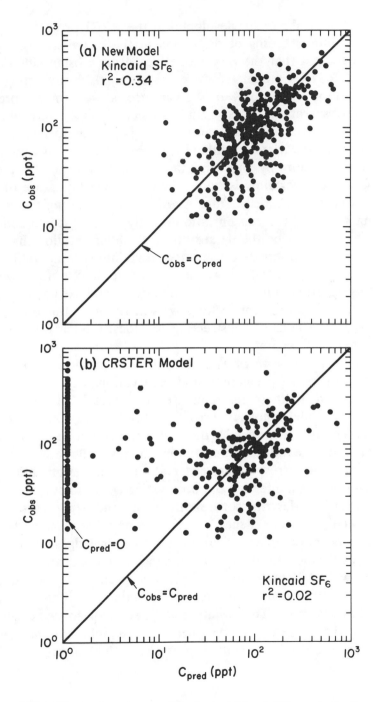

Figure 6.13. Observed versus predicted ground-level SF_6 concentrations at the Kincaid power plant for (a) a new model, and (b) the CRSTER model.

model (Weil et al., 1986), developed in the EPRI project, is based on recent understanding of dispersion in the convective boundary layer. It is seen that the new model performs substantially better than the CRSTER model. However, r^2 is only 0.34, suggesting that we still cannot explain 66% of the variance between model predictions and observations. This result is typical of those obtained from other studies. The modeling community is beginning to suspect that this unexplained variance is inherent in the sense that it cannot be reduced much more by improving the physics of air quality models. This inherent uncertainty is caused by the stochastic nature of turbulence that controls dispersion. Another way of stating this is that we cannot predict the concentration exactly because it is practically impossible to describe the details of the turbulent motion that governs the concentration. Because this uncertainty is comparable with the predicted concentration, we cannot ignore it. We need to estimate this uncertainty in order to both evaluate and use an air quality model. In the subsequent sections we will reiterate this point, and also suggest methods of constructing statistical models for inherent uncertainty.

Model evaluation revolves around the examination of the residuals between model predictions and observations. The four factors that contribute to these residuals are 1) errors in model physics, 2) errors in model inputs, 3) errors in observations, and 4) "inherent" uncertainty. In principle, the first three components of model uncertainty can be reduced by examining the residuals between model predictions and corresponding observations. This examination will depend upon the model and the observations used to evaluate it. In a later section, I describe some methods of conducting a systematic residual analysis. Here, I shall concentrate on inherent uncertainty.

The concept of "inherent" uncertainty acknowledges the fact that we cannot predict the details of the stochastic concentration field. The best we can hope to do is to be able to predict the average over a large number of concentrations corresponding to different realizations of an ensemble. The deviation between the predicted ensemble average and any one observed realization is the inherent uncertainty, whose variance can be formally expressed as (Venkatram, 1979; Fox, 1984)

$$\sigma_c^2 = \langle (C - \langle C \rangle)^2 \rangle , \qquad (6.95)$$

where C is the concentration and angle brackets refer to an ensemble average. In order to convert Eq. (6.95) into an operational definition we need to define an ensemble. The textbook (Lumley and Panofsky, 1964) definition is "a set of experiments corresponding to fixed external conditions." In order to use this definition we need to relate the term "external conditions" to measurements that are normally made for modeling applications. One way of doing so is to use the model inputs to define the ensemble.

The convenience afforded by this definition can be best shown by writing the observed concentration C_0 as follows:

$$C_0 = C_0(\alpha, \beta) \ . \tag{6.96}$$

In Eq. (6.96), α refers to known model inputs—wind speed, mixed-layer height, etc.; β refers to the unresolved turbulent quantities that are inaccessible to us. In principle we can conduct a set of experiments in which the values of α are fixed. In these experiments, although α is fixed, β can take on any value. This implies that C_0 will vary from experiment to experiment. Any one observation belonging to this ensemble can always be written as

$$C_0(\alpha, \beta) = \overline{C_0(\alpha, \beta)}^\beta + c(\alpha, \beta) \ . \tag{6.97}$$

The first term on the right hand side of the equation is an average over all the observations corresponding to different values of the unknown set β. Since this ensemble average is over all possible values of β, it can only be a function of the model inputs α, and we can write

$$C_0(\alpha, \beta) = C_p(\alpha) + c(\alpha, \beta) \ . \tag{6.98}$$

Our definition of the ensemble leads naturally to the conclusion that the model prediction $C_p(\alpha)$ is the ensemble average. The deviation $c(\alpha, \beta)$ is a stochastic variable because it is a function of the unknown set of variables β.

In principle, the stochastic component $c(\alpha, \beta)$ can be reduced by increasing the set α of model inputs to include more of β. However, the nature of turbulence places limitations on this process, and we have to accept $c(\alpha, \beta)$ as inherent to the system being considered.

The idealized experiment that defines our ensemble is clearly impossible in practice. We cannot keep α fixed. Therefore, we cannot determine the statistics of the inherent uncertainty $c(\alpha,\beta)$ directly from observations. On the other hand, it might be possible to use indirect means, which, as we show later, depends upon a model for the statistics of $c(\alpha,\beta)$.

With the recent advent of supercomputers it is now possible to simulate some of the important features of turbulence. The computer can now be used as a numerical laboratory in which we can conduct experiments that would be impossible in the real world (Wyngaard, 1984). This suggests that direct estimates of σ_c should be possible in the very near future. In the past few years, the Langevin equation has been used to simulate dispersion (see Sawford, 1985). Although it has been applied primarily to idealized problems of theoretical interest, it has the potential for providing insight into the physics of inherent uncertainty in practical problems.

The preceding discussion points out that inherent uncertainty has to be explicitly accounted for in evaluating models. This point becomes rather clear if one thinks of modeling as an attempt to explain observations. This suggests that a model prediction has to contain information on the deterministic component $C_p(\alpha)$ as well as the stochastic component $c(\alpha,\beta)$ before it can be compared with an observation. In other words, an air quality model has to provide estimates of both $C_p(\alpha)$ and the statistics of $c(\alpha,\beta)$. In a later section, I suggest possible methods of evaluating this "expanded" model. It is worthwhile first to describe one possible method of estimating σ_c. This will illustrate some of the concepts presented in this section, and also provide estimates of the expected magnitude of σ_c.

Concentration fluctuations in the convective PBL

The modeling approach taken here is based on the assumption that most of the variation of the observations $C_0(\alpha,\beta)$ about the ensemble mean $C_p(\alpha)$ can be explained through the variation of a subset $\overline{\beta}$ of β. We also assume that we do know the form of $\overline{C}_0(\alpha,\overline{\beta})$, which in a sense represents a pseudo-observation. This implies that

$$c(\alpha,\beta) \simeq \overline{C}_0(\alpha,\overline{\beta}) - C_p(\alpha) , \qquad (6.99)$$

where

$$C_p(\alpha) = \overline{\overline{C}_0(\alpha,\overline{\beta})}^{\overline{\beta}} . \qquad (6.100)$$

Equation (6.100) says that the ensemble average $C_p(\alpha)$ is the average over all possible values of the pseudo-observation $\overline{C}_0(\alpha, \overline{\beta})$ for fixed values of the known inputs α. These ideas will be illustrated through an example.

In our example, the pseudo-concentration is assumed to be given by an expression suggested by Venkatram (1983). This formulation yields results that compare well with those from the water tank experiments of Willis and Deardorff (1978). Then, the "observed" cross-wind integrated concentration is

$$\overline{C}^y = \frac{2Q}{x} P(w|z); \quad w = w_s, \quad z = H_s . \tag{6.101}$$

In Eq. (6.101), Q is the emission rate from a point source with stack height H_s, x is the downwind distance to the receptor, and $P(w \mid z)$ is the probability density function of vertical velocity fluctuations at z. The velocity w_s is given by

$$w_s = -U H_s / x \tag{6.102}$$

and corresponds to the value required to bring plume material from the source to the receptor in a straight line. This physical picture is consistent with the assumption of infinite time scales in the convective boundary layer.

Next we will assume that we cannot predict the value of $\overline{C}^y$ because the precise form of $P(w|z)$ is unknown for a given realization. Although $P(w|z)$ is positively skewed (Lamb, 1982), the experience of Weil and Brower (1984) suggests that adequate estimates of concentration can be obtained with a Gaussian probability density function. We can then write

$$P(w|z) = \frac{1}{\sqrt{2\pi}\sigma_w} \exp\left[-\frac{(w - w_0)^2}{2\sigma_w^2}\right] . \tag{6.103}$$

In Eq. (6.103), σ_w is taken to be a time-averaged value obtained from measurements.

The mean vertical velocity w_0 is taken to be unknown for a given realization. This results in the variation of concentrations belonging to the ensemble defined by

$$\alpha = (Q, H_s, x, \sigma_w, U, T) \tag{6.104a}$$

$$\bar{\beta} = (w_0) . \tag{6.104b}$$

In Eq. (6.104a), T is the averaging time corresponding to σ_w and U. In the analysis to follow, it will be convenient to deal with $\ln \bar{C}^y$ given by

$$\ln \bar{C}^y = \ln \left(\sqrt{\frac{2}{\pi} \frac{Q}{\sigma_w x}} \right) - \frac{(w_s - w_0)^2}{2\sigma_w^2} . \tag{6.105}$$

The expression for the geometric mean concentration $\bar{C}_g^y$ follows readily from Eq. (6.105):

$$\bar{C}_g^y = \sqrt{\frac{2}{\pi} \frac{Q}{\sigma_w x}} \exp \left(-\frac{\langle w_0^2 \rangle}{2\sigma_w^2} \right) \exp \left(-\frac{w_s^2}{2\sigma_w^2} \right) , \tag{6.106}$$

where the angle brackets refer to an ensemble average. To estimate $\langle w_0^2 \rangle$, let us assume that the records of the vertical velocity $w(t)$, corresponding to the different realizations of the ensemble, are derived from a stationary time series with an integral time scale T_w. Then we can show that for $T \gg T_w$ (Lumley and Panofsky, 1964),

$$\langle w_0^2 \rangle = 2\sigma_w^2 T_w / T . \tag{6.107}$$

Note that σ_w^2 is not the ensemble variance in the sense that it corresponds to infinite averaging time. In our discussion, σ_w^2 constrains the energy in each of the realizations of the ensemble,

$$\sigma_w^2 = \frac{1}{T} \int_0^T w^2(t) dt . \tag{6.108}$$

With Eq. (6.107), and using $w_s = -UH_s/x$ and $\sigma_z = \sigma_w x/U$, Eq. (6.106) becomes

$$\bar{C}_g^y = \sqrt{\frac{2}{\pi} \frac{Q}{U\sigma_z}} \exp \left(-\frac{T_w}{T} \right) \exp \left(-\frac{H_s^2}{2\sigma_z^2} \right) . \tag{6.109}$$

Equation (6.109) is essentially the Gaussian formulation in common use.

Because w_0 is a time average of $w(t)$, we can invoke the Central Limit Theorem and assume that w_0 is normally distributed. Then, it is easy to show that the standard deviation of concentration "fluctuations" is given by (Venkatram, 1984):

$$\sigma[\ln C(x,y)] = \left[\frac{2T_w}{T}\left(\frac{T_w}{T} + \frac{H_s^2}{\sigma_z^2}\right) + \frac{2T_v}{T}\left(\frac{T_v}{T} + \frac{y^2}{\sigma_y^2}\right)\right]^{1/2}.$$

(6.110)

In Eq. (6.110), T_v is the integral time scale for the lateral velocity fluctuation $v(t)$ and $\sigma_y = \sigma_v x/U$.

We can specialize Eq. (6.110) for the convective boundary layer by taking $\sigma_z = 0.6w_*x/U$ (Panofsky, 1978) and $T_w \cong 1.5h/U$ (Kaimal et al., 1976), where h is the mixed-layer height and w_* is the convective velocity scale (Ch. 1). Here, we have assumed that these expressions are exact and that the release height $H_s > 0.1h$. Then, the σ for the centerline concentration is

$$\sigma(\ln C) \cong 3\left(\frac{h}{UT}\right)^{1/2}\frac{\overline{H}_s}{X}; \quad X = w_*x/z_iU, \overline{H}_s = H_s/h \quad (6.111a)$$

$$s_c \equiv \exp(\sigma) = \exp[3(h/UT)^{1/2}\overline{H}_s/X]. \quad (6.111b)$$

Equation (6.101) assumes that the concentration is determined by the flux of serially released particles through an elemental area (see Venkatram, 1984 for details). This implies that the averaging time $T \gg T_w$. Therefore, the proposed formulation for s_c is not applicable for T less than or comparable to T_w. The variance of instantaneous concentrations has to be modeled using methods such as that suggested by Durbin (1980).

Typical results

To get an idea of the magnitude of s_c, we have plotted Eq. (6.111b) for $h = 1000$ m, $U = 5$ m s^{-1}, and $T = 3600$ s (1 h). Figure 6.14 shows the variation of s_c with the nondimensional distance X; $\overline{H}_s$ is the normalized source height. Because our formulation does not account for the limitation of mixing by the inversion lid, results beyond $X = 1$ (especially for $\overline{H}_s = 0.5$) should be viewed with caution.

To interpret Fig. 6.14, let us assume that the concentrations are lognormally distributed. (The actual distribution, consistent with the assumed normality of w_0, is not lognormal.) Then s_c^2 is approximately equal to the 95% confidence interval for the ratio C_0/C_p. For example, $s_c = 2$ indicates that we expect 95% of the observations to lie within a factor of $4(= s_c^2)$ of the prediction.

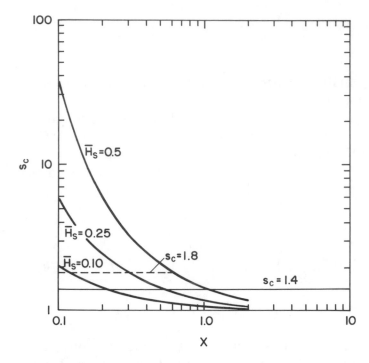

Figure 6.14. Variation of the geometric standard deviation of concentration fluctu-ations s_c with $X \equiv w_* x / z_i U$.

As one would expect, s_c increases sharply with $\overline{H}_s$, and falls off rapidly with X. If we insist that a "good" model should predict within a factor of 2 of the observations 95% of the time, the required s_c is $\cong 1.4$. We see that for $\overline{H}_s = 0.1$, the uncertainty meets this requirement for $X > 0.52$, and the corresponding X for $\overline{H}_s = 0.5$ is roughly 1.0.

It is easily shown that at the location at which the model predicts the maximum concentration, $s_c \cong 1.8$ for all source heights. This means that we should expect 25% of the observations to lie outside a factor of 2 of the *maximum* predicted concentration. Clearly, this should be accounted for in regulatory applications of models. Note that the preceding results depend on the assumed values of the time scales of the vertical and horizontal velocity fluctuations.

We have assumed that w_0 is the only unknown variable that is responsible for the variance of concentrations. We can readily extend our analysis to a case in which the ensemble is specified in terms of a measured velocity U, which differs from the "true" transport velocity by an expected error ε_u. This error could be the difference between the estimated and the actual wind speed in the mixed layer. Then it is easy to show that

$$\sigma[\ln C(x,0)] = \frac{H_s}{\sigma_z} \left[\frac{2T_w}{T} + \left(\frac{\varepsilon_u}{U}\right)^2 \left(\frac{H_s}{\sigma_z}\right)^2\right]^{1/2} . \qquad (6.112)$$

If we take $\varepsilon_u/U = 0.1$, we find that except for large H_s/σ_z (close to the source), the first term in the parentheses is much larger than the second.

6.4.4. The Analysis of Residuals

The results of the previous section clearly show that "inherent" uncertainty is a function of model inputs [see Eqs. (6.111) and (6.112)]. One would suspect that other components of model uncertainty are also functions of model inputs. This suggests that each residual between model prediction and observation is drawn from a different ensemble, each of which is defined by a different *value* of the model input set α. Because the residual includes more than the inherent uncertainty $c(\alpha,\beta)$ let us denote it by ϵ. Now we can express ϵ formally as

$$\epsilon(\alpha,\beta) = c(\alpha,\beta) + i(\Delta\alpha) + f(\alpha) . \qquad (6.113)$$

In Eq. (6.113), $i(\Delta\alpha)$ refers to the error associated with model input error $\Delta\alpha$ while $f(\alpha)$ refers to errors caused by model formulation. Both $c(\alpha,\beta)$ and $i(\Delta\alpha)$ are expected to be random in the sense that they can take on arbitrary negative and positive values. On the other hand, the model formulation error $f(\alpha)$ has to be systematic.

These properties of model errors can be used to examine model performance. We can obtain a visual indication of model performance by plotting the residuals $\epsilon(\alpha,\beta)$ against chosen inputs. If $f(\alpha)$ is close to zero, the residuals should be uniformly distributed about zero as indicated in the top plot of Fig. 6.15. Notice that the variance of the residuals can be a function of the model inputs α. The bottom part of Fig. 6.15 shows residual behavior that might indicate a problem with the formulation of the model.

A careful analysis of residual plots can provide very useful information on model performance. Residuals should be plotted against

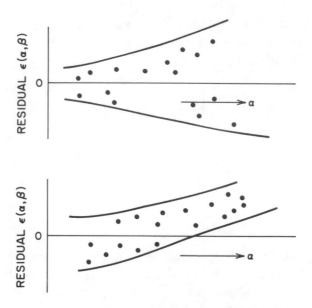

Figure 6.15. Examples of plots of residuals between model predictions and corresponding observations. α refers to a chosen model input. Solid lines indicate envelopes for points on the plots. Top plot shows acceptable residual behavior; bottom plot indicates a problem with the model with increasing α.

most of the important model input variables. This usually allows us to isolate the variable that is incorrectly accounted for in the model.

Because residuals are functions of model inputs, statistics computed from them can provide misleading information on model performance. Then, how do we quantify model performance? One possible approach is to estimate $\epsilon(\alpha,\beta)$ by modeling the terms $c(\alpha,\beta)$ and $i(\Delta\alpha)$ in Eq. (6.113). This is equivalent to simulating the observations because

$$C_0(\alpha,\beta) = C_p(\alpha) + \epsilon(\alpha,\beta) . \qquad (6.114)$$

Note that we cannot predict $C_0(\alpha,\beta)$ exactly because β is unknown for the particular realization corresponding to the observation. However, we can generate possible sets of observations corresponding to the model inputs by choosing appropriate values of β and model input errors $\Delta\alpha$. A Monte-Carlo method can be used to choose the values required to compute $c(\alpha,\beta)$ and $i(\Delta\alpha)$.

Once we have generated these sets of pseudo-observations, we can then test whether the available observations belong to this set.

We can design statistical tests to test the hypothesis that the "predicted" observations are indistinguishable from the "observed" observations. This is the basic idea behind the test proposed by Lewellen et al. (1985). They generated probability density functions (p.d.f.) using predicted residuals. The p.d.f. of the observed residuals was then checked to see whether it could belong to the predicted p.d.f.'s This is only one possible way of using predicted residuals.

The proposed method of evaluating a model might suggest that we can make a model acceptable simply by predicting a large inherent uncertainty $c(\alpha,\beta)$. However, this impression is wrong if we view the air quality model as consisting of two components: a deterministic component $C_p(\alpha)$, and a statistical component $c(\alpha,\beta)$. (The objective of model improvement is to reduce the statistical component relative to the deterministic component.) The prediction from a model has to be stated in terms of $C_p(\alpha)$ and its associated uncertainty $c(\alpha,\beta)$.

References

Carson, D. J., and F. B. Smith, 1974: Thermodynamic model for the development of a convectively unstable boundary layer. *Advances in Geophysics*, **18A**, Academic Press, New York, 111–124.

Deardorff, J. W., and G. W. Willis, 1982: Ground-level concentrations due to fumigation into an entraining mixed layer. *Atmos. Environ.*, **16**, 1159–1170.

Durbin, P. A., 1980: A stochastic model of two-particle dispersion and concentration fluctuations in homogeneous turbulence. *J. Fluid Mech.*, **100**, 279–302.

Fox, D. G., 1981: Judging air quality model performance. *Bull. Amer. Meteor. Soc.*, **62**, 599–609.

Fox, D. G., 1984: Uncertainty in air quality modeling. *Bull. Amer. Meteor. Soc.*, **65**, 27–36.

Hunt, J. C. R., 1982: Diffusion in the stable boundary layer. *Atmospheric Turbulence and Air Pollution Modelling*, F. T. M. Nieuwstadt and H. van Dop, Eds., Reidel, Dordrecht, 231–274.

Hunt, J. C. R., R. E. Britter, and J. S. Puttock, 1979: Mathematical models of dispersion of air pollution around building and hills. *Mathematical Modelling of Turbulent Diffusion in the Environment*, Academic Press, New York, 145–200.

Kaimal, J. C., J. C. Wyngaard, D. A. Haugen, O. R. Coté, Y. Izumi, S. J. Caughey, and C. J. Readings, 1976: Turbulence structure in the convective boundary layer. *J. Atmos. Sci.*, **33**, 2152–2169.

Kerman, B. R., R. E. Mickle, R. V. Portelli, N. B. Trivett, and P. K. Misra, 1982: The Nanticoke shoreline diffusion experiment, June 1978–II. Internal boundary layer structure. *Atmos. Environ.*, **16**, 423–437.

Kitaigorodskii, S. A., 1973: *The Physics of Air-Sea Interaction.* Israel Program for Scientific Translations, Jerusalem, 236 pp.

Lamb, R. G., 1978: A numerical simulation of dispersion from an elevated point source in the convective planetary boundary layer. *Atmos. Environ.*, **12**, 1297–1304.

Lamb, R. G., 1982: Diffusion in the convective boundary layer. *Atmospheric Turbulence and Air Pollution Modelling*, F. T. M. Nieuwstadt and H. van Dop, Eds., Reidel, Dordrecht, 159–229.

Lenschow, D. H., J. C. Wyngaard, and W. T. Pennell, 1980: Mean-field and second-moment budgets in a baroclinic convective boundary layer. *J. Atmos. Sci.*, **37**, 1313–1326.

Lewellen, W. R., R. I. Sykes, and S. F. Parker, 1985: An evaluation technique which uses the prediction of both concentration mean and variance. *Proceedings of the DOE/AMS Air Pollution Evaluation Workshop*, October 1984, Savannah River Lab, Report No. DP–1701–1, Sec. 2, 1–24.

Louis, J. F., 1979: A parametric model of vertical eddy fluxes in the atmosphere. *Bound.-Layer Meteor.*, **17**, 187–207.

Lumley, J. L., and H. A. Panofsky, 1964: *The Structure of Atmospheric Turbulence*. Wiley Interscience, New York, 239 pp.

Lyons, W. A., and H. S. Cole, 1973: Fumigation and trapping on the shores of Lake Michigan during stable onshore flow. *J. Appl. Meteor.*, **12**, 494–510.

Misra, P. K., 1980: Dispersion from tall stacks into a shoreline environment. *Atmos. Environ.*, **14**, 393–397.

Misra, P. K., and S. Onlock, 1982: Modeling continuous fumigation of Nanticoke generating station plume. *Atmos. Environ.*, **15**, 479–489.

Nieuwstadt, F. T. M., and H. van Dop, Eds., 1982: *Atmospheric Turbulence and Air Pollution Modelling*. Reidel, Dordrecht, 358 pp.

Panofsky, H. A., 1978: Matching in the convective boundary layer. *J. Atmos. Sci.*, **35**, 272–276.

Panofsky, H. A., and J. Dutton, 1984: *Atmospheric Turbulence: Models and Methods for Engineering Applications*. Wiley, New York, 397 pp.

Plate, E. J., 1971: *Aerodynamic Characteristics of Atmospheric Boundary Layers*. U. S. Atomic Energy Commission, TID–25465, 190 pp.

Raynor, G. S., P. Michael, R. M. Brown, and S. Sethuraman, 1975: Studies of atmospheric diffusion from a nearshore oceanic site. *J. Appl. Meteor.*, **14**, 1080–1094.

Sawford, B. L., 1985: Lagrangian statistical similarity of concentration mean and fluctuation fields. *J. Climate Appl. Meteor.*, **24**, 1152–1166.

Schiermeier, F. A., 1984: Scientific assessment document on status of complex terrain models for EPA regulatory applications. EPA Report No. EPA–600/3–84–103.

Smith, S. D., 1980: Wind stress and heat flux over the ocean in gale-force winds. *J. Phys. Ocean.*, **10**, 709–726.

Snyder, W. H., R. S. Thompson, R. E. Eskridge, R. E. Lawson, Jr., I. P. Castro, J. T. Lee, J. C. R. Hunt, and Y. Ogawa, 1983: The Structure of Strongly Stratified Flow over Hills—Dividing Streamline Concept. Appendix A to EPA–600/3–83–015, U. S. Environmental Protection Agency, Research Triangle Park, NC, 320–375.

Strimaitis, D. G., T. F. Lavery, A. Venkatram, D. C. DiCristofaro, B. R. Greene, and B. A. Egan, 1985: EPA Complex Terrain Model Development. Fourth Milestone Report—1984. EPA Report No. EPA–600/3–84/110.

Tennekes, H., 1973: A model for the dynamics of the inversion above a convective boundary layer. *J. Atmos. Sci.*, **30**, 558–567.

Thompson, R. S., and W. H. Snyder, 1984: Dispersion from a source upwind of a three-dimensional hill of moderate slope. EPA report included in Strimaitis et al. (1985).

van Dop, H., R. Steenkist, and F. T. M. Nieuwstadt, 1979: Revised estimates for continuous shoreline fumigation. *J. Appl. Met.*, **18**, 133–137.

Venkatram, A., 1977a: A model of internal boundary-layer development. *Bound.-Layer Meteor.*, **11**, 419–437.

Venkatram, A., 1977b: Internal boundary layer development and fumigation. *Atmos. Environ.*, **11**, 479–482.

Venkatram, A., 1979: The expected deviation of observed concentrations from predicted ensemble means. *Atmos. Environ.*, **13**, 1547–1550.

Venkatram, A., 1980: Estimating the Monin-Obukhov length in the stable boundary layer for dispersion calculations. *Bound.-Layer Meteor.*, **19**, 481–485.

Venkatram, A., 1983: On dispersion in the convective boundary layer. *Atmos. Environ.*, **17**, 529–533.

Venkatram, A., 1984: The uncertainty in estimating dispersion in the convective boundary layer. *Atmos. Environ.*, **18**, 307–310.

Venkatram, A., D. Strimaitis, and D. DiCristofaro, 1984: A semi-empirical model to estimate vertical dispersion of elevated releases in the stable boundary layer. *Atmos. Environ.*, **18**, 823–928.

Weil, J. C., and R. P. Brower, 1984: An updated Gaussian plume model for tall stacks. *J. Air Pollut. Control. Assoc.*, **34**, 818–827.

Weil, J. C., L. A. Corio, and R. P. Brower, 1986: Dispersion of buoyant plumes in the convective boundary layer. *5th Joint Conference on Applications of Air Pollution Meteorology*, Amer. Meteor. Soc., Boston, 335–338.

Weisman, B., 1976: On the criteria for the occurrence of fumigation inland from a large lake—a reply. *Atmos. Environ.*, **12**, 172–173.

Willis, G. E., and J. W. Deardorff, 1978: A laboratory study of dispersion from an elevated source within a modeled convective planetary boundary layer. *Atmos. Environ.*, **12**, 1305–1311.

Wyngaard, J. C., Ed., 1984: *Large-Eddy Simulation: Guidelines for its Application to Planetary Boundary Layer Research*. Available from DTIC, AD–A146381, 122 pp.

CHAPTER 7

Concentration Fluctuations in Dispersing Plumes

R. Ian Sykes

7.1. Introduction

One may argue about whether the Navier-Stokes equations of fluid motion are deterministic, or even about their relevance to atmospheric motion, but at a practical level one is forced to accept that it is impossible to make sufficiently detailed measurements to specify the entire velocity field. We therefore treat the flow as a random, or turbulent, field and attempt to make our predictions by means of statistical measures such as fluctuation correlations or probability distributions. The dispersion of a scalar contaminant in this stochastic velocity field is also a random process, and the present recognition of this aspect of dispersion is evinced by the inclusion of this chapter in a monograph on atmospheric dispersion.

Any measurement in a random flow will involve a random component. It is the object of the research under discussion here to quantify this component, and relate it to measures of randomness in the velocity field. This is an important aspect of almost any application of diffusion modeling, since we are always trying to predict a real event, i.e., the concentration which some type of sampler would measure. This is a random variable, so information about the probability distribution is valuable, even if it tells us that the ensemble mean is an extremely likely measurement. Environmental impact assessment, or short-time-average problems such as toxic or flammable vapor dispersion all require information about probability distributions, because we need to know the range of possible concentrations that could actually be encountered at a given point. Model evaluation is also an area needing such information, since we are generally comparing a prediction of the ensemble mean with an observation of a random variable.

The conservation equation for a scalar contaminant in a fixed coordinate system $x = (x,y,z)$ with the z-axis oriented vertically, i.e., in the opposite direction to the gravity vector, can be written

$$\frac{\partial c}{\partial t} + (u \cdot \nabla)c = D\nabla^2 c , \qquad (7.1)$$

where $u(x,t)$ is the velocity field, D the molecular diffusivity, and c the scalar concentration. The formal statistical approach requires specification of an ensemble of velocity fields, from which we observe a particular realization with a certain probability distribution. It is clear that different realizations of the velocity field will produce a different evolution of the scalar field by means of Eq. (7.1). We may thus formally relate the probability distribution of c to that of u. Some theoretical work has proceeded along this line, e.g., Pope (1981), but the route is extremely complex and requires a great deal of computational effort, in addition to several empirical modeling assumptions.

The definition of the ensemble of velocity fields is a very difficult problem for the atmospheric flows we consider here. The situation is much easier in the laboratory, where either a statistically steady flow can be developed so that long-time averages converge to ensemble averages, or the experiment can be repeated many times with fixed external parameters to give a true ensemble. In the atmosphere, we cannot repeat the experiment and the flow is never completely steady because there are velocity variations on a wide range of time scales.

Lamb (1984) discussed the philosophy of the ensemble, and related it to the observations by defining the set of all possible velocity fields as those that match the observed velocity at the observation points. This allows a very large range of velocity fields, since values at points other than observation points are arbitrary. Lamb suggested reducing this range by assuming that the probability of a certain state depends on the spectrum of that state in such a way that "unusual" spectra are highly unlikely. This is an appealing formal approach, but clearly has practical difficulties; we shall present a less formal, but more practical definition so that our discussion of concentration fluctuations will have some relevance to the real world.

The lack of an obvious ensemble in our problem forces us to recognize the importance of the time and space scales of interest to us. We must specify the range over which we wish to predict the diffusion, L say, which then defines a time scale of interest, $T = L/U$, where U is the wind speed and T the travel time from the source out to a distance L. Following Lamb, we must also consider what observations of the flow field are available to us; let us assume that we measure the velocity on a network with a spacing characterized by a length L_0. If we have only one observation at the source, we can set $L_0 = L$ since our range of interest will always fall between observations in such a network. Neglecting inhomogeneities in the surface conditions for the moment, we can define velocity fluctuations on scales shorter than L_0 to be the turbulent component, and those on scales larger than L_0 to be the deterministic or mean-flow component. This is because our observation network can resolve the larger-scale variations, but not the smaller ones. The information on the smaller scales must come from the time series at one of our observation points, so we relate the length scale L_0 to a time scale T_0 by the relation $T_0 = L_0/U$. We then filter our observed time series for the velocity to remove fluctuations on time scales shorter than T_0. Turbulent fluctuations can be computed by time filtering the square of the deviation from the mean. Thus, we define

$$\bar{u}(t) = \frac{1}{T_0} \int_{-\infty}^{\infty} w(t')u(t + t')dt' ,$$

where $u(t)$ is the observed velocity variation, and w is a weighting filter satisfying

$$\int_{-\infty}^{\infty} w(t')dt' = T_0 .$$

As an example, we have employed the triangular filter

$$w(t) = \begin{cases} (1 - |t|/T_0), & |t| < T_0 \\ 0, & |t| \geq T_0 , \end{cases}$$

and Fig. 7.1 shows the results of applying this to a time series to give $\bar{u}(t)$, and $\overline{u'^2}(t) = \overline{[u(t) - \bar{u}(t)]^2}$.

We are now in a position to discuss the fluctuations in the concentration field using either the idealized laboratory situation or our less rigorous atmospheric ensemble of velocity fields. Following Chatwin and Sullivan (1979), we first consider the case with $D = 0$, i.e., no molecular diffusivity, because this sheds some light on the nature of the real problem. Chatwin and Sullivan consider relative diffusion

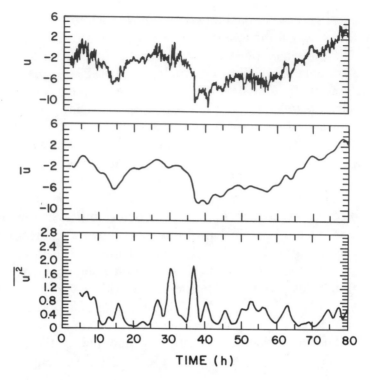

Figure 7.1. Mean velocity obtained from triangular time filter.

of a cloud of pollutant about its center of mass, but a more relevant problem for us is the diffusion in a uniform stream from a continuous source, i.e., a plume. The transport equation for the fluctuating concentration field c is

$$\frac{\partial c}{\partial t} + (U_0 + u')\frac{\partial c}{\partial x} + v'\frac{\partial c}{\partial y} + w'\frac{\partial c}{\partial z} = S , \qquad (7.2)$$

where U_0 is the mean speed, and u', v', and w' are zero-mean random fields. S represents the source term. If we use the overbar to represent the ensemble mean and assume a steady mean flow and source, then we obtain

$$U_0\frac{\partial \bar{c}}{\partial x} = -\frac{\partial \overline{u'c'}}{\partial x} - \frac{\partial}{\partial y}\overline{v'c'} - \frac{\partial}{\partial z}\overline{w'c'} - S , \qquad (7.3)$$

where the prime denotes a fluctuation from the mean. Multiplying Eq. (7.2) by c' and averaging gives

$$U_0 \frac{\partial \overline{c'^2}}{\partial x} = -2\overline{u'c'}\frac{\partial \overline{c}}{\partial x} - 2\overline{v'c'}\frac{\partial \overline{c}}{\partial y} - 2\overline{w'c'}\frac{\partial \overline{c}}{\partial x} - \frac{\partial}{\partial x}\overline{u'c'^2} - \frac{\partial}{\partial y}\overline{v'c'^2} - \frac{\partial}{\partial z}\overline{w'c'^2}.$$

(7.4)

Combining Eqs. (7.3) and (7.4) gives

$$U_0 \frac{\partial}{\partial x}(\overline{c'^2} + \overline{c}^2) = 2S\overline{c} - \frac{\partial}{\partial x}\overline{u'c'^2} - \frac{\partial}{\partial y}\overline{v'c'^2} - \frac{\partial}{\partial z}\overline{w'c'^2}.$$ (7.5)

If we now make the assumption that the streamwise transport term $\partial(\overline{u'c'^2})/\partial x$ is negligible, then the equations are hyperbolic with characteristics in the positive x-direction so that the source can be represented as an initial condition at $x = 0$, and subsequently set to zero, i.e.,

$$U_0 \frac{\partial}{\partial x}(\overline{c'^2} + \overline{c}^2) = -\frac{\partial}{\partial y}\overline{v'c'^2} - \frac{\partial}{\partial z}\overline{w'c'^2}.$$ (7.6)

Now integrating Eq. (7.6) in the transverse (y, z)-plane gives

$$U_0 \frac{\partial}{\partial x}\int\int(\overline{c'^2} + \overline{c}^2)dy\,dz = 0$$ (7.7)

if $c \to 0$ as $y^2 + z^2 \to \infty$, i.e. unbounded flow.

Equation (7.7) demonstrates conservation of total c^2-stuff in the plume, and the following simple analysis shows that in the zero-diffusivity case, all the c^2-stuff is converted into the fluctuating component.

We have $\int\int(\overline{c'^2} + \overline{c}^2)dy\,dz$ = constant, and if the source is a Gaussian with spread σ_0 in the y and z-directions and total mass Q, then the constant is $Q^2/4\pi\sigma_0^2$ if $\overline{c'^2} = 0$ initially. At any subsequent time,

$$\int\int \overline{c}^2 dy\,dz = \frac{Q^2}{4\pi\sigma_y\sigma_z}.$$

if $\overline{c}$ spreads as a Gaussian with second moments of σ_y^2 and σ_z^2. The Gaussian assumption is quite realistic, and is theoretically correct for short distances from the source if the velocity fluctuations have a Gaussian probability distribution. Thus $\int\int \overline{c}^2 dy\,dz \to 0$ as $x \to \infty$ since $\sigma_y, \sigma_z \sim x^{1/2}$ using Taylor's (1921) theory. Hence $\int\int \overline{c'^2} dy\,dz \to Q^2/4\pi\sigma_0^2$ as $x \to \infty$ and all the c^2-stuff is converted into $\overline{c'^2}$-stuff. In fact, it will be seen that $\int\int \overline{c}^2 dy\,dz \ll \int\int \overline{c'^2}$ as soon as $\sigma_y\sigma_z \gg$

σ_0^2. Thus, the majority of the conversion has taken place when the plume has tripled its initial size. This result also shows that discussion of concentration variance without specification of the source size is meaningless, at least if molecular diffusion is negligible.

Chatwin and Sullivan discussed relative diffusion about the center of mass of a cloud, but the arguments are virtually identical here. They also show that the distance-neighbor function,

$$P(y,z) = Q^{-2} \int \int c(y',z')c(y' + y, z' + z)dy'dz',$$

must have a double structure with much larger magnitude for $|y|$, $|z| = O(\sigma_0)$. The result for our plume is

$$P(y,z) = O(\sigma_0^{-2}), \qquad y^2 + z^2 = O(\sigma_0)$$

$$P(y,z) = O(\sigma^{-2}), \qquad y^2 + z^2 = O(\sigma),$$

where we have assumed $\sigma_y = \sigma_z = \sigma$. P is clearly related to the spatial correlation function for the cloud; the two-scale structure shows us that correlation falls off very rapidly on the scale of the initial source size, but also has an extensive tail on the scale σ. Chatwin and Sullivan argued that $\bar{c}$ and $\overline{c'^2}$ should also exhibit this two-scale structure, but this does not seem likely; it is certainly *not* true in the absolute diffusion case considered here.

If we include molecular diffusion in the preceding analysis, Eq. (7.7) becomes

$$U_0 \frac{\partial}{\partial x} \int \int (\overline{c'^2} + \bar{c}^2) dy\, dz = -2D \int \int \left[\overline{\left(\frac{\partial c}{\partial y}\right)^2} + \overline{\left(\frac{\partial c}{\partial z}\right)^2} \right] dy\, dz .$$

$$(7.8)$$

We now have monotonic decrease of c^2-stuff, since the right-hand side of Eq. (7.8) is always negative. Equation (7.8) shows that molecular dissipation will be negligible until the concentration field has been sufficiently stretched and distorted by the velocity field to produce high gradients. However, our knowledge of the details of those small-scale distortion mechanisms is too limited to allow us to say much directly from Eq. (7.8). We therefore turn now to some conceptual modeling ideas and some known results.

7.2. Modeling Concepts

7.2.1. Gifford's Fluctuating-Plume Model

The earliest model, and one of the most useful, is the fluctuating-plume model of Gifford (1959). Observation of a small source in a turbulent flow, e.g., a cigarette plume out of doors, clearly shows that in its early stage the plume is coherent and narrow, and meanders from side to side as the large eddies sweep past the source; see Fig. 7.2 for a schematic example. Gifford used this idea to produce a simple model of a meandering plume by defining an instantaneous Gaussian plume with spread σ_i, whose centroid location (Y, Z) is a

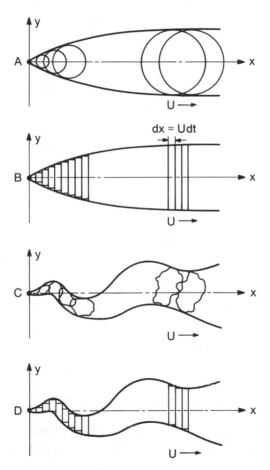

Figure 7.2. Schematic view of plume models: A, puff model, steady; B, plume model, steady; C, fluctuating plume represented as puffs; and D, fluctuating plume with spreading disks. (From Gifford, 1959.)

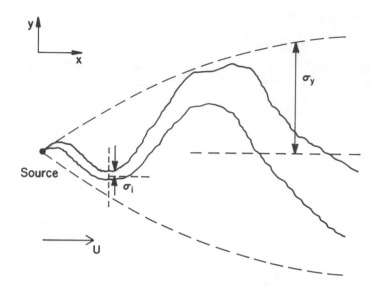

Figure 7.3. Schematic representation of Gifford's meandering-plume model.

normal random variable with spreads σ_y, σ_z, (Fig. 7.3). We therefore have a realization of the plume with centroid at (Y,Z) as

$$c(y,z;Y,Z) = \frac{Q}{2\pi U\sigma_i^2} \exp\left[-\frac{(y-Y)^2}{2\sigma_i^2} - \frac{(z-Z)^2}{2\sigma_i^2}\right] \qquad (7.9)$$

and ensemble averages are obtained by averaging over all possible centroid positions (Y,Z) using the appropriate probability density function. Thus,

$$\overline{\phi}(y,z) = \int\int \phi(y,z;Y,Z)P(Y,Z)dY\,dZ \qquad (7.10)$$

for any variable ϕ. Gifford assumed

$$P(Y,Z) = \frac{1}{2\pi\sigma_y\sigma_z} \exp\left(-\frac{Y^2}{2\sigma_y^2} - \frac{Z^2}{2\sigma_z^2}\right)$$

i.e., independent Gaussian random variables. We assume $\sigma_y = \sigma_z$ in order to simplify the algebra considerably, and obtain

$$\overline{c}(y,z) = \frac{Q}{2\pi U(\sigma_i^2 + \sigma_y^2)} \exp\left[-\frac{y^2 + z^2}{2(\sigma_i^2 + \sigma_y^2)}\right] \qquad (7.11)$$

$$\overline{c^2}(y,z) = \frac{Q^2}{(2\pi U)^2 \sigma_i^2 (2\sigma_y^2 + \sigma_i^2)} \exp\left(-\frac{y^2 + z^2}{2\sigma_y^2 + \sigma_i^2}\right) \qquad (7.12)$$

Then $\overline{c'^2} = \overline{c^2} - \overline{c}^2$.

There are several restrictive assumptions in Gifford's model; the most serious is the neglect of fluctuations in the instantaneous plume. Equation (7.9) shows that the distribution about the centroid is fixed. The model is thus most applicable in the early stages of diffusion, but as we shall see below, it is capable of predicting the meander component of the fluctuations reasonably accurately. The dominance of the meandering effect requires that the turbulent eddies be much larger than the instantaneous plume, so that the plume is advected as a whole, and in this situation Gifford's model is very useful.

The principal results from Gifford's model are that

$$\frac{\overline{c'^2}(0,0)}{\overline{c}^2(0,0)} = \frac{\sigma_y^2}{\sigma_i^2(2\sigma_y^2 + \sigma_i^2)} \qquad (7.13)$$

and

$$\frac{\overline{c'^2}(y,z)}{\overline{c}^2(y,z)} \sim \exp\left[\frac{(y^2 + z^2)\sigma_y^2}{(\sigma_i^2 + \sigma_y^2)(\sigma_i^2 + 2\sigma_y^2)}\right] . \qquad (7.14)$$

That is, as the ratio gets large, the relative intensity of the fluctuations increases as the square of the ratio of the ensemble spread to the instantaneous spread. Furthermore, the intensity increases as an inverse Gaussian in the tails of the plume. Both of these predictions have been verified in laboratory experiments, as we shall see later.

Gifford's model, being very simple and explicit, affords a very clear view of the meandering fluctuations. It therefore provides a sound basis for studying the effect of time averaging on the variance of the concentration fluctuations. It is clear that a time-averaged measurement of the concentration should reduce the variance by smoothing out the short-time variations. We would like to know the effect of time averaging in model evaluation exercises, for example, where a comparison is made between a time-averaged sample and a predicted ensemble mean, or perhaps in toxic exposure where an integrated dose is important. If we define a time-averaged quantity

$$\phi(t,S) = \frac{1}{S} \int_{t-S/2}^{t+S/2} \phi(t')dt' , \qquad (7.15)$$

where S is the sampling or averaging time, then Tennekes and Lumley (1972) show that the time-averaged variance can be related to the autocorrelation function by

$$\overline{c'^2}(S) = \frac{2\overline{c'^2}(0)}{S} \int_0^S \left(1 - \frac{t'}{S}\right) \rho(t')dt' , \qquad (7.16)$$

where $\rho(t') = \frac{\overline{c'(t)c'(t+t')}}{\overline{c'^2}}$, and $\overline{c'^2}(0) = \overline{c'^2}$.

In Eq. (7.16) we have suppressed the explicit dependence on t since we are considering a stationary process. Venkatram (1979) followed this procedure and postulated an exponential form for $\rho(t')$ with integral timescale proportional to the Eulerian timescale T_E of the fluctuating velocity field. However, the two-scale nature of Gifford's model gives a more complex correlation if one assumes that the centroid position has an exponential autocorrelation with integral scale T_E, as in Sykes (1984). This is an obvious extension of Gifford's model, since the centroid position is the random variable; this is moved around by the velocity field and therefore might be expected to have the same timescale. Since the concentration is defined explicitly in terms of the centroid positions by Eq. (7.9), the autocorrelation function can be calculated and gives

$$\rho(\tau) = \frac{\sigma_i^2 \left(\sigma_i^2 + 2\sigma_y^2\right) \exp(-2\tau/T_E)}{\left[(\sigma_i^2 + \sigma_y^2)^2 - \sigma_y^4 \exp(-2\tau/T_E)\right]} . \qquad (7.17)$$

The two-scale nature of Gifford's model can be seen in Eq. (7.17). If $\sigma_i \ll \sigma_y$, then

$$\rho(\tau) \sim \left(1 + \frac{\sigma_y^2 \tau}{\sigma_i^2 T_E}\right)^{-1} , \tau \ll T_E$$

$$\rho(\tau) \sim \frac{2\sigma_i^2}{\sigma_y^2} \exp\left(-\frac{2\tau}{T_E}\right) , \tau = O(T_E) . \qquad (7.18)$$

Equation (7.18) is similar to Chatwin and Sullivan's result for the distance-neighbor function insofar as $\rho = O(1)$ for $\tau < \sigma_i^2 T_E/\sigma_y^2$, but $\rho = O(\sigma_i^2/\sigma_y^2)$ for $\tau = O(T_E)$.

Equation (7.17) defines an integral timescale for the concentration fluctuation,

$$T_C = \int_0^\infty \rho(\tau)\,dt \approx T_E \frac{\sigma_i^2}{\sigma_y^2} \ln\left[\frac{\sigma_y^2}{2\sigma_i^2}\right] \tag{7.19}$$

for $\sigma_i^2 \ll \sigma_y^2$.

If the autocorrelation is exponential, Venkatram (1979) showed that

$$\frac{\overline{c'^2}(S)}{\overline{c'^2}} = \frac{2}{\beta^2}(\beta - 1 + e^{-\beta}), \tag{7.20}$$

where $\beta = S/T_c$. Figure 7.4 shows the reduction in variance due to time averaging for various ratios of σ_i^2/σ_y^2 and compares estimates using Eqs. (7.20) and (7.19) with the full result from Eq. (7.17). Note that for $\sigma_i^2 \ll \sigma_y^2$ and $S \gg T_c$,

$$\overline{c'^2}(S) \approx \frac{T_E}{S} \ln\left(1 + 2\frac{\overline{c_0'^2}}{\overline{c}_0^2}\right) \overline{c}_0^2, \tag{7.21}$$

where the subscript zero refers to plume centerline, and we have used the result of Eq. (7.13) to express σ_y^2/σ_i^2 in terms of the relative fluctuation intensity. Equation (7.21) says that if we average a highly intermittent plume for times longer than the Eulerian timescale of the velocity fluctuations, we find only a weak logarithmic dependence on the actual variance. This is because a long time average samples

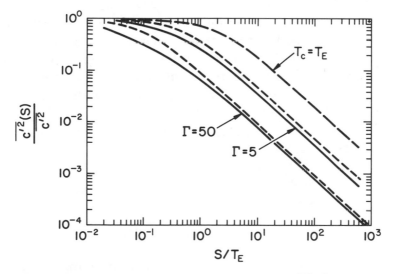

Figure 7.4. Variance reduction due to time averaging. $\Gamma = \overline{c'^2}/\overline{c}^2 \simeq \sigma_y^2/2\sigma_i^2$. (From Sykes et al., 1984).

the whole of the instantaneous plume and therefore integrates its entire mass, which is a constant, and therefore most of the detail of the instantaneous plume is lost.

7.2.2. Stochastic Two-Particle Models

Durbin (1980) presented a model of concentration fluctuations based on the idea of treating particle diffusion as a random-flight process. This technique has been developed and used extensively to predict the mean concentration, where each particle can be treated independently. Durbin considered turbulent dispersion in one dimension only, and defined

$$\overline{c^2}(z,t) = \int\int_{-\infty}^{\infty} P_2(z,t;z_1',z_2')S(z_1')S(z_2')dz_1'dz_2' \,, \qquad (7.22)$$

where $S(z')$ is the source distribution at the initial time, and $P_2(z,t;z_1',z_2')$ is the joint probability that the pair of particles at the point z and time t came from the points z_1', z_2' at the initial time. Durbin argued that the two-particle distribution is necessary for modeling molecular-diffusion effects. He calculated the solution by tracing particle pairs, which are initialized very close together at the receptor, back to the source time. The particles need to be distinct; otherwise, we have only one particle and it can be shown that this definition conserves total c^2-stuff which means no molecular diffusion, as we showed in the discussion of Chatwin and Sullivan's results. Durbin showed that the results from Eq. (7.22) are independent of the particle-pair separation at the receptor (z,t) provided it is small enough, and that this represents a smearing due to instrumental response since it can be shown to be the outer limit of the spatial correlation function (Fig. 7.5).

The actual equations for the motion of the particle pair are written in center-of-mass coordinates as

$$\frac{dZ}{dt} = [\alpha(\Delta) + \beta(\Delta)]U^{(1)}(t)$$
$$\frac{d\Delta}{dt} = [\alpha(\Delta) - \beta(\Delta)]U^{(2)}(t) \qquad (7.23)$$

where Z is the center of mass, Δ is the particle separation, and $U^{(1)}$, $U^{(2)}$ are independent stationary Gaussian Markov processes with zero

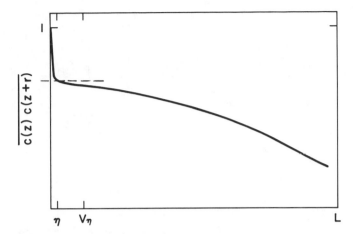

Figure 7.5. Schematic representation of the spatial correlation for concentration fluctuations. η represents the Kolmogorov scale, and the dashed line shows the effect of averaging over the scale V_η. (From Durbin, 1980.)

mean and variance σ_w^2 obtained from the Langevin equation

$$dU = -\frac{U}{\tau_L}dt + \sigma_w \left(\frac{2}{\tau_L}\right)^{1/2} dW_t .$$

Here τ_L is the Lagrangian correlation time, and dW_t is a Gaussian white noise process. The Langevin equation has been used extensively to model single-particle diffusion and is discussed elsewhere, e.g. Gifford (1982). α and β are correlations that Durbin defined as

$$\alpha^2 = \frac{1}{2}\left\{1 - [2R(\Delta) - R^2(\Delta)]^{1/2}\right\}$$

$$\beta^2 = 1 - \alpha^2$$

$$\text{and } R(\Delta) = \left(\frac{\Delta^2}{L^2 + \Delta^2}\right)^{1/3} .$$

He presented arguments that justify this choice as a representation of the inertial range structure function with L as the integral scale.

Durbin obtained numerical and analytic solutions to Eq. (7.23), and these were extended by Sawford (1983). For a line source Durbin and Sawford showed that $\overline{c'^2}/\overline{c}^2$ tends to a constant value on the centerline of the plume at large downstream distances in homogeneous

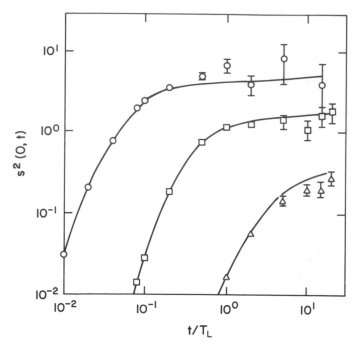

Figure 7.6. Centerline relative fluctuation intensities from Sawford's (1983) two-particle model. Symbols: $\bigcirc$, $\sigma_o^2/L^2 = 5 \times 10^{-5}$; , $\sigma_o^2/L^2 = 5 \times 10^{-3}$; $\triangle$, $\sigma_o^2/L^2 = 5 \times 10^{-1}$.

stationary turbulence. This is illustrated in Fig. 7.6, which shows the centerline intensity

$$s^2(0,t) = \frac{\overline{c'^2}(0,t)}{\overline{c}^2(0,t)} ,$$

where t measures the time from release, i.e., $t = x/U$, and T_L is the Lagrangian integral timescale for the velocity fluctuations. The three curves show results for three ratios of σ_0/L, and confirm the analytic result that $s^2 \to A_0(L/\sigma_0)^{1/3}$, where A_0 is a constant depending on the source shape; σ_0 is proportional to the source size. For a rectangular source, Durbin calculated $A_0 = 0.9 \cdot 2^{1/3}$. Figure 7.7 shows the transverse shape of the concentration variances, and demonstrates that they collapse to the same Gaussian as the mean concentration. This result is consistent with Gifford's meandering plume, and again gives relative fluctuation intensity increasing in the tails of the plume, i.e., $s^2(y,t) \sim \exp(y^2/2\sigma_y^2)$ as $y \to \infty$. Note that the two-particle model predicts that the ultimate relative

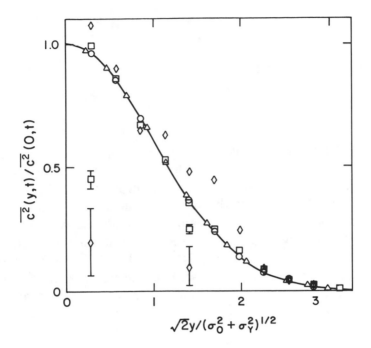

Figure 7.7. Transverse shape of the $\overline{c^2}$ -profile at different downstream locations normalized by the mean plume width σ_y. (From Sawford, 1983.)

fluctuation intensity far downstream from a line source in homogeneous turbulence goes to a constant on the centerline, the constant being proportional to $(L/\sigma_0)^{1/3}$. That is, the fluctuations never "forget" the source. This prediction is very difficult to verify and the late-time behavior is still an open question.

A major obstacle to the practical use of the two-particle model is the number of particle pairs needed to build a complete distribution of $\overline{c'^2}$. The number increases dramatically with the number of dimensions in the problem, and published results to date have been restricted to one dimension. The model is, however, very valuable as a research tool and provides insight into the nature of the fluctuations.

7.2.3. Closure Models

Csanady (1967) was one of the first to apply ensemble closure to concentration fluctuations. The most important aspect of this problem, which distinguishes it from other turbulent flows such as pipe flow or boundary layer flow, is the fluctuation dissipation term

$$\epsilon_c = D\overline{\frac{\partial c'}{\partial x_i}\frac{\partial c'}{\partial x_i}} \, ,$$

which is usually modeled as

$$\epsilon_c = \frac{\overline{c'^2}}{\tau_c} \, , \tag{7.24}$$

which can be thought of as defining the dissipation time scale τ_c. Csanady proposed to model τ_c in such a way as to obtain a self-similar solution for the plume in homogeneous turbulence. This requires that τ_c increase linearly with distance from the source. A self-similar solution may be valid far downstream from the source, where we have very little laboratory data or sound theory, but it does not model the early meander phase of a plume. Sykes et al. (1984, 1986) proposed an equation for a length scale describing the instantaneous scale of the plume, i.e., the scale of the relative or two-particle dispersion. This model can be written

$$\frac{d\Lambda_c}{dt} \propto q_c \, , \tag{7.25}$$

where $q_c = q_B(\Lambda_c/\Lambda_B)^{1/3}$, q_B^2 is the background turbulence intensity, i.e., $\overline{u'^2} + \overline{v'^2} + \overline{w'^2}$, and Λ_B is the background turbulence length scale. Thus q_c represents energy at the scale Λ_c, assuming a Kolmogorov inertial-range spectrum. The dissipation time scale τ_c is then modeled as Λ_c/q_c; this is an inertial range time scale appropriate when $\Lambda_c \ll \Lambda_B$. As the plume grows this model must be modified, but at present we do not have sufficient information to fix the parameterization. The solution to Eq. (7.25) gives $\Lambda_c \propto t^{3/2}$, which agrees with the relative diffusion law of Richardson (1926). The closure model gives results similar to those of the meandering plume and two-particle models for short times, including the source size-dependence and the transverse Gaussian shape.

7.3. Experimental Data

The best data on concentration fluctuations come from controlled laboratory experiments, where the ensemble is well defined and unambiguous measurements are possible. Although there have been relatively few studies of passive plumes in a turbulent flow, the most comprehensive study is undoubtedly that of Fackrell and Robins

(1982a). They measured fluctuating concentrations from a variety of source sizes in a wind-tunnel turbulent boundary layer. Figures 7.8 to 7.10 show some of the major features of their plume.

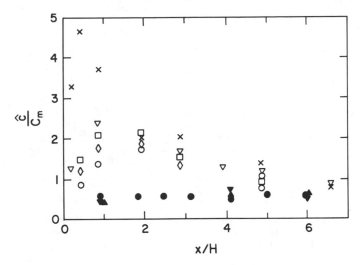

Figure 7.8. Dependence of fluctuation intensity on source characteristics from Fackrell and Robins (1982a). $\hat{c}$ is maximum r.m.s. fluctuation at any downstream location, x; C_m is maximum mean concentration. Symbols for ground level releases: ▲, source diameter d_s = 3 mm; ▼, d_s = 9 mm; ●, d_s = 15 mm. Elevated releases (~ 20 cm): ×, d_s = 3 mm; ▽, d_s = 9 mm; □, d_s = 15 mm; ◊, d_s = 25 mm; and ○, d_s = 35 mm.

Figure 7.8 shows the maximum root-mean-square (r.m.s) fluctuation at a downstream x-station divided by the maximum mean concentration at the same x-location; the maxima are not always at the same location in the transverse plane. The curves are for different source diameters and both elevated and surface releases. There is a clear tendency for larger values to occur at earlier locations with smaller elevated sources. The surface releases are independent of source size, however. Vertical profiles of the mean concentration and the fluctuation variance on the plume centerline are shown in Figs. 7.9 and 7.10 respectively. These have self-similar profiles for the surface release, and an elevated maximum moving down onto the surface for $\bar{c}$ but staying aloft for $\overline{c'^2}$. Note that for short times the elevated release shows similarly shaped profiles of $\bar{c}$ and $\overline{c'^2}$; this is also true for transverse profiles.

Fackrell and Robins (1982b) used Gifford's conceptual model of a meandering plume to predict the fluctuation variance in their

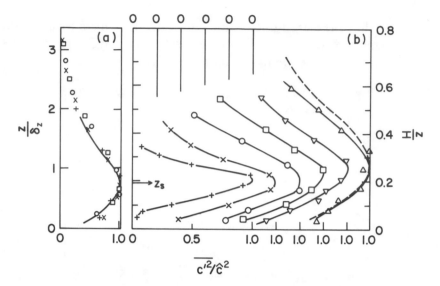

Figure 7.9. Mean concentration profiles in the vertical, from Fackrell and Robins, (1982a). (a) Ground-level releases; δ_z is half-width. (b) Elevated releases; $x/H =$ 0.96, 1.92, 2.88, 3.83, 4.79, 6.52.

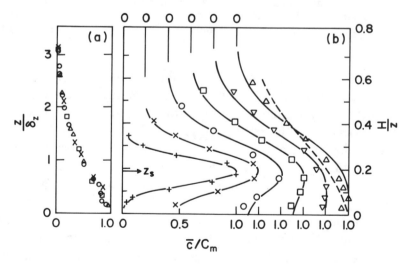

Figure 7.10. Concentration fluctuation variance profiles at same locations as Figure 7.9.

elevated releases. Some extra work was required here, since Gifford specified the distributions in terms of instantaneous and ensemble plume widths, σ_i and σ_y, but did not give equations for σ_i and σ_y themselves. Fackrell and Robins used formulations due to Hay and Pasquill (1959) and Smith and Hay (1961) for the ensemble and relative diffusion in terms of the velocity spectra, which were actually measured. Their results for the relative fluctuation intensity are shown in Fig. 7.11, where it can be seen that this simple model accounts for a good deal of the experimental variation. The details of the calculation are somewhat complicated, but the general picture can be derived from simpler parameterizations as follows.

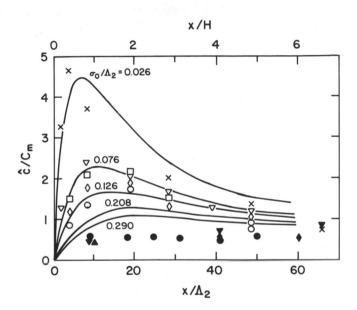

Figure 7.11. Prediction of relative fluctuation intensity using Gifford's model. Symbols as in Fig. 7.8; only elevated sources are predicted. From Fackrell and Robins (1982b).

In the early stage of the meandering plume, we assume that the ensemble spread is growing linearly, i.e.,

$$\sigma_y = \sigma_0 + \frac{\sigma_v x}{U} , \qquad (7.26)$$

where σ_v is the r.m.s. velocity fluctuation. We also assume

$$\sigma_i = \left(\sigma_0^{2/3} + \alpha\sigma_v \frac{x\Lambda_B^{-1/3}}{U}\right)^{3/2}, \qquad (7.27)$$

where Λ_B is the background scale. Equation (7.27) is the $t^{3/2}$ relative diffusion law, with α an $O(1)$ constant. Gifford's model then gives centerline relative fluctuation

$$\frac{\hat{c}}{C_m} = \frac{\sigma_y^2}{\sigma_i(2\sigma_y^2 + \sigma_i^2)^{1/2}}.$$

If we simplify the algebra by assuming $\sigma_y \gg \sigma_0$, and also $\sigma_y \gg \sigma_i$, then

$$\frac{\hat{c}}{C_m} \simeq \frac{\sigma_y}{\sqrt{2}(\sigma_i)} \simeq \frac{\sigma_v x/U}{\sqrt{2}(\sigma_0^{2/3} + \alpha\sigma_v x\Lambda_B^{-1/3}/U)^{3/2}} \qquad (7.28)$$

$$\frac{\hat{c}}{C_m} = 0 \text{ at } x = 0 \text{ and } \frac{\hat{c}}{C_m} \sim x^{-1/2} \text{ as } x \to \infty;$$

thus $\hat{c}/C_m$ has a maximum value. Differentiating Eq. 7.28 with respect to x shows that the maximum occurs at

$$x = \frac{2}{\alpha} \frac{U}{\sigma_v} \sigma_0^{2/3} \Lambda_B^{1/3}, \qquad (7.29)$$

and takes the value

$$\frac{\hat{c}}{C_m} = \frac{2}{3\alpha} (\Lambda_B/\sigma_0)^{1/3}. \qquad (7.30)$$

If we take the scale of the turbulence to be the "mixing length", which is kz near the surface, with $k = 0.4$ being von Kármán's constant, we can estimate α from the Fackrell-Robins data. If we assume σ_0 equal to the source radius, then the 9 mm diameter source, released at height 23 cm has $(\sigma_0/\Lambda_B) = 4.5/92 = 0.05$. This release has $\hat{c}/C_m = 2.2$ at its maximum value; therefore, $\alpha = 2/3 \cdot 1/2.2(0.05)^{1/3} = 0.11$. The downstream distance predicted from Eq. (7.29) is $2H$, where H is the boundary-layer depth (1.2 m in the experiment); we have used the experimental value of $\sigma_v/U \approx 0.1$. This is in good agreement with the data, and the power-law predictions also give good agreement for other source sizes.

The concept of a meandering plume is valid only if the turbulent eddies are significantly larger than the plume itself, for only then can the plume be carried to and fro as a coherent entity by the random velocity field. This is evident in the surface-release data of Fackrell and Robins which show that the fluctuation variance is *independent* of the source size (Fig. 7.8). We can explain this by observing that the plume is always as big as the turbulent eddies in the surface-release case, since the turbulence scale goes to zero at the surface. Thus, as the plume grows it always "sees" turbulence of its own size, and thus exhibits self-similar behavior. We should add a note of caution for atmospheric modeling here; in the convective boundary layer, there are large-scale horizontal velocity fluctuations close to the ground (see Kaimal, 1978 and Ch. 1); therefore, it is possible to have horizontal meander even in a surface release in such a boundary layer.

Other studies of concentration fluctuations are much less complete. Deardorff and Willis (1984) measured surface-concentration fluctuations from an elevated source in a convectively mixed layer. Their relative intensity is shown as a function of nondimensional downstream distance $X = xw_*/Uh$, where w_* is the convective scaling velocity, and h the mixed layer depth, in Figs. 7.12 and 7.13 for the nonbuoyant and buoyant sources respectively. The mean concentrations on the plume centerline are shown in Fig. 7.14. It can be seen that significant surface impact is accompanied by very large vari-

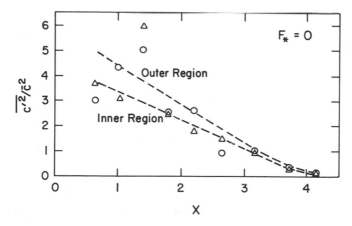

Figure 7.12. Relative fluctuation intensity at ground level in a convectively mixed layer, from Deardorff and Willis (1984); the source is elevated but not buoyant. $\triangle$ represents average over $y < \sigma_y/2$; o represents average over $\sigma_y/2 < y < \sigma_y$.

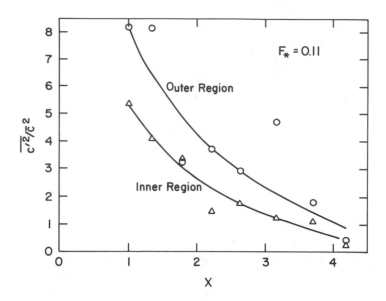

Figure 7.13. As in 7.12 but for a buoyant source.

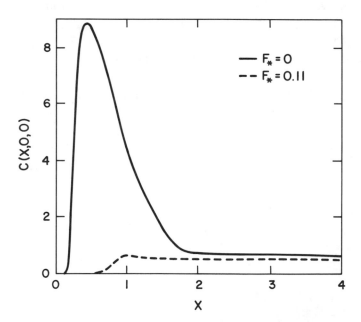

Figure 7.14. Mean centerline concentrations at ground level. From Deardorff and Willis (1984).

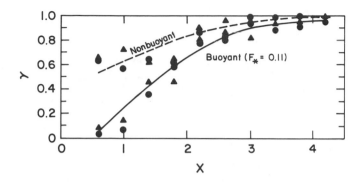

Figure 7.15. Intermittency γ from the ground-level measurements of Deardorff and Willis (1984). γ is the fraction of time with non-zero concentration; symbols as in Figure 7.12.

ance, indicating that the plume is highly intermittent. The measured intermittency is shown in Fig. 7.15 where γ is the fraction of time with measured concentrations greater than zero. Buoyancy and momentum in the source jet complicate the dynamics significantly, and it is therefore difficult to draw quantitative conclusions from these data.

Hanna (1984) reviewed some of the atmospheric data on concentration fluctuations, most of which are for short travel distance. There generally are insufficient data to apply any of the models confidently, but Hanna used "plausible" estimates, when data were unavailable, to show that the simple meandering plume model and its variations can make reasonable predictions of the concentration variance.

My final topic is the probability distribution for the concentration field. Again there are several experimental sources. Figure 7.16 shows p.d.f.'s from Fackrell and Robins (1982a). The p.d.f.'s are at different heights on the plume centerline, and basically show the changing shape as $\sigma_c/\bar{c}$ increases from a value less than 1 at $z/H = 0.06$ to a value in excess of 3 at $z/H = 0.54$. Here, we have defined $\sigma_c^2 = \overline{c'^2}$. The lowest values show a shape similar to Gaussian but obviously distorted by the requirement that c always be positive. We might expect the distribution to tend to Gaussian as $\sigma_c/\bar{c} \to 0$, in analogy with the central limit theorem, if we imagine that the variance is reduced by some sort of averaging process. However, at large values of $\sigma_c/\bar{c}$, Fackrell and Robins, and many other workers, obtain p.d.f.'s that are close to exponential together with a large fraction of

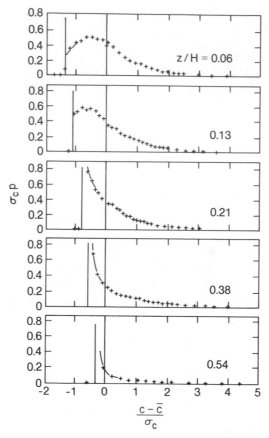

Figure 7.16. Probability density functions from Fackrell and Robins (1982a).

observations close to zero. This is consistent with the notion of an intermittent, meandering plume where a good deal of time is spent outside the plume where concentrations are small. Deardorff and Willis (1984) tried to fit a log-normal distribution to the non-zero part of the p.d.f., using Csanady's (1973) argument that a turbulent eddy can be considered to dilute the concentration by a certain factor. However, the data did not fit a simple log-normal, and they did not compare with an exponential distribution.

Lewellen and Sykes (1986) analyzed lidar observations of atmospheric plumes to obtain estimates of the p.d.f.; their results are summarized in Fig. 7.17. In general, there were about 10 lidar cross-sections of the plume in each hour. These were power plant plumes from stacks of about 200 m, and the cross sections were 2–20 km

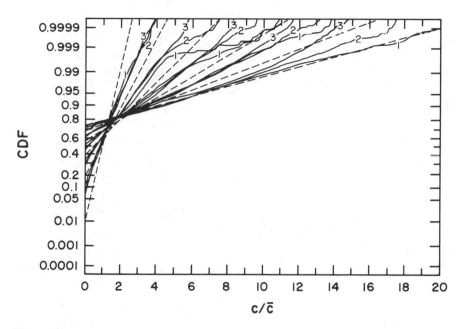

Figure 7.17. Observed distributions of concentration from lidar cross sections. Data are in bins using $\sigma_c/\bar{c}$ with boundaries 0.4, 0.8, 1.2, 1.6, 2.0, 2.4. Dashed lines show truncated Gaussians. From Lewellen and Sykes (1986).

downstream. The 10 samples were not enough to obtain good statistics, so the sample was increased to 100 by assuming that the instantaneous plume shape and the centroid position were independent so that each shape could be assigned to each centroid. This then allows a reasonable estimate of $\bar{c}$ and σ_c to be obtained. The second assumption is that the p.d.f. depends only on these two parameters, i.e., the mean and the variance. Over 100 h of data were then combined in ranges of $\sigma_c/\bar{c}$ to obtain the distribution functions shown in Fig. 7.17. The three different data sets, two from Illinois and one from Tennessee, collapse together and fall neatly between the straight lines. The latter represent the truncated Gaussian distributions, which are obtained by replacing all the non-realizable negative tail of a Gaussian by a delta function at the origin, i.e., the intermittent zeroes. This idealized distribution is Gaussian at small $\sigma_c/\bar{c}$ and is exponential plus intermittent zeroes at large $\sigma_c/\bar{c}$, and therefore spans the known range sensibly. It also fits the atmospheric observations remarkably well.

7.4. Applications

We close with an attempt to provide some idea of practical applications of the research on concentration fluctuations. It should be clear that our knowledge is still very limited and the applications far from developed as yet. My own research efforts have concentrated on model validation, and therefore this is my principal topic here.

Lewellen and Sykes (1983) outlined a framework for model evaluation that uses concentration fluctuation information. We argued that in order to assess the accuracy of a model prediction, one needs information about the probability distribution of the concentration, since we are sampling a random variable. Under the assumption that the probability distribution is a function of the mean and standard deviation only, the p.d.f. is obtainable for any averaging time provided we know how to reduce the variance to account for the averaging. Some experimental data to support the dependence of the p.d.f. on the first two moments were given in the previous section, and some estimates of the time-averaging effect were made using the meandering plume model. Given one of the models discussed in Sec. 7.2 for predicting the variance of the concentration fluctuations, we have, in principle, the necessary ingredients for model evaluation.

The important feature precluding the standard statistical tests in the case of atmospheric dispersion modeling is the fact that each sample is essentially a single realization from an unknown probability distribution. We are not allowed to resample, and thus determine the distribution, because the distribution is changing in both space and time, and therefore any other sample comes from a different distribution. This is the main problem behind any determination of the variance in atmospheric measurements. We therefore conclude that the variance must be part of the modeling prediction, and the model evaluation must test the consistency of the data with the predicted probability distribution.

A specific method of comparison was proposed by Lewellen et al. (1985) at a DOE-AMS Model Evaluation Workshop. The idea is to compare the differences between the observations and the predicted mean values with the differences we would expect on the basis of the p.d.f. prediction. This means that the model must make an accurate prediction of both the mean and the variance (and hence the p.d.f.) in order to compare favorably. Before describing the procedure for the comparison, the philosophy of the approach deserves a few words.

It may appear a major change to demand that a model predict the p.d.f. before it can be evaluated, and it is clearly not achievable in the near future. However, it is virtually impossible to conclude anything by examining quantities such as r.m.s. error of a model if one has no information about the inherent uncertainty in the samples. In the previous section, we showed that relative fluctuation intensities, $\sigma_c/\bar{c}$, of 5 or more on plume centerline have been observed in the wind tunnel. The ratio depends on the relative size of the source and the turbulence length scale, and much larger values are possible in the atmosphere. Furthermore, the relative intensity increases in the edges of the plume. Thus, with turbulence time scales of order 10 min, and a 1-h time average, we can still find $\sigma_c/\bar{c}$ significantly greater than unity for parts of a plume with a significant impact. Shorter time averages would exhibit even larger intensities, provided we are discussing dispersion ranges greater than the boundary layer depth so that the boundary layer eddies must be counted as random fluctuations. We note that the intensities are largest for small passive sources in a deep boundary layer; the intensities would be reduced in a buoyant source because the internally generated turbulence would mix the early plume rapidly to increase the effective source size. In any event, it is clear that in general, even with 1-h averages on the 10-km scale, there can be significant inherent uncertainty in sampler measurements. If we are to determine whether the discrepancies between prediction and observation are simply a manifestation of this uncertainty, we need a quantitative assessment of the variation.

The method suggested by Lewellen et al. is to compare the distribution of the actual residuals (i.e., the difference between the observation and the predicted mean) with the expected distribution of residuals obtained by randomly selecting a sample from the predicted probability distribution at each sampler location. An example of such a comparison is shown in Fig. 7.18. The solid line represents the cumulative distribution of residuals, $d_i = \bar{c}_i - O_i$, where i denotes a sampler, O_i is the observation, and $\bar{c}_i$ is the predicted mean; i.e., at any value of d, the ordinate represents the fraction of observations with $d_i \leq d$. If the model were to predict the observations exactly, i.e., $d_i = 0$ for each i, then the line would jump from 0 to 1 at $d = 0$, but we know this will not happen in practice and we do not even expect it for a "perfect" model, since the observations are random samples. The dashed lines in Fig. 7.18 are an estimate of the range of expected distributions of residuals. These curves were obtained by generating

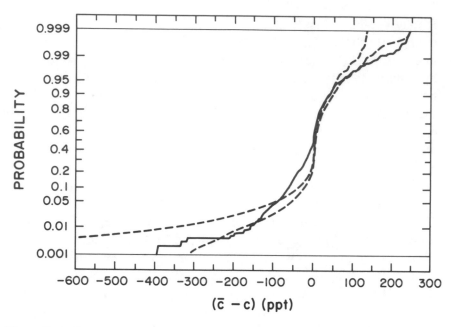

Figure 7.18. Example residual plot. ——, observed residuals; ––, 95% confidence
limits on expected residuals from model prediction.

150 realizations and calculating the spread of the distributions. One
realization is generated by making a random selection at each sampler location from the appropriate predicted p.d.f.; note that these
selections are independent, so we are neglecting space and time correlations in the data. Our realization will thus give a realization of
the cumulative residual distribution. We compute 150 such distributions and then plot a percentile as a dashed line to obtain confidence
limits on our expected distribution. In fact, we calculate the mean
and the standard deviation at each value of d, and the dashed lines
represent mean plus or minus 2 standard deviations; this was done
in the interests of efficiency and economy in the computational procedure. In any event, this procedure gives in principle a comparison
between the observed values and the predicted values in a realistic
sense. It is evident that the comparison will depend on both the predicted mean and the predicted variance, but a successful comparison
will require a consistent prediction of both. It is not in general possible to improve the comparison by increasing the predicted variance
so that the observation falls within the wider range of predictions,
because if there is a consistent error, then the observations will not
span the expected range of the predictions. The effect of increasing

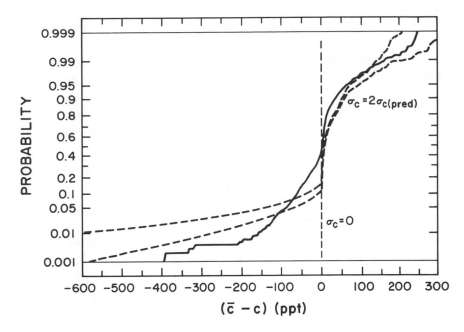

Figure 7.19. Effect of different fluctuation level prediction on expected residual distribution.

or decreasing the predicted variance is shown in Fig. 7.19. It will be seen that the dashed lines move toward or outward from the d-origin as σ_c is decreased or increased respectively. However, this occurs for both under-predictions and over-predictions, so a consistent under-prediction error could not be disguised by this technique. As with all statistical measures, a large amount of data is required to obtain confident estimates of model errors.

Venkatram (1982) proposed a fundamentally different evaluation procedure incorporating a different notion of concentration fluctuations. Briefly, his idea is to define the ensemble in terms of model input parameters, i.e., to reject the concept of a definable atmospheric ensemble. This has some merit, in view of the difficulty we experienced in the definition at the start of this chapter, which left the ensemble still somewhat vaguely defined as a filtered flow field. Venkatram's approach would consider two flows to be members of the same ensemble for a given model if both result in the same inputs for the model. This is the approach taken by Hanna (1982) to determine the inherent variability of an atmospheric dispersion data set. Working with a simple Gaussian plume model, he grouped the data into bins of wind speed, direction, and stability and found significant variability within the bins. The major drawback within this approach is that it is virtually impossible to work with any but the very simplest models, since the dimension of the input parameter space quickly gets out of hand, and every observation then represents a different ensemble. Venkatram's approach is clearly related to the filtering approach insofar as the filtering depends on the inputs being used by the model, i.e., the spacing of the wind observations and the spatial scales they are assumed to represent.

The AMS Workshop on Updating Applied Diffusion Models in Clearwater, Florida, 1984 (Weil, 1985) recommended that natural uncertainty estimates should ultimately become an integral part of air quality predictions, but concluded that the models are not sufficiently developed to warrant routine application at present. However, it is clear that an estimate of the probability distribution of the concentration is a much sounder basis for decision making than we now have. Furthermore, assessment of flammability or toxicity on the basis of ensemble-average concentrations can be seriously in error. These effects depend on short temporal- and spatial-scale fluctuations and thus the variance is essential for these predictions. At present, though, our modeling ability is only just developing, and we have a long way to go before we are able to make reliable assessments in general atmospheric conditions for arbitrary sources.

References

Chatwin, P. C., and P. J. Sullivan, 1979: The relative diffusion of a cloud of passive contaminant in incompressible turbulent flow. *J. Fluid Mech.*, **91**, 337–355.

Csanady, G. T., 1967: Concentration fluctuations in turbulent diffusion. *J. Atmos Sci.*, **24**, 21–28.

Csanady, G. T., 1973: *Turbulent Diffusion in the Environment*. Reidel, Dordrecht, 248 pp.

Deardorff, J. W., and G. E. Willis, 1984: Ground level concentration fluctuations from a buoyant and a non-buoyant source within a laboratory convectively mixed layer. *Atmos Environ.*, **18**, 1297–1309.

Durbin, P. A., 1980: A stochastic model of two-particle dispersion and concentration fluctuations in homogeneous turbulence. *J. Fluid Mech.*, **100**, 279–302.

Fackrell, J. E., and A. G. Robins, 1982a: Concentration fluctuations and fluxes in plumes from point sources in a turbulent boundary layer. *J. Fluid Mech.*, **117**, 1–26.

Fackrell, J. E., and A. G. Robins, 1982b: The effects of source size on concentration fluctuations in plumes. *Bound.-Layer Meteor.*, **22**, 335–350.

Gifford, F. A., 1959: Statistical properties of a plume dispersion model. *Adv. Geophys.*, **6**, 117–138.

Gifford, F. A., 1982: Horizontal diffusion in the atmosphere: A Lagrangian-dynamical theory. *Atmos. Environ.*, **16**, 505–512.

Hanna, S. R., 1982: Natural variability of observed hourly SO_2 and CO concentrations in St. Louis. *Atmos. Environ.*, **16**, 1435–1440.

Hanna, S. R., 1984: Concentration fluctuations in a smoke plume. *Atmos. Environ.*, **18**, 1091–1106.

Hay, J. S., and F. Pasquill, 1959: Diffusion from a continuous source in relation to the spectrum and scale of turbulence. *Adv. Geophys.*, **6**, 345–365.

Kaimal, J. C., 1978: Horizontal velocity spectra in an unstable surface layer. *J. Atmos. Sci.*, **35**, 18–24.

Lamb, R. G., 1984: Air pollution models as descriptors of cause-effect relationships. *Atmos. Environ.*, **18**, 591–600.

Lewellen, W. S., and R. I. Sykes, 1983: On the use of concentration variance as a measure of natural uncertainty in observed concentration samples. *6th Symposium on Turbulence and Diffusion*, Amer. Meteor. Soc., Boston, 47–50.

Lewellen, W. S., and R. I. Sykes, 1986: Analysis of concentration fluctuations from lidar observations of atmospheric plumes. *J. Climate Appl. Meteor.*, **85**, 1145–1154.

Lewellen, W. S., R. I. Sykes, and S. F. Parker, 1985: An evaluation technique which uses the prediction of both concentration mean and variance. *Proceedings of the DOE/AMS Air Pollution Model Evaluation Workshop*, October 1984, Savannah River Lab. Report No. DP–1701–1.

Pope, S. B., 1981: Transport equation for the joint PDF of velocity and scalars in turbulent flow, *Phys. Fluids*, **24**, 588–596.

Richardson, L. F., 1926: Atmospheric diffusion shown on a distance-neighbor graph. *Proc. Roy. Soc. A*, **110**, 709–737.

Sawford, B. L., 1983: The effect of Gaussian particle-pair distribution functions in the statistical theory of concentration fluctuations in homogeneous turbulence. *Quart. J. Roy. Meteor. Soc.*, **109**, 339–356.

Smith, F. B., and J. S. Hay, 1961: The expansion of clusters of particles in the atmosphere. *Quart. J. Roy. Meteor.*, **87**, 89–91.

Sykes, R. I., 1984: The variance in time-averaged samples from an intermittent plume. *Atmos. Environ.*, **18**, 121–123.

Sykes, R. I., W. S. Lewellen, and S. F. Parker, 1984: A turbulent-transport model for concentration fluctuations and fluxes. *J. Fluid Mech.*, **139**, 193–218.

Sykes, R. I., W. S. Lewellen, and S. F. Parker, 1986: A Gaussian plume model of atmospheric dispersion based on second-order closure. *J. Climate Appl. Meteor.*, **25**, 322–331.

Taylor, G. I., 1921: Diffusion by continuous movements. *Proc. London Math. Soc.*, **20**, 196–211.

Tennekes, H., and J. L. Lumley, 1972: *A First Course in Turbulence*. MIT Press, Cambridge, 300 pp.

Venkatram, A., 1979: The expected deviation of observed concentrations from predicted ensemble means. *Atmos. Environ.*, **13**, 1567–1569.

Venkatram, A., 1982: A framework for evaluating air quality models. *Bound.-Layer Meteor.*, **24**, 371–385.

Weil, J. C., 1985: Updating applied diffusion models. *J. Climate Appl. Meteor.*, **24**, 1111–1130.

CHAPTER 8

Concentration Fluctuations Within a Laboratory Convectively Mixed Layer

James W. Deardorff
Glen E. Willis

8.1. Introduction

The unstably stratified boundary layer is a relatively simple entity in which to try to understand and model dispersion. The depth of vertical mixing (h) in this case is usually clearly defined by the height above which stable stratification commences. The turbulence intensities responsible for the dispersion are rather well known functions of height relative to h (Caughey, 1982) if the convective velocity scale w_* is known. Also, the wind speed and direction can then usually be treated as independent of height above the surface layer and within the boundary layer if the latter is not too baroclinic.

The simple convectively mixed layer can then be studied in the laboratory, when simulating atmospheric cases for which $h >> -L$, where L is the Obukhov length; i.e., cases for which buoyant convection, rather than mean wind shear, supplies most of the turbulence energy. One proviso, however, is that the turbulence Reynolds number (hw_*/ν where ν is the kinematic viscosity) should be of order 1000 or larger. This is satisfied for most scales of turbulence within a laboratory tank of the size used in this study (Willis and Deardorff, 1974), but not for scales comparable with the exit diameter of the laboratory stack. Hence, we must be alert to any indications of insufficient mixing on scales smaller than those to which the laboratory inertial subrange extends.

Mixed-layer dispersion studies performed in the laboratory so far include the investigation of the mean-concentration fields downstream of a nonbuoyant source at different heights within the mixed layer (summarized in Deardorff, 1985; Poreh and Cermak, 1985), and the concentration fluctuations near the surface from a buoyant and a nonbuoyant source (Deardorff and Willis, 1984; Deardorff, 1985). The present analysis extends the latter study to all heights within the mixed layer, for a nonbuoyant and a buoyant source, and examines the concentration probability distribution more comprehensively. However, the present sources possessed substantial momentum jets whose effects cannot be ignored even at moderate distances downstream.

The concentration fluctuations are of special interest, since peak values may exceed the mean by an order of magnitude or more. Their interpretation will require some knowledge of the intermittency of the fluctuations, which then will also be examined.

8.2. Experimental Analysis

Data were collected within the convection tank of Willis and Deardorff (1974) from experiments involving the emission of fluorescent dye from a stack translating across the floor of the tank at uniform speed U. A vertical y-z plane near the center of the tank was illuminated with a planar-spread beam of blue laser light directed downward from above. The origin of effluent travel time ($t = 0$) was taken as the instant the stack crossed through the vertical plane of light. The thickness of the plane of illumination was about 2 mm, which may be compared with an average boundary layer height of about 20 cm. Only dye within the plane of light was visible; it fluoresced with a yellow-green color, permitting all the original blue color reflected elsewhere within the tank to be filtered out of the camera's view. To a very high degree of approximation, the intensity of scattered fluoresced light was directly proportional to the concentration of the dye. Color photographs of this y-z plane were taken every 5 to 10 s after the $t = 0$ time. Since the downstream distance x is given by Ut, the quantity $X = (w_*/U)(x/h) = (w_*/h)t$ represents a nondimensionalized (with mixed-layer scales) effluent travel time during which a parcel of dye was under the influence of turbulent mixing in the environment. Hence, we use X as the dimensionless downstream coordinate.

Photographs were also taken of 11 calibration bottles, containing a 2^{10} range of known concentration of fluorescent dye, placed within the tank shortly before each experiment. Absolute values of the concentrations of the diffusing dye were then obtained by comparison against these known standards, subject to the calibration adjustments to be described.

Photographs were subsequently analyzed by projecting plume slides onto the plane of a roving photomultiplier tube with an aperture of 0.007 h. It was traversed, under computer control, along rows of z = constant and y incremented; the coordinates and light intensities were fed onto digital tape and later converted to dimensionless coordinates and concentrations.

Saturation of the film often occurred in regions close to the source. In order to prevent unrealistically excessive values of the dimensionless concentration

$$C = cUh^2/S$$

from occurring in these regions, where c is the actual concentration and S is the effluent source strength (mass emitted per unit time), values of C were suppressed whenever they exceeded a value of N, where N was set between 10 and 20. The suppression correction was

$$C_{\text{revised}} = N + \log_e(C - N + 1) \quad (C > N) \tag{8.1}$$

(The value of C of the undiluted stack effluent ranged between 100 and 200; its well-mixed centerline value was about 0.8 at X = 1.5.) Although this correction permits variations in C even under conditions of near film saturation, it tends to damp real fluctuations occurring in these regions quite strongly where $X < 0.5$ or so.

For the opposite extreme of very weak concentrations, spurious noise appearing as small values of C but outside the areas of apparent plume fragments was removed by estimating its average intensity, plus two standard deviations along the y = constant edges of the first photographic film in any one experiment. This mean value was then subtracted away from the entire field of concentrations in the experiment, and any negative values set to zero. In so doing, small values of truly non-zero concentrations were probably lost at the peripheries of the main plume for the larger values of X. Thus, the lateral standard deviations of the concentration σ_y that could be deduced from the data became increasingly too small as X increased beyond a value of order 1, and consequently we do not deal with any statistics depending strongly on σ_y. This uncertainty on the low side is estimated to be on the order of 0.05 in C, and also prevents our examining intermittency accurately when a small threshold is used.

Finally, the data had to be recalibrated frame by frame so that the total mass of effluent within each photograph (y-z plane) should not deviate too greatly from its average value stemming from mass conservation:

$$\int_0^\infty \int_{-\infty}^\infty <c> U \, dy \, dz = <S> , \qquad (8.2)$$

where the angular brackets represent the ensemble average. The difficulty here is that Eq. (8.2) does not hold precisely if the ensemble-averaging operation is omitted, and in any one data frame some deviation from the true average is to be expected. It was assumed that such deviations, when genuine, are due primarily to σ_u/U, where σ_u is the root-mean-square x-component flow variation at the stack, although the vagaries of diffusion along x would also be a factor.

From inspections of photographs of blue dye observed looking in the crosswind direction, from the study of Hukari (1984) where intensity was roughly proportional to the y-integrated concentration, it was estimated that the integral I, given by

$$I = \int_0^\infty \int_{-\infty}^\infty (cUh^2/S) d(y/h) d(z/h)$$

could deviate from unity by amounts specified by δI, where

$$\delta I = (\sigma_u/U) \exp(-0.4X) \qquad (8.3)$$

and with σ_u/U given by

$$\sigma_u/U = 0.3 . \qquad (8.4)$$

The above relations were assumed independent of effluent buoyancy as measured by

$$F_* = w_s r_s^2 (g/\rho_0)(\rho_0 - \rho_s)/(w_*^2 Uh) , \qquad (8.5)$$

where w_s is the stack-exit vertical velocity, r_s the stack radius, ρ_s the initial effluent density, and ρ_0 the ambient fluid density.

The final recalibration of concentrations consisted of multiplying C by a particular factor R for any given photographic frame, where R is given by

$$R = [1 + A(I_i/\bar{I}_i - 1)]/I_i = R(X) , \tag{8.6}$$

$$A = \delta I / [\overline{(I_i/\bar{I}_i - 1)^2}]^{1/2} = A(X) , \tag{8.7}$$

I_i is the value of I of a given photo or realization (for a given value of X in an experiment), and $\bar{I}_i$ is the mean measured value of I_i in the ensemble of like experiments at a given X. The dependence of A upon X is through the dependence of δI in Eq. (8.3).

The use of Eq. (8.6) is perhaps the simplest method of adjusting concentrations in particular y-z realizations in such a manner as to cause the mean y-z integrated dimensionless concentration over an ensemble of experiments at given X to have unit value; to cause the variance from the mean at given X to have a desired small value determined by Eq. (8.3); and to prevent the generation of any negative concentrations in the process. The method reverts to no correction whenever the mean has unit value and the measured variance has the value prescribed by Eq. (8.3).

Tests conducted using Eqs. (8.3), (8.4), (8.6), and (8.7) with δI altered by a factor of 2 indicated that $\sigma_c/\bar{c}$, where σ_c is the root-mean-square variation in c about the mean $\bar{c}$, was affected at only the 5% level. This relatively small effect occurred even though individual values of R could be as large as 3 or as small as 1/3. Presumably, this result means that the concentration fluctuation was determined more by the internal variations of concentration than it was by plume meandering between realizations.

Two cases are studied here, one of neutral or near-neutral effluent buoyancy ($F_* = 0$ or 0.03) and one of buoyant effluent with $F_* = 0.26$. The stack radius was 1 mm for $F_* = 0.03$, and 2 mm for $F_* = 0$ and 0.26. Values of parameters that were essentially constant for all the experiments reported here are $w_s = 9.7$ cm s^{-1}, $U = 2.0$ cm s^{-1}, $w_* = 0.9$ cm s^{-1}, stack height (z_s) = 3.1 cm, stability aloft = 0.63 K cm^{-1}, number of experiments per case = 7 to 9, and entrainment rate for the boundary layer as a whole about 0.014 w_*.

8.3. Patterns of Mean Concentration

Patterns of the crosswind-integrated mean dimensionless concentration $\overline{C}^y$, defined by

$$\overline{C}^y = \int_{-\infty}^{\infty} \overline{C} \, dy/h \; ,$$

are shown for the $X - z/h$ plane in Fig. 8.1 for the case $F_* = 0$, and in Fig. 8.2 for the case $F_* = 0.03$. In both cases, the effect of the momentum jet in causing a strong initial plume rise is evident. A measure of this effect, relative to h, is given by

$$\ell_m^* = (4w_s r_s^2/U)^{1/2}/h \; ,$$

which had a value of about 0.055 for $F_* = 0$ and a value of about 0.028 for $F_* = 0.03$. A value of 0.02 is typical of atmospheric stacks. Hence this study will overemphasize somewhat the effect of the momentum jet for most applications.

The most interesting aspect of the figures is the close similarity of the pattern, especially that of Fig. 8.1, to that of the elevated nonbuoyant source without a momentum jet investigated by Willis and Deardorff (1981) and shown again here in Fig. 8.3. Figure 8.1 appears as a weak, diffuse version of Fig. 8.3, with an elevated source

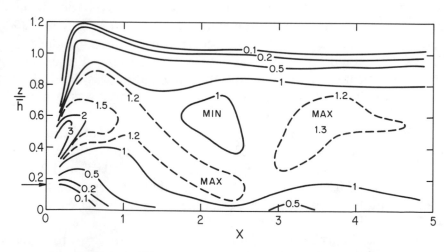

Figure 8.1. Contours of $\overline{C}^y$ for the case in which $F_* = 0$ and there is a strong source momentum jet. Source height in this and subsequent figures is indicated by arrow at lower left.

(due to plume rise from the momentum jet) displaced to about $X = 0.7$ where it has become quite diffuse vertically. The slower rate of descent of the maximum in this case than in Fig. 8.3 is presumably caused by a continuing influence of the momentum jet even after maximum plume rise is reached.

Differences between the contours of Figs. 8.1 and 8.2, which are relatively minor, are probably associated more with sampling error than with the effect of the weakly buoyant release of Fig. 8.2.

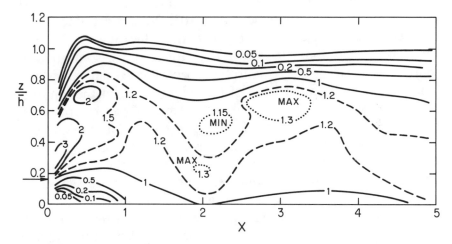

Figure 8.2. Contours of $\overline{C}^y$ for the case in which $F_* = 0.03$ and there is a strong source momentum jet.

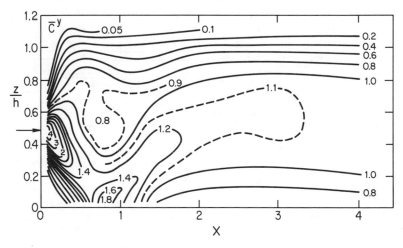

Figure 8.3. Contours of $\overline{C}^y$ for the case in which $F_* = 0.0$ and there is no source momentum jet, but source height is elevated. From Willis and Deardorff (1981).

Values of $\overline{C}$ at $y = 0$ are obtainable from $\overline{C}^y$ through use of

$$\overline{C}(y = 0) = \overline{C}^y/(\sqrt{2\pi}\sigma_y/h) . \qquad (8.8)$$

Estimates of σ_y/h for use in Eq. (8.8) may be obtained from Willis and Deardorff (1976), Briggs (1984), and Poreh and Cermak (1985).

8.4. Patterns of $\sigma_c/\overline{c}$ and of Intermittency

The pattern of $s = \sigma_c/\overline{c}$ near the centerline ($|y| < \sigma_y/2$) is shown in Fig. 8.4 for the case $F_* = 0.03$. Values are interpolated between data collected in zones of vertical thickness $\Delta z/h = 0.2$ and data spaced about 0.4 apart in X for $X < 1.2$ and about 0.7 apart in X at larger X. As is to be expected, maximal values occur at the upper and lower fringes of the "plume," i.e., at the top of the boundary layer and near the surface at small X before appreciable concentrations have descended to ground level. The minimal value within the momentum jet for $X < 0.5$ is smaller than anticipated, and will be seen to have been associated with minimal plume meandering. The minimal near-surface values at large X, which are significantly smaller than unity, will be seen to occur where the intermittency γ is essentially 1 (detectable amounts of concentration always present), and to indicate that the exponential probability distribution function (p.d.f.) is not a good fit there. The exponential p.d.f. requires that $s = 1$ when $\gamma = 1$. Although s may decline even below 0.5 at

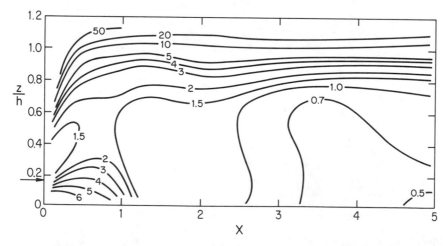

Figure 8.4. Contours of $\sigma_c/\overline{c}$ for $|y| \le 1/2\sigma_y$ and for $F_* = 0.03$.

still larger X, it cannot reach zero, even near the centerline ($y = 0$), because of the mixed-layer entrainment process, which creates concentration fluctuations as long as the air above the boundary layer is purer than the air within.

The data were too coarse in the vertical to disclose how s behaved as $z \to 0$. However, visual observations during the experiments, as well as inspection of the photographs, disclosed no significant tendency for s to approach zero as $z \to 0$. Hence the contours of Fig. 8.4 are drawn to reflect these observations. This behavior is probably due to the large lateral length scale of the horizontal turbulence even as $z \to 0$, as suggested by Deardorff and Willis (1985, Fig. 13), which permits in-plume fluctuations to exist there even though the length scale of the turbulent vertical velocity may approach zero as $z \to 0$. This behavior contrasts with Fackrell and Robins' (1982) finding, for concentration fluctuations in neutral shear-flow turbulence in the wind tunnel, that $s \to 0$ as $z \to 0$.

Although any inspection of the raw photographic data discloses the existence of very small scales of concentration fluctuations, most of their intensity, along y, appears to reside on scales greater than a few tenths of σ_y. This conclusion stems from an analysis for the case $F_* = 0.03$ in which the instantaneous fluctuations of c from the mean were smoothed along y over successively greater increments (Δy), and the reduced intensitites ($\tilde{s}$) were calculated. Results are shown in Fig. 8.5, where the ordinate values approach 1 rather discontinuously as $\Delta y \to 0$. It should be mentioned that in this and the previous analysis for s, $\overline{C}(y)$ was obtained by assuming Gaussianity along y and utilizing Eq. (8.6) along with measured values of $\overline{C}^y$ and of σ_y/h. The normalization of σ_y in the figure is with σ_y values from Willis and Deardorff (1976).

Figure 8.5 indicates an increasingly rapid dropoff in $\tilde{s}^2$ with Δy, the dropoff being less rapid at very small X where meandering may have been significant. Figure 8.5(a) also shows the behavior of the smoothed intensity for a quantity obeying a first-order Markov process with correlation 0.8 between successive data points. Such a process has an exponential autocorrelation and a -2 spectral slope, and for the figure the smoothing calculations were performed in the same discrete manner as for the concentration data. The similarity with the experimental data is close, the more rapid falloff in the experimental case at larger values of Δy indicating that the "red" spectrum probably existing in the experiments reached a peak at moderate

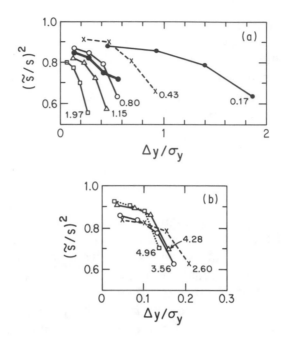

Figure 8.5. Square of the ratio of smoothed concentration-fluctuation strength relative to its point-value strength s, versus the smoothing interval Δy, normalized by σ_y, for $F_* = 0.03$; the parameter is X. (a) For values of X up to about 2; (b) for values of X between 2 and 5. Heavy line in (a) is from a Monte-Carlo first-order model.

values of Δy and then declined at larger scales, unlike a strictly -2 spectral decay. The overall spectral shape is thus deduced to be very much like that observed by Fackrell and Robins (1982) where a short inertial subrange was observed. Here, this subrange would be very short indeed, because the inertial subrange observed in the turbulent velocity fluctuations (Deardorff and Willis, 1985) generally occupied less than a decade of wavenumber variation. In the present case the usual type of spectrum analysis could not be performed since continuous data along a dimension of homogeneity were not available, unlike measurements taken at a fixed point within a mean wind.

The variation of σ_c^2 with $|y|/\sigma_y$ is shown in Fig. 8.6 at heights near the center of the plume (at $z/h = 0.3$ for $X = 0.17$ and at $z/h = 0.5$ for all larger X). They are normalized by the inner value, which represents σ_c^2 in the interval $0 < |y| < \sigma_y/2$. The falloff with $|y|$ closely resembles that reported by Csanady (1967, Fig. 2) and Fackrell and Robins (1982). The latter note that the profile is broader than the Gaussian shape based on σ_y deduced from $\bar{c}$, and that there may be a

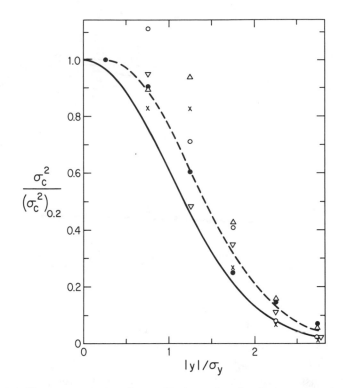

Figure 8.6. The concentration-fluctuation intensity σ_c^2, relative to its value at $y/\sigma_y = 0.2$, versus y/σ_y, for $F_* = 0.03$ and $z/h = 0.5$, except $z/h = 0.3$ at $X = 0.17$. Symbols: solid circle, $X = 0.17$; $\times$, $X = 0.43$; triangle with point up, $X = 0.80$; open circle, $X = 1.15$; triangle with point down, $X = 1.96$. Solid curve is the Gaussian distribution assumed for $\overline{C}$; dashed curve is the same Gaussian displaced 1/4 abscissa unit to the right.

slight off-center peak near $|y|/\sigma_y = 0.4$, since gradient production of σ_c^2 operates only away from the plume centerline. The present data are too scattered and of insufficient resolution to confirm this expectation. It is noted here only that a fair fit to the data is obtained by shifting the Gaussian shape for $\overline{c}$ about 0.25 y/σ_y units to the right.

The intermittency for the case $F_* = 0.03$ is shown in Fig. 8.7, also for the same near-centerline region of the x-z plane as in Fig. 8.4. Here, the intermittency γ is defined as the fraction of occurrence of C exceeding a threshold value within a given volume of space ($\Delta y \leq \pm\sigma_y/2$, $\Delta z = 0.2h$, and $\Delta x = 0.01h$). For Fig. 8.7 the threshold is $C_T = 0.01$.

As may be expected, γ behaves inversely to s, approaching zero at the edges of the "plume." However, the value of γ within the

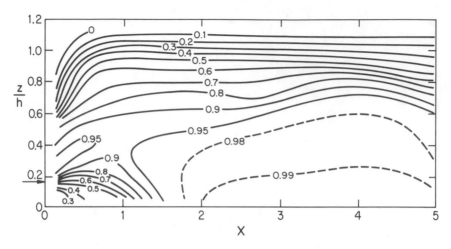

Figure 8.7. Contours of the intermittency γ in the $z/h - X$ plane for the case $F_* = 0.03$ and for the near-centerline region $|y| \leq 1/2\sigma_y$. The threshold is $C_T = 0.01$.

momentum jet (up to 0.95) is surprisingly large at such small X, and has caused us to speculate that the disturbances caused by both the momentum jet and its surrounding compensating downdraft (which must encompass a considerably larger area) tend to shield the effluent from scales of lateral motion smaller than about 0.5 h. This would cause less meandering than would be expected in the atmosphere where ℓ_m^* may be smaller than the values used in the laboratory experiments, and where meandering motions of scales larger than that of boundary-layer turbulence are frequently present.

We may question how dependent γ may be upon the threshold value C_T, especially for the present experiments in which small values of C had marginal significance. If we select $C_T = 0.01\overline{C}$ instead of $C_T = 0.01$, we obtain the γ values of Fig. 8.8, which do not differ very much from those of Fig. 8.7. However, if we choose $C_T = 0.1\overline{C}$ we obtain the γ values of Fig. 8.9, which are substantially decreased, at least in the vicinity of $z/h = 0.5$ and $X = 1$. The greatest decrease is 20%. This choice for C_T is not overly large, since peak values are many times greater than $\overline{C}$. Hence, we may conclude that the use of γ in theory and analysis should be avoided if possible, when uncertainties of up to 20% cannot be tolerated.

The falloff of γ with increasing X for $X > 4$ in Figs. 8.7–8.9 is believed to be a spurious effect of dilution causing values of C here and there to fall below the measurement threshold level.

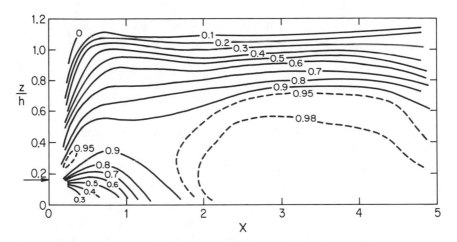

Figure 8.8. Contours of γ as in Fig. 8.7, except that $C_T = 0.01\overline{C}$.

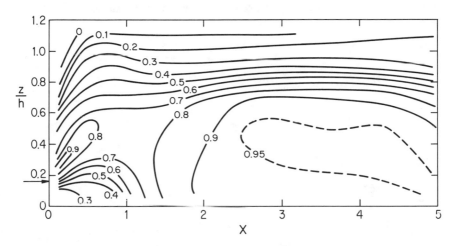

Figure 8.9. Contours of γ as in Fig. 8.7, except that $C_T = 0.1\overline{C}$.

Now, an in-plume concentration fluctuation c_p' has been defined as

$$c_p' = c - \bar{c}_p \quad c > c_T, \text{ and } c_T \text{ small },$$

where

$$\bar{c}_p = \bar{c}/\gamma$$

is the mean c of observations for which $c > c_T$. We may then use the identity derived by Wilson et al. (1985) to relate the in-plume concentration-fluctuation strength, $s_p = (\overline{c_p'^2}/\bar{c}_p^2)^{1/2}$, to the total concentration-fluctuation strength, $s = (\overline{c'^2}/\bar{c}^2)^{1/2}$. This identity is

$$s_p/s = [\gamma(1 + s^2) - 1]^{1/2}/s . \qquad (8.9)$$

The ratio s_p/s is shown within the x-z plane near the $y = 0$ centerline for our $F_* = 0.03$ case in Fig. 8.10. In obtaining the ratio, the γ values of Fig. 8 were used. The pattern closely resembles the distribution of $\gamma^{1/2}$ (see Fig. 8.8), and indicates that near $y = 0$ the in-plume fluctuations greatly dominate over the meandering contribution to s except at the plume fringes. The dominance is even greater if σ_{cp}/σ_c is examined. In the atmosphere's boundary layer, in-plume dominance as pronounced as this is not expected at small X unless all mesoscale and synoptic scales of motion are filtered out, requiring that the x axis be redefined every 20 minutes or so. In the laboratory tank, the stack towing line uniquely defines the x axis.

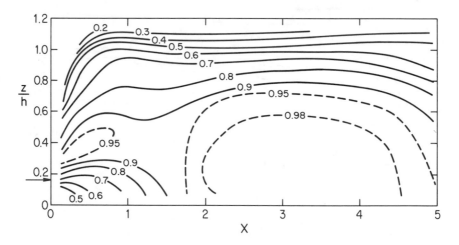

Figure 8.10. Contours of s_p/s using the intermittency data of Fig. 8.8, where $s = \sigma_c/\bar{c}$ and s_p is the same statistic using only in-plume data for which $C > C_T$.

8.5. The Probability Distribution

8.5.1. Neutrally buoyant effluent

Probability distributions in this case were obtained from concentration data within the zone $0 < z/h < 0.2$ and $|y|/\sigma_y < 1$, and include only those values above the $C = 0.01$ threshold. (Inspection of data collected within zones for which $|y|/\sigma_y < 1/2$ showed no systematic differences from those for $1/2 < |y|/\sigma_y < 1$, so that both data sets were averaged together for presentation here.) Probability distributions, $P_i(\ln C_p)$, for the case $F_* = 0$, are shown in Fig. 8.11 in histogram form for values of X increasing downward. P_i is defined such that

$$\sum_{i=1}^{17} P_i(\ln C_p) = 1$$

and is related to the continuous probability density functions $P(\ln C_p)$ and $P(C_p)$ by

$$P_i(\ln C_p) = P(\ln C_p)d(\ln C_p) = P(C_p)dC_p . \qquad (8.10)$$

Here, subscript p refers to the fact that only in-plume values are treated (for concentrations lying above a threshold); thus, $P(\ln C_p)$ and $P(C_p)$ are the conditional p.d.f.'s $P(\ln C_p) = P(\ln C; C > C_T)$, $P(C_p) = P(C; C > C_T)$. From Eq. (8.10) it follows that

$$P_i(\ln C_p) = (\Delta C/C_p)P(\ln C_p) = C_p(\Delta C/C_p)P(C_p) , \qquad (8.11)$$

where $\Delta C/C_p = 0.4605$ for the data of this paper.

Although the total number of data points, N, constituting each histogram of Fig. 8.11 is much smaller than is desirable, for $X \geq 1.46$ the distributions appear to follow the expected tendency, as X increases, of becoming more peaked as the fluid becomes more well mixed.

The solid curves in Fig. 8.11 represent the Gamma distribution usually expressed as

$$P(C_p) = \frac{n^n}{\overline{C}_p\Gamma(n)}(C_p/\overline{C}_p)^{n-1}\exp(-nC_p/\overline{C}_p) , \qquad (8.12)$$

where

$$n = \bar{c}_p^2/\sigma_{cp}^2 = s_p^{-2} . \qquad (8.13)$$

Here $\bar{c}_p$ is the mean dimensional concentration of the data entering the histogram (which excludes any zeros), $\overline{C}_p$ is its dimensionless counterpart, σ_{cp} is the r.m.s. fluctuation for these data, and $\Gamma(n)$ is the Gamma function. For $n = 1$, this Gamma p.d.f. becomes synonomous with the exponential distribution that has been supported by the studies of Barry (1977) and Hanna (1984a). The

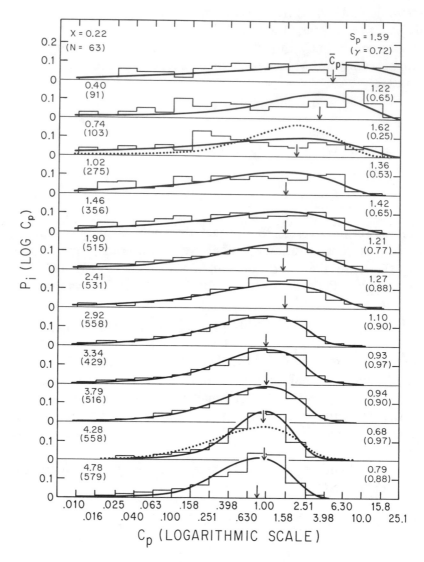

Figure 8.11. Probability of occurrence, $P_i(\ln C_p)$, within the logarithmically spaced intervals shown on the abscissa for various X for the case $F_* = 0$, $0 \leq z/h \leq 0.2$, and $|y| < \sigma_y$. Exceptions are that at $X = 0.22$ and 0.40, z/h lay between $0.\overline{2}$ and 0.4. The histogram ordinates with separate scales are from the data of this study. Numbers at upper left of each histogram denote X, at lower left denote N (number of samples within each histogram), at upper right denote s_p, and at lower right (in parentheses) denote γ. The γ values are average of γ for $|y| < \sigma_y/2$ and γ for $\sigma_y/2 < |y| < \sigma_y$. Arrows on abscissa indicate $\overline{C}_p$ obtained from the summation of $C_p P_i$. The solid curves are the Gamma p.d.f.'s based upon the given s_p and $\overline{C}_p$ values, from Eqs. (8.12) and (8.13). Dotted curves are the exponential distribution.

actual Gamma p.d.f. curves based on $\ln C_p$ and plotted in Fig. 8.11, from Eq. (8.11) and Eq. (8.12), are

$$P_i(\ln C_p) = 0.46 \frac{n^n}{\Gamma(n)} (C_p/\overline{C}_p)^n \exp(-n C_p/\overline{C}_p) .\qquad (8.14)$$

In earlier studies (Netterville, 1979; Deardorff and Willis, 1984; Deardorff, 1985; Wilson and Simms, 1985), as well as here, it has been found that the log-normal p.d.f. is not satisfactory. The arguments of Csanady (1973) which yielded the log-normal p.d.f. may be in error for several possible reasons:

(a) All parcels emanating from the stack were assumed to mix with essentially clean air—no allowance was made for dilute parcels to mix with nearby parcels of comparable or much higher concentration even though γ may have approached 1.

(b) No allowance was made for the fact that the volume rate at which ambient air entrains into the plume is much greater farther downwind that close to the stack.

(c) The number of dilution events experienced by particular effluent parcels traveling from the stack exit to the point of measurement was assumed to have a Gaussian distribution even though a Poisson distribution would actually be expected.

Hence, lack of agreement between observations and the log-normal p.d.f. should not be surprising.

For $X > 1.4$ or so, the Gamma distributions of Fig. 8.11 fit the data quite well. In obtaining $\overline{C}_p$ and s_p for these fits, the observed distributions were used; i.e.

$$\overline{C}_p = \sum_{i=1}^{17} C_p P_i(\ln C_p)$$

$$\overline{C_p^2} = \sum_{i=1}^{17} C_p^2 P_i(\ln C_p) .$$

The observed distributions higher in the mixed layer ($0.4 \leq z/h \leq 0.6$) for the same case of neutrally buoyant effluent are shown in Fig. 8.12. The fitted Gamma p.d.f.'s are again good approximations for the same range of X.

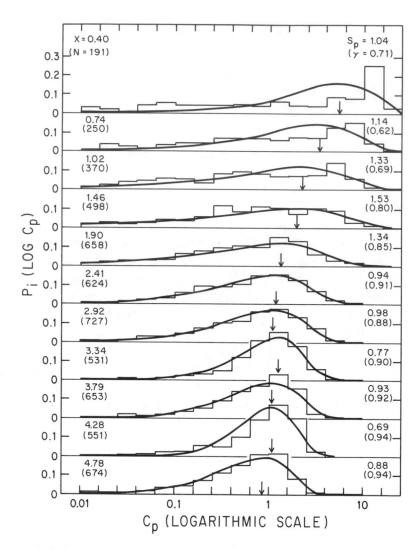

Figure 8.12. Probability of occurrence, $P_i(\ln C_p)$, versus $\ln C_p$, as in Fig. 8.11, except that $0.4 \le z/h \le 0.6$ at all X values.

However, when the Gamma distributions were similarly compared against our observed p.d.f.'s for the case $F_* = 0.03$, results were less favorable for $X > 2$. Although the agreement was good for $C > 2\overline{C}$, peak probabilities of the Gamma p.d.f.'s were up to 30% too small, and at somewhat smaller values of C were correspondingly too large. Since we doubt that the small amount of plume buoyancy involved, for $F_* = 0.03$, could have been responsible, the results cast some doubt on the reliability of the agreement with the Gamma p.d.f. found for the $F_* = 0$ case.

The Gamma p.d.f.'s may be tested against those measured by Fackrell and Robins (1982) for wind-tunnel boundary layer shear turbulence downwind of an elevated source. They presented $\sigma_c P(c)$ versus $(c - \bar{c})/\sigma_c$, which then causes Eq. (8.12) to take on the form

$$\sigma_c P(c) = \left[n^{n/2}/\Gamma(n) \right] c''^{n-1} \exp(-n^{1/2} c'') \qquad (8.15)$$

$$c'' = (c - \bar{c})/\sigma_c + n^{1/2} .$$

Their values of $\sigma_c/\bar{c}$ were here estimated from their Fig. 8 and the abscissa values for which $c = 0$, assuming $\gamma = 1$. The results are shown in Fig. 8.13 at a substantial downwind distance of 4.8 boundary layer heights. The Gamma p.d.f. agrees rather well with their measured distributions, especially for c equal to, or exceeding, the mean.

The Gamma p.d.f. was also tested against Netterville's (1979) observed distributions and found to perform well—much better than the log-normal p.d.f.

The first test of the Gamma p.d.f. in atmospheric pollution data was apparently made by Bencala and Seinfeld (1976), who reported that it closely resembled the log-normal distribution. In view of later findings by others that contradict this conclusion, it is not clear how to reconcile their results. Possibly it was because they did not normalize the Gamma p.d.f. relative to $\bar{c}$ and σ_c.

Berger et al. (1982) were apparently the first to point out that the Gamma p.d.f. gives a better agreement with field observations than does the log-normal p.d.f. They made use of 24-h averaged SO_2 data from 12 stations in Belgium over a 3-year period. However, they also did not normalize the Gamma p.d.f. relative to $\bar{c}$ and σ_c, but used a method of "maximum likelihood."

Wilson and Simms (1985) tested the Gamma p.d.f. in the form of Eq. (8.12). They noted that it definitely gives better predictions than the log-normal p.d.f., although in a theoretical analysis they predicted that near the source the p.d.f. should appear Gaussian, and farther from the source, exponential.

In addition to the Gamma p.d.f., the following variable-exponent power distribution was tested here:

$$P(C_p) = \frac{\Gamma(2n)}{n\Gamma^2(n)\overline{C}_p} \exp\left\{-\left[\frac{(2n)}{(n)}\frac{C_p}{\overline{C}_p}\right]^{1/n}\right\} \tag{8.16}$$

where n is determined from

$$\frac{\Gamma(3n)\Gamma(n)}{\Gamma^2(2n)} = s_p + 1 .$$

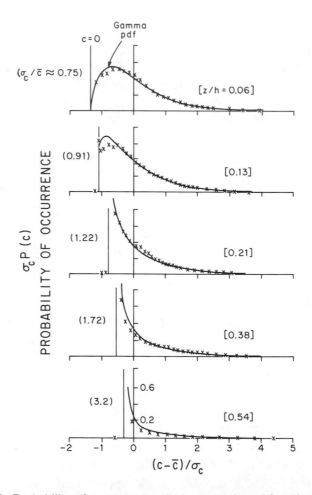

Figure 8.13. Probability of occurrence, $\sigma_c P(c)$, versus $(c - \bar{c})/\sigma_c$ (linear scale) for the data of Fackrell and Robins (1982) for which $x/h = 4.79$, $y/h = 0$, and h is the boundary-layer height. Their data are indicated by xxx; the smooth curves are the Gamma p.d.f.'s assuming $\gamma \simeq 1$. Values in parentheses at left give estimates of $\sigma_c/\bar{c}$. Values within square brackets at right give z/h. Source height was $0.19h$.

After converting this distribution to the observed $P(\ln C_p)$ form, the fits of Eq. (8.16) to the present data were found to be not as good as for the Gamma p.d.f.

A distribution that has been favored by Lewellen and Sykes (1985) is the truncated Gaussian distribution (with c linear on the abscissa, not logarithmic). For this p.d.f., the negative tail to the Gaussian p.d.f. that would exist is removed, and the deleted fraction of occurrences is considered to represent zero values associated with intermittency ($\gamma < 1$). In comparing this truncated Gaussian p.d.f. with the observed ones, however, we find that it greatly underestimates the probability of occurrence of large concentrations, the ones usually of most interest. The right-hand tail tends to be more exponential-like than Gaussian. Also, in comparing the truncated Gaussian p.d.f. against the distributions from Fackrell and Robins (1982), the truncation procedure appears to cause a substantial underestimation of γ. However, it is a true two-parameter p.d.f., unlike the log-normal and Gamma p.d.f.'s, which require three parameters: γ, $\bar{c}$, and σ_c. In this sense, the truncated Gaussian distribution can be said to perform surprisingly well.

The exponential p.d.f., as applied to $P_i(\ln C_p)$ of this study, is shown in Fig. 8.11 for $X = 0.74$ and 4.28. Since on this type of plot its maximum always remains the same ($e^{-1}\Delta c/c_p$), occurring at $c_p = \bar{c}_p$, it cannot even qualitatively represent the effect of the tendency of progression toward a well-mixed state. This, of course, is a consequence of its also being a two-parameter p.d.f. (only γ and $\bar{c}$ are required to be known), with σ_c implied to equal $\bar{c}$. The fact that the exponential p.d.f. has been found to perform well in some instances is here attributed to $\sigma_{cp} \simeq \bar{c}_p$ in those instances. The same occurs for the present data of Fig. 8.11 for X between about 2.9 and 3.3. However, if s_p does not lie between about 0.9 and 1.25, the exponential distribution deviates too far from the observed to seem satisfactory.

Since the Gamma p.d.f. appears to hold well for large values of $c_p/\bar{c}_p$, for X not too small, a graph of the probability that any particular value of $c_p/\bar{c}_p$ will be exceeded has been prepared here, making use of the tables of the incomplete Gamma function (Pearson, 1946). This graph is shown in Fig. 8.14 for values of $s_p = \sigma_{cp}/\bar{c}_p$ ranging from 0.2 to 2. It is seen that if the exponential p.d.f., corresponding to $s_p = 1$, were used as an approximation when s_p actually equaled 1.5, for example, the probability that $c_p/\bar{c}_p$ would exceed 11.5 would be predicted to be 100 times too small.

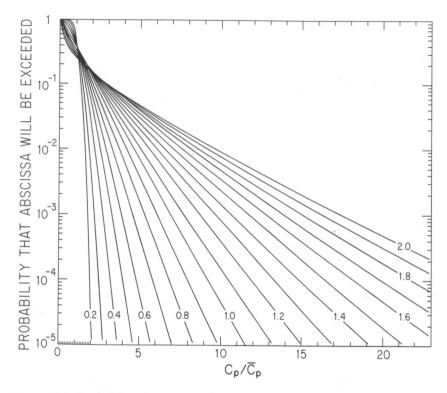

Figure 8.14. Probability that any particular value of $c_p/\bar{c}_p$ on abscissa will be exceeded, based on the Gamma distribution. Numbers near ends of lines indicate values of $s_p = \sigma_{cp}/\bar{c}_p$. Only the $s_p = 1$ line is perfectly straight.

Since Fig. 8.14 treats only the probabilities associated with $\gamma = 1$, a correction needs to be applied if $\gamma < 1$. As noted by Wilson and Simms (1985), for example,

$$P(c) = (1 - \gamma)\delta(c) + \gamma P(c_p)$$

where $\delta(c)$ is the delta function and $P(c_p)$ is the conditional p.d.f. $P(c; c > c_T)$, represented here by the Gamma distribution. Hence, the exceedance probabilities of Fig. 8.14 need to be multiplied by the intermittency γ and $\bar{c}_p$, s_p converted to $\bar{c}$, s, if $\gamma < 1$. The figure is designed for use when c exceeds (1 or 2)$\bar{c}$, and therefore exceeds any intermittency threshold.

If $\bar{c}$ and σ_c are known, and if γ is also known or can be assumed to be nearly unity, a graph such as Fig. 8.14 will be useful for estimating what fraction of the time, downwind of a pollutant source, a particular value of c will be exceeded. However, it cannot answer questions like "How long must one wait in exposure to a pollutant before there is an even chance that a fluctuation in c exceeding a particular value will actually occur?" Such questions involve knowledge of space and time scales of the concentrations, for which the reader is directed to studies of Wilson and Simms (1985) and Hanna (1984a, b).

8.5.2. Probability distributions for buoyant effluent

The case examined here represents a strongly buoyant effluent with $F_* = 0.26$. The pattern of $\overline{C}^y$ for this case is shown in Fig. 8.15, where a strong maximum persists in the upper mixed layer and resists downward mixing and entrainment. This represents a continuing source of only partially diluted effluent, supplied to the middle and lower mixed layer from above, for $X > 1$ or 2. In addition, contaminant is supplied through the usual route of downstream advection of that which found its way into the mixed layer during initial plume

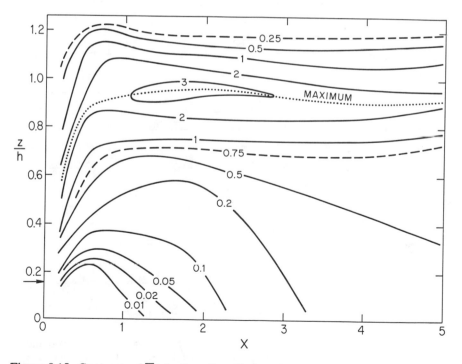

Figure 8.15. Contours of $\overline{C}^y$ in the $z/h - X$ plane for the case $F_* = 0.26$. Effects of effluent buoyancy mask the effects of initial source momentum jet.

rise, and of that which had entrained downward for small values of X. In essence this constitutes two different sources of effluent. In addition, if we should examine the p.d.f. near the top of the mixed layer, within a zone ($0.8 \leq z/h \leq 1.0$) where turbulent and nonturbulent fluid are interspersed, we might expect to observe one distribution with larger c collected within the nonturbulent fluid superimposed on another distribution with smaller c collected within the turbulent fluid. With these complications, the p.d.f.'s to be presented (Fig. 8.16 for $0.4 \leq z/h \leq 0.6$ and Fig. 8.17 for $0.8 \leq z/h \leq 1.0$) might not be expected to follow such a simple (Gamma) distribution as for the nonbuoyant source.

Figs. 8.16 and 8.17 indeed fail to obey the Gamma p.d.f. or any other simple distribution, although for $z/h = 0.5$ the log-normal

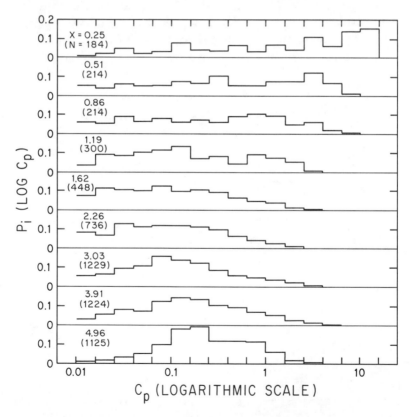

Figure 8.16. Probability of occurrence, $P_i(\ln C_p)$, within logarithmically spaced intervals of abscissa at various X for the case $F_* = 0.26$, $0.4 \leq z/h \leq 0.6$, and $|y| < \sigma_y$. Numbers at upper left of each histogram denote X; below them are the number of samples N per histogram.

p.d.f. may apply for X near 3 or 4 (Fig. 8.16). At $z/h = 0.9$, the simplest shaped histogram, for $X = 1.62$ in Fig. 8.17, has the corresponding Gamma p.d.f. displayed for comparison, and the differences even then are seen to be substantial.

It is concluded here that the Gamma p.d.f. applies only to the single point source whose effluent parcels all experience the same, nearly homogeneous, turbulence in traveling from source to receptor, and for X not "too small." These conditions are not met by a buoyant effluent that becomes partially trapped, temporarily, at the top of a mixed layer where the turbulence level is much reduced.

Although it is not clear from the imperfect data of Figs. 8.11 and 8.12 that the Gamma p.d.f. is unsatisfactory for small X, that conclusion is suggested. It might then be conjectured that the Gamma p.d.f.

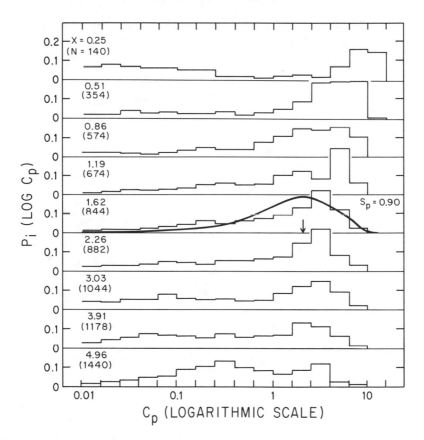

Figure 8.17. Probability of occurrence, $P_i(\ln C_p)$ as in Fig. 8.16 except that $0.8 \leq z/h \leq 1.0$. Solid curve for $X = 1.62$ is the Gamma p.d.f.

is, for the nonbuoyant effluent, only an equilibrium p.d.f. gradually approached with continued mixing. That is, once a Gamma p.d.f. is approximately established, further random mixing amongst themselves of parcels constituting that distribution may merely result in another Gamma p.d.f. farther downwind having smaller σ_c. An initial p.d.f. different from the Gamma p.d.f. might then eventually approach the Gamma distribution as mixing under quasi-homogeneous conditions progresses. The degree of intermittency may not affect the argument if all parcels have equal chance of mixing with ambient fluid, since by this effect all values of c (and $\bar{c}$ and σ_c) would tend to be diminished by the same factor. However, none of this speculation has yet been proved.

Acknowledgments

This study was supported by research Grant No. R810904 of the Environmental Protection Agency, and by Grant No. ATM–8318847 of the Division of Atmospheric Sciences of the National Science Foundation. Helpful ideas stemmed from corrspondence with Dave Wilson and discussions with Ian Sykes. P.d.f. analyses by graduate student K. Gurer are appreciated, as is the photographic analysis by J. Hollingsworth.

References

Barry, P. J., 1977: Stochastic properties of atmospheric diffusivity. *Sulphur and its Inorganic Derivatives in the Canadian Environment*, National Research Council of Canada, 313–358.

Bencala, K. E., and J. H. Seinfeld, 1976: On frequency distributions of air pollutant concentrations. *Atmos. Environ.*, **10**, 941–950.

Berger, A., Melice, J. L., and C. L. Demuth, 1982: Statistical distribution of daily and high atmospheric SO$_2$ concentrations. *Atmos. Environ.*, **16**, 2863–2877.

Briggs, G. A., 1984: Plume rise and buoyancy effects. *Atmospheric Science and Power Production*, D. Randerson, Ed., U. S. Dept. of Energy DOE/TIC–27601, NTIS DE84005177, 327–366.

Caughey, S. J., 1982: Observed characteristics of the atmospheric boundary layer. *Atmospheric Turbulence and Air Pollution Modelling*, F. T. M. Nieuwstadt and H. van Dop, Eds., Reidel, Dordrecht, 107–156.

Csanady, G. T., 1967: Concentration fluctuations in turbulent diffusion. *J. Atmos. Sci.*, **24**, 21–28.

Csanady, G. T., 1973: *Turbulent Diffusion in the Environment*. Reidel, Boston, 248 pp.

Deardorff, J. W., 1985: Laboratory experiments on diffusion: The use of convective mixed-layer scaling. *J. Climate Appl. Meteor.*, **24**, 1143–1151.

Deardorff, J. W., and G. E. Willis, 1984: Groundlevel concentration fluctuations from a buoyant and a non-buoyant source within a laboratory convective planetary boundary layer. *Atmos. Environ.*, **18**, 1297–1309.

Deardorff, J. W., and G. E. Willis, 1985: Further results from a laboratory model of the convective planetary boundary layer. *Bound.-Layer Meteor.*, **32**, 205–236.

Fackrell, J. E., and A. G. Robins, 1982: Concentration fluctuations and fluxes in plumes from point sources in a turbulent boundary layer. *J. Fluid Mech.*, **117**, 1–26.

Hanna, S. R., 1984a: The exponential probability distribution function and concentration fluctuations in smoke plumes. *Bound.-Layer Meteor.*, **29**, 361–375.

Hanna, S. R., 1984b: Concentration fluctuations in a smoke plume. *Atmos. Environ.*, **18**, 1019–1106.

Hukari, N. F., 1984: Photographic analysis of buoyant stack plumes in a laboratory model of the turbulent mixed layer. M. S. Thesis, Dept. of Atmospheric Sciences, Oregon State University, 89 pp.

Lewellen, W. S., and R. I. Sykes, 1985: Analysis of concentration fluctuations from lidar observations of atmospheric plumes. *7th Symposium on Turbulence and Diffusion*, Amer. Meteor. Soc., Boston, 327–330.

Netterville, D. D. J., 1979: *Concentration fluctuations in plumes.* Syncrude Environ. Research Monograph, 1979-4, 288 pp.

Pearson, K., 1946: *Tables of the Incomplete Gamma Function.* Cambridge University Press, 164 pp.

Poreh, M., and J. E. Cermak, 1985: Study of neutrally buoyant plumes in a convective boundary layer with mean velocity and shear. *7th Symposium on Turbulence and Diffusion*, Amer. Meteor. Soc., Boston, 327–330.

Willis, G. E., and J. W. Deardorff, 1974: A laboratory model of the unstable planetary boundary layer. *J. Atmos. Sci.*, **31**, 1297–1307.

Willis, G. E., and J. W. Deardorff, 1976: A laboratory model of diffusion into the convective planetary boundary layer. *Quart. J. Roy. Meteor. Soc.*, **102**, 427–445.

Willis, G. E., and J. W. Deardorff, 1981: A laboratory study of dispersion from a source in the middle of the convectively mixed layer. *Atmos. Environ.*, **15**, 109–117.

Wilson, D. J., and B. W. Simms, 1985: Exposure time effects on concentration fluctuations in plumes. Dept. of Mechanical Engineering, University of Alberta, Edmonton, Canada, Report No. 47, 162 pp.

Wilson, D. J., A. G. Robins, and J. E. Fackrell, 1985: Intermittency and conditionally averaged concentration fluctuation statistics in plumes. *Atmos. Environ.*, **19**, 1053–1064.

Index